T0257638

# Applications of Remote Sensing

# Applications of Remote Sensing

Edited by **Matt Weilberg**

New York

Published by Callisto Reference,
106 Park Avenue, Suite 200,
New York, NY 10016, USA
www.callistoreference.com

**Applications of Remote Sensing**
Edited by Matt Weilberg

International Standard Book Number: 978-1-63239-075-2 (Hardback)

Printed in the United States of America.

# Contents

# Preface

Remote sensing has majorly profited almost all areas of human activity and development. Remote sensing provides a common platform to physical, natural and social activities for interaction and advancement. This book discusses the impacts of remote sensing on various areas of science, human activity and technology by presenting a selected number of high quality contributions related to various remote sensing applications organized under two sections: Land Cover and Climate & Atmosphere. The book includes contributions of prominent experts and researchers, who possess vast knowledge and years of experience in this field.

This book is the end result of constructive efforts and intensive research done by experts in this field. The aim of this book is to enlighten the readers with recent information in this area of research. The information provided in this profound book would serve as a valuable reference to students and researchers in this field.

At the end, I would like to thank all the authors for devoting their precious time and providing their valuable contribution to this book. I would also like to express my gratitude to my fellow colleagues who encouraged me throughout the process.

Editor

# Section 1

# Land Cover

# Narrowband Vegetation Indices for Estimating Boreal Forest Leaf Area Index

Ellen Eigemeier, Janne Heiskanen, Miina Rautiainen, Matti Mõttus,
Veli-Heikki Vesanto, Titta Majasalmi and Pauline Stenberg
*University of Helsinki*
*Finland*

## 1. Introduction

### 1.1 Leaf area index

The green photosynthesizing leaf area of a canopy is an important characteristic of the status of the vegetation in terms of its health and production potential. At stand level, the amount of leaf area in a canopy is represented by a variable called the leaf area index (LAI), which is one of the key biophysical parameters in the global monitoring and mapping of vegetation by satellite remote sensing (Morisette et al., 2006). In this paper we adopt the, by now widely accepted, definition of LAI as the hemi-surface or half of the total surface area of all leaves or needles in the vegetation canopy divided by the horizontal ground area below the canopy. The definition is in line with the original definition of LAI, formulated for flat and (assumedly) infinitely thin leaves (Watson, 1947), as the one-sided leaf area per unit ground area. For coniferous canopies, the question arose on how to define the "one-sided" area of non-flat needles. While projected needle area formerly often has been used erroneously as a synonym to one-sided flat leaf area, it is now commonly accepted that the hemi-surface needle area represents the logical counterpart to the one-sided area of flat leaves (e.g. Chen & Black, 1992; Stenberg, 2006).

LAI controls many biological and physical processes, driving the exchange of matter and energy flow. Because LAI responds rapidly to different stress factors and changes in climatic conditions, monitoring of LAI yields a dynamic indicator of forest status and health. The link between forest productivity and LAI, in turn, lies in that LAI is the main determinant of the fraction of incoming photosynthetically active radiation absorbed by the canopy (fAPAR). The absorbed photosynthetically active radiation (APAR) quantifies the energy available for net primary production (NPP) and is thus a critical variable in NPP and carbon flux models. NPP is related to APAR by the light-use-efficiency originally introduced by Monteith (1977) for agricultural crops.

Traditionally, ground-based measurements of LAI have typically involved destructive sampling and determination of allometric relationships, e.g. between leaf area and the basal area of stem and/or branches carrying the leaves (the pipe model theory) (Shinozaki et al., 1964; Waring et al., 1982). However, such "direct methods" are quite laborious and indirect measurements of LAI using optical instruments are today the preferred choice (Welles &

Cohen, 1996; Jonckheere et al., 2004). They provide inverse estimates of LAI based on the fraction of gaps through the canopy in different directions, which can be measured using devices such as the LAI-2000 Plant Canopy Analyzer (LI-COR, 1992) or hemispherical photography. A vast body of classical literature exists on the dependency between LAI and canopy gap fraction underlying these techniques (e.g. Wilson, 1965; Miller, 1967; Nilson, 1971; Lang, 1986). In short, the inversion methods rely upon the assumption that leaves are randomly distributed in the canopy, in which case Beer's law can be applied to plant canopies (Monsi & Saeki, 1953). However, as the organization of leaves (needles) in forest canopies is typically more aggregated ("clumped") than predicted by a purely random distribution, the technique causes underestimation of LAI, especially in coniferous stands (e.g. Smith et al., 1993; Stenberg et al., 1994). Instead of the true LAI, the inversion of gap fraction data without correction for clumping yields the quantity commonly referred to as the "effective leaf area index" (Black et al., 1991).

Monitoring LAI in a spatially continuous mode and on a regular basis is possible only using remote sensing. Estimation of LAI from optical satellite images is considered feasible because LAI is closely linked to the spectral reflectance of plant canopies in the shortwave solar radiation range (Myneni et al., 1997). The physical relationships between canopy spectral reflectances and LAI form the basis of retrieval algorithms used in current Earth observation programs (e.g. MODIS, CYCLOPES, GLOBCARBON products) for mapping LAI at global scales. They produce bi-weekly and monthly vegetation maps that are widely used by biologists, natural resources managers, and climate modelers, e.g. to track seasonal fluctuations in vegetation or changes in land use. The arrival of narrowband reflectance data (also known as hyperspectral or imaging spectroscopy data) opens up new possibilities for satellite-derived estimation/monitoring of variables connected to the status and structure of vegetation, including LAI.

## 1.2 Spectral properties of boreal forests

The boreal forest zone, which spreads through Fennoscandia, Russia, Canada and Alaska, is the largest unbroken forest zone in the world and accounts for approximately one fourth of the world's forests. The boreal zone is a major store of carbon and thus plays an important role in determining global albedo and climate.

The reflectance spectra of coniferous forests (even if they have the same leaf area) are very distinct from similar broadleaved forests. The reasons for the special spectral behaviour of coniferous forests are versatile, yet primarily related to their structural, not optical, properties. Firstly, a high level of within-shoot scattering of conifers was originally noted nearly four decades ago (Norman & Jarvis, 1975). More recently, Landsat ETM+ data and a forest reflectance model were used to show that the low near infrared (NIR) reflectances observed in coniferous areas can largely be explained simply by within-shoot scattering (Rautiainen & Stenberg, 2005). Secondly, absorption by coniferous needles is higher than that by broadleaved species (Roberts et al., 2004; Williams, 1991), a phenomenon which can partly contribute to the lower reflectances of conifer-dominated areas. Other explanations include, for example, that the tree crown surface of coniferous stands is more heterogeneous than in broadleaved stands (Häme, 1991; Schull et al., 2011). In other words, when surface roughness (i.e. crown-level clumping) increases, the shaded area within the canopy increases, thus leading to lower reflectances. Overall, these results highlight the importance of various geometric properties as the main reason for the reflectance differences between broadleaved and coniferous stands.

Remote sensing of the biophysical properties, such as LAI, of a boreal coniferous forest canopy layer is further complicated by the often dominating role of the understory in the spectral signal (Rautiainen et al., 2011; Rautiainen et al., 2007; Eriksson et al., 2006; Eklundh et al., 2001; Chen & Cihlar, 1996; Spanner et al,. 1990). Coniferous forests that are regularly treated according to forest management practices tend to have relatively clumped and open canopies. Thus, the role of the understory vegetation in forming boreal forest reflectance cannot be neglected (Pisek et al., 2011).

## 1.3 Vegetation indices in LAI estimation

Canopy biophysical variables, such as LAI, can be estimated from remotely sensed data by two types of algorithms: empirical models and methods that use physically-based radiative transfer (RT) models. In empirical algorithms, the estimation is based on statistical relationships modelled between concurrent ground reference measurements and surface reflectance data. These relationships are typically expressed in the form of vegetation indices (VI). VIs include various combinations of spectral bands designed to maximize the sensitivity to vegetation characteristics while minimizing it to atmospheric conditions, background, view and solar angles (Baret & Guyot, 1991; Myneni et al., 1995). Operational LAI algorithms at global-scale typically make use of RT models, but the empirical models usually outperform them in more localized applications.

The design of a VI that is optimally correlated with a particular vegetation property requires good physical understanding of the factors affecting the spectral signal reflected from vegetation. The sensitivity of a VI to a vegetation characteristic is typically maximized by including bands with high sensitivity (e.g. high absorption) to the monitored entity and bands mostly unaffected by the same entity. The simplest forms of VIs are simple differences ($R_{B1}$–$R_{B2}$), ratios ($R_{B1}/R_{B2}$) and normalized differences [($R_{B1}$-$R_{B2}$)/($R_{B1}$+$R_{B2}$)] of the reflectances of two spectral bands ($R_{B1}$, $R_{B2}$). (In Table 2 we give examples of common VIs used in this study.) The most apparent characteristic of the green vegetation spectrum is the pronounced difference between the red and NIR reflectances, the so called red-edge around 700 nm. For example, the normalized difference vegetation index (NDVI) utilizes this difference and has been shown to correlate with many interrelated vegetation attributes, such as chlorophyll content, LAI, fractional cover, fAPAR and productivity.

The most commonly used VIs were designed for broadband sensors (one spectral band spans about 50 nm or more) having red and NIR bands, such as NOAA AVHRR and Landsat MSS (e.g. Tucker, 1979). However, the basic VIs in red and NIR spectral range suffer from three well-known problems in LAI estimation: (1) they are not sensitive to LAI over its natural range but tend to saturate already at moderate levels of LAI, (2) they are sensitive to canopy background variability, and (3) the VI-LAI relationships are dependent on the vegetation type. These VIs are also sensitive to atmospheric noise and correction.

The saturation of NDVI occurs typically at LAI levels of 2 to 6 depending on the vegetation type and environmental conditions (e.g. Sellers, 1985; Myneni et al., 1997). In general, NDVI saturates as the fractional cover of vegetation approaches one, although LAI still increases (e.g. Carlson & Ripley, 1998). Over conifer-dominated boreal forests, NDVI varies typically in a narrow range and shows poor relationships with canopy LAI (Chen & Cihlar, 1996; Stenberg et al., 2004). The reason for this is the green understory, which results in a non-contrasting background in the visible part of the spectrum (Nilson & Peterson, 1994; Myneni et al., 1997).

Many modifications of basic VIs have been suggested to give better sensitivity to LAI. Typical modifications use other visible bands than red (e.g. the green vegetation index, GNDVI, Gitelson et al., 1996), try to reduce soil effects based on the soil line concept (e.g. the soil adjusted vegetation index, SAVI, Huete, 1988), or include short wave infrared (SWIR) bands. Many modifications also attempt to reduce atmospheric effects (e.g. the enhanced vegetation index, EVI, Huete et al., 2002). The soil line is based on the observation that soil reflectances fall in a line in the red-NIR spectral space (e.g. Huete, 1988). Many VIs utilize the parameterized soil line in their calculation, but these VIs have not been successful in boreal forests as bare soil is rarely visible (e.g. Chen, 1996).

The sensitivity of shortwave infrared (SWIR) reflectance to forest biophysical variables has been recognized for a long time (e.g. Butera, 1986; Horler & Ahern, 1986) and several VIs utilizing the SWIR band have been designed. Rock et al. (1986) showed that the moisture stress index (MSI), i.e. the ratio of SWIR reflectance to NIR reflectance, was an indicator of forest damage. Later, the ratio has commonly been referred to as the infrared simple ratio (ISR, Chen et al., 2002; Fernandes et al., 2003). The SWIR reflectance has also been used for adjusting NDVI (Nemani et al., 1993) and SR (Brown et al., 2000). The reduced simple ratio (RSR) has been used specifically for estimating LAI (Brown et al., 2000; Stenberg et al., 2004) and has been employed also in regional and global-scale operational algorithms (Chen et al., 2002; Deng et al., 2006). RSR seems to reduce the sensitivity to the type and amount of understory vegetation, because background reflectance varies less in SWIR than in visible and NIR (Brown et al., 2000; Chen et al., 2002). RSR has also some capability to unify coniferous and broadleaved forest types, which reduces the need for land cover type specific LAI algorithms. However, in comparison to ISR, the use of red band makes RSR sensitive to atmospheric effects (Fernandes et al., 2003). However, although inclusion of SWIR reflectance increases the sensitivity of VIs to LAI, these indices also have a tendency to saturate at high levels of LAI (e.g. Brown et al., 2000; Heiskanen et al., 2011).

Imaging spectroscopy provides much narrower spectral bands than typical multispectral sensors. Due to the more detailed sampling of the vegetation spectra, such data can detect specific absorption features of vegetation and therefore improve the estimation of vegetation biochemical properties. For example, the SPOT 5 HRG sensors capture a spectral range from 500 nm to 1750 nm with four broad bands, in comparison to Hyperion's 242 (10 nm wide) bands between 400 nm and 2500 nm. At the canopy scale, the contents of biochemical components and LAI are highly inter-related (e.g. Asner, 1998; Roberts et al., 2004). Therefore, imaging spectroscopy could potentially improve LAI estimates. Furthermore, there is potentially complementary information outside the typical spectral bands of broadband sensors.

One way to utilize imaging spectroscopy data is to calculate narrow-band VIs in a similar fashion as for broadband data but using narrower bands. The aim is to improve the sensitivity of the VI to a specific vegetation biochemical property. For example, Ustin et al. (2009) give a comprehensive review on VIs used as indicators of plant pigments (chlorophyll, carotenoids and anthocyanin). The methods of estimating the non-pigment biochemical composition of vegetation (water, nitrogen, cellulose and lignin), on the other hand, are reviewed by Kokaly et al. (2009). Many of the developed indices have been designed to work at leaf level and do not necessarily upscale to canopy level, because of the high sensitivity to canopy structure, background, solar and view geometry. Another approach is to find iteratively the simple combinations of bands that give the best correlation with empirical data (e.g. Mutanga & Skidmore, 2004; Schlerf et al., 2005).

Most chlorophyll indices exploit the information in the red edge around 700 nm (Ustin et al., 2009). Imaging spectroscopy data also enables the estimation of the red edge position (REP), which is particularly sensitive to changes in chlorophyll content (e.g. Dawson & Curran, 1998). Water indices, on the other hand, utilize the water absorbing regions in the SWIR region of the spectrum (e.g. Gao, 1996; Zarco-Tejada et al., 2003). Those indices seem particularly interesting for LAI estimation considering the importance of the SWIR spectral region in estimating LAI using broadband indices.

There is growing evidence that imaging spectroscopy data can improve LAI estimates in comparison to broadband data by reducing the saturation effects. Depending on the vegetation type and range of LAI, different types of VIs have been found useful. However, the red edge indices have been most effective in estimating LAI of crops (Wu et al., 2010), grasslands (Mutanga & Skidmore, 2004) and thicket shrubs (Brantley et al., 2011). On the other hand, indices based on NIR and SWIR bands have been successful in broadleaved (le Maire et al., 2008) and coniferous forests (Gong et al., 2003; Schlerf et al., 2005; Pu et al., 2008). The importance of the SWIR spectral region in estimating boreal forest LAI has also been emphasized by multivariate regression analysis (e.g. Lee et al., 2004). However, broadband sensors can also have advantages over narrowband sensors in LAI estimation, for example, by being less sensitive to noise due to the sensor, atmosphere and background (e.g. Broge & Leblanc, 2000). Although there are case studies from different biomes, the performance of narrowband VIs has been poorly assessed over European boreal forests.

## 2. Case study

### 2.1 Aims

The aim of the study is to establish the extent to which vegetation indices can be used to measure variation in LAI based on a test site in southern boreal forest in Finland. We explore different VIs in LAI estimation during full leaf development. We compare the performance of narrowband VIs to traditional broadband VIs. The objective is to identify VIs, which are least sensitive to species composition and, on the other hand, perform well in coniferous stands.

### 2.2 Materials and methods

### 2.2.1 Study area

The study area, Hyytiälä, is located in the southern boreal zone in central Finland (61° 50'N, 24°17'E) and has an annual mean temperature of 3°C and precipitation of 700 mm. Dominant tree species in the Hyytiälä forest area are Norway spruce (*Picea abies* (L.) Karst), Scots pine (*Pinus sylvestris* L.) and Silver birch (*Betula pendula* Roth). Understory vegetation, on the other hand, is composed of two layers: an upper understory layer (low dwarf shrubs or seedlings, graminoids, herbaceous species) and a ground layer (mosses, lichens). The growing season typically begins in early May and senescence in late August. We measured twenty stands from the Hyytiälä forest area in July 2010 (see Section 2.2.2, Table 1). The stands represented different species compositions that are typical to the southern boreal forest zone in Finland.

| Site | Vegetation | Site type | Tree height, m | Basal area, m2/ha | LAI |
|------|-----------|-----------|----------------|-------------------|-----|
| A4 | Pine | mesic | 15.8 | 20.4 | 1.77 |
| A5 | pine, understory broadleaf | mesic | 18.6 | 24.3 | 2.67 |
| B2 | spruce, understory birch | mesic | 7.5 | 10 | 2.64 |
| D3 | pine, understory spruce & birch | sub-xeric | 17.8 | 20.5 | 2.37 |
| D4 | spruce, 25% birch | mesic | 16.5 | 27.5 | 3.72 |
| E1 | birch, spruce understory | mesic | 19.1 | 10.7 | 2.58 |
| E5 | 50% spruce, 50% birch | mesic | 23.1 | 27.2 | 4.12 |
| E6 | 50% spruce, 40% birch, 10% pine | mesic | 10.2 | 22.2 | 3.34 |
| E7 | Spruce | mesic | 13.3 | 31.7 | 3.91 |
| F1 | birch, spruce understory | mesic | 13.8 | 20.9 | 3.37 |
| G4 | spruce, 15% birch, 10% pine | herb-rich | 15.5 | 29.1 | 4.57 |
| H3 | Birch | herb-rich | 14.9 | 10.7 | 2.63 |
| H5 | Birch | herb-rich | 14.1 | 20.6 | 2.77 |
| I4 | birch, understory pine, spruce seedlings | mesic | 2.4 | 4 | 2.61 |
| T | Spruce | mesic | 24.6 | 56 | 3.43 |
| U16 | Birch | mesic | 14 | 21 | 2.69 |
| U17 | birch, 10% spruce | herb-rich | 11.7 | 27 | 3.35 |
| U18 | 65% pine, 25% spruce, 10% birch | sub-xeric | 16.5 | 26 | 3.45 |
| U26 | 20% pine, 70% spruce, 10% birch | mesic | 16.8 | 24.9 | 2.43 |
| U27 | 5% pine, 90% spruce, 5% birch | mesic | 15.2 | 20.9 | 2.63 |

[pine = Scots pine, spruce = Norway spruce, birch = Silver birch]

Table 1. Study stands.

## 2.2.2 Ground reference measurements

The LAI-2000 Plant Canopy Analyzer (PCA) is one of the most commonly used optical devices to measure LAI. The PCA's optical sensor includes five concentric rings of different zenith angles ($\theta$) (together covering almost a full hemisphere), which measure diffuse sky radiation between 320-490 nm (LI-COR, 1992). Measurements by the PCA performed below and above the canopy yield canopy transmittances, $T(\theta)$, for each ring. Finally, LAI is calculated by numerical approximation of the integral (Miller, 1967):

$$LAI = -2 \int_0^{\pi/2} \ln[T(\theta)]\cos\theta\sin\theta d\theta \tag{1}$$

There are four fundamental assumptions behind the LAI calculation method: 1. leaves (needles) are optically black in the measured wavelengths (implying that canopy transmittance closely corresponds to canopy gap fraction), 2. leaves (needles) are randomly distributed inside the canopy volume, 3. leaves (needles) are small compared to the area of view of the PCA's rings, and 4. leaves (needles) are azimuthally randomly oriented. The LAI estimate produced by Eq. 1 is commonly called effective LAI as the foliage elements are not randomly organized but typically clumped (or grouped) together, which causes the estimate produced by the PCA to be smaller than the "true" LAI (Chen et al., 1991; Deblonde et al., 1994).

The LAI measurements can be done either with one or two PCA instruments. One PCA is used for small plants such as crops, but for taller plants (e.g. trees), two units are necessary. When only one instrument is used, the measurement is at first taken below and then above the canopy. If two instruments are used, one instrument remains above the canopy and the other one below the canopy. The use of two instruments is preferable since data are logged nearly simultaneously with both sensors. The LAI estimate is calculated by combining below and above canopy data. The measurements should be conducted under diffuse light conditions; for example, when the sky has a full cloud cover or the sun angle is low (less than 16 degrees). The radius of the sample plot should be at least three times the dominant tree height as the PCA instrument has a relatively large opening angle.

In this study, the ground reference LAI (Table 1) was acquired by operating two LAI-2000 PCA instruments simultaneously. The instruments were intercalibrated before measurements were performed. The reference sensor was located above the forest canopy and set at a 15-second logging interval, while the other sensor was used inside the forest. The sampling scheme was a 'VALERI-cross' (Validation of Land European Remote Sensing Instruments, VALERI) which consists of two perpendicular 6-point transects. The distance between two measurement points was four meters, so that the sampling scheme corresponded roughly to a 20 m x 20 m plot. Measurement height was kept constant at 0.7 meters.

### 2.2.3 Satellite data

In this study, we used narrowband spectral data obtained from a Hyperion satellite image. Hyperion is a narrowband imaging spectrometer aboard the National Aeronautics and Space Administration (NASA) Earth Observer-1 (EO-1) satellite launched in 2000. Hyperion captures data in the 'pushbroom' manner in 7.7 km wide strips using 242 spectral bands. The spectral range of Hyperion is 356-2577 nm with each band covering a nominal spectral range of 10 nm. Each pixel in a Hyperion image corresponds to an area of 30 m x 30 m on the ground. During an acquisition, a scene with a length of either 42 km or 185 km is recorded. Hyperion is in a repetitive, circular, sun-synchronous, near-polar orbit at an altitude of 705.3 km measured at the equator. Thus, it can image almost any point on Earth and it flies over all locations at approximately the same local time. The nominal revisit time is 16 days, but due to the possibility of tilting the sensor, the potential revisit frequency is higher. The scene used in this study was captured on 03 July 2010, and was provided courtesy of the U.S. Geological Survey (USGS) Earth Explorer service.

Out of the potential 242 spectral bands, several lack illumination (due to the absorption in the atmosphere or a decrease of incident solar spectral irradiance in the longer infrared wavelengths) or have a very low spectral response. This leaves the user with 198 usable spectral bands: bands 8-57 in the visible and NIR (wavelengths 436-926 nm) and bands 77-224 in SWIR (wavelengths 933-2406 nm) (Pearlman et al., 2003). Hyperion images have several known deficiencies which can be corrected using algorithms given in scientific literature. Firstly, Hyperion suffers from systematic striping in along-track direction of the image. The stripes are characteristic to all pushbroom sensors. Instruments belonging to this broad class have a different receiving element for each image line. Hyperion has thus 256 radiation-sensitive elements for each spectral band, each seeing a separate 30 m strip of the ground, thus producing the 7.7 km wide image. The striping can be broadly divided into two classes, completely missing lines (due to non-functioning receiving elements) and actual stripes (arising from slightly different sensitivities of the 256 receivers). We removed the actual striping using Spectral Moment Matching (SpecMM), outlined by Sun et al. (2008), which uses the average and standard deviation statistics between highly correlated bands to remove stripes. Next, the missing lines containing no information were identified and corrected using the values from spatially adjacent pixels using local destriping (Goodenough et al., 2003). The results of the destriping can be seen in Figure 1.

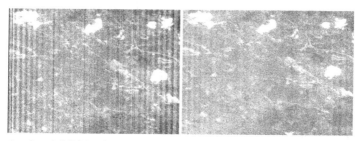

Fig. 1. Hyperion band 8 (436nm) uncorrected image (left), and corrected using Spectral Moment Matching and local destriping (right).

The second known defect in Hyperion imagery is a shift in the wavelength of each column in the across track direction from the band central wavelength. This shift, known as spectral smile, is also characteristic to pushbroom sensors and is a result of different optical paths leading to the different receiving elements. The shift is a function of wavelength and the position of the receiving element in the receiving array. As is the case for most instruments, the "smile" manifests itself in Hyperion imagery as a "frown", with the wavelengths of the columns near the edges of each band shifting negatively from the bands average wavelength (Figure 2). The smile was corrected using the pre-launch laboratory measured spectral shift (Barry, 2001). We used interpolation to bring each individual pixel to a common central wavelength based on the pre-launch calibration measurements.

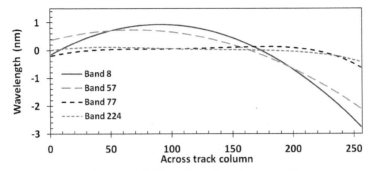

Fig. 2. Laboratory measured spectral shift of Hyperion (Barry, 2001).

The signal received by the Hyperion instrument consists of the photons scattered by the atmosphere as well as the ground surface. To study surface reflectance, the influence of the atmosphere needs to be eliminated in a process commonly known as atmospheric correction. We performed this correction using an algorithm known as Fast Line-of-sight Atmospheric Analysis of Spectral Hypercubes (FLAASH, Matthew et al, 2000). FLAASH is an absolute atmospheric correction that incorporates the MODTRAN4 radiation transfer code to model the scattering and transmission properties of the atmosphere at the time of image capture (San & Suzen, 2010). The FLAASH algorithm is incorporated into the ITT Visual Information Solutions (ITT VIS) ENVI software. For processing, FLAASH requires an input value for visibility to estimate atmospheric aerosol levels, in addition to basic geographic and temporal details about the scene. The visibility can be recalculated by FLAASH, using a ratio between dark pixels at 600 nm and 2100 nm. However, a more accurate estimate of visibility was achieved using ground based optical measurements from a weather station in the area.

The final processing stage is to resample the image pixels into a geographic coordinate system, known as geocorrection. This was done using a polynomial transformation to a vector base map from the National Land Survey of Finland. The Hyytiälä area contains numerous roads, providing a large number of easily identifiable potential ground control points (GCPs) at intersections. Around 20 GCPs were selected, with a root mean square error of 0.4 pixels being achieved. Bilinear interpolation was chosen for resampling the image pixels due to the better geometric accuracy over nearest neighbour.

The final product is a geocorrected image of the surface hemispherical-directional reflectance factors (HDRF) of the Hyytiälä area. To validate the atmospheric correction, we compared the HDRF to a field measured reflectance factor. A soccer field of about 130 m by 60 m in the area was sampled during the summer of 2010 every two to three weeks using an ASD handheld portable spectroradiometer covering a spectral range from 325-1075 nm. The sampling was done using a transect approach with 42 measurements at around 1 meter intervals. The final hemispherical-conical reflectance factor (HCRF) used for the comparison is an average of the transect representing the average for the whole field. While no ground measurements fell on the exact date of the Hyperion image, the ground measured spectra was interpolated to dates between two measurements. After interpolation the ground measured HCRF was binned into corresponding Hyperion bands using the spectral response of each band.

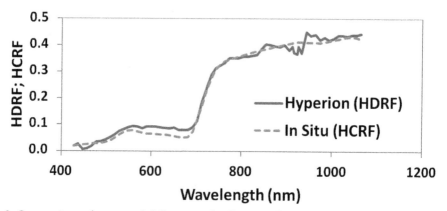

Fig. 3. Comparison of a soccer field's spectral reflectance factors from in situ radiometric measurements and corrected Hyperion data.

Overall, there is a very good correlation between the field measured reflectance and the fully processed Hyperion reflectance (Fig. 3). An overall RMSE of 1.8% is achieved, which gives us confidence in the validity of the pre-processing and atmospheric correction. However, as the *in situ* spectrum is considerably smoother than the one measured from the satellite, a considerable amount of noise is also present in the satellite-derived HDRF.

### 2.2.4 Vegetation indices and statistical analysis

First, we studied how HDRFs in single bands are correlated with LAI. Next, in order to evaluate narrow-band VIs for estimating LAI, we did regression analyses between various VIs and LAI. We used two approaches to select narrowband indices: 1) We made a literature survey for narrow-band VIs that have been designed to estimate foliage biochemical components. (A collection of VIs showing the highest $R^2$ with LAI are shown in Table 2.) 2) We calculated all the possible Ratio Indices (RI) and Normalized Difference Indices (NDI) of Hyperion bands and correlated them with LAI. In other words, the first approach also contains VIs combining several bands and the second approach aims to identify the simple two-band VIs that best correlate with LAI.

To facilitate the comparison of narrowband VIs with broadband indices, we calculated synthetic HDRFs based on Landsat 7 ETM+ bands. The HDRFs were calculated according to Jupp et al. (2002) using the ETM+ spectral sensitivity functions, and Hyperion's central wavelengths and bandwidths. Four broadband indices were calculated for comparison, SR, NDVI, ISR and RSR (Table 2). All these indices have been used for LAI estimation in various biomes. SR and NDVI were included for reference, and ISR and RSR because they have shown best performance over conifer-dominated boreal forests (see 1.3).

We analyzed the data both by grouping all the sample plots together and separately for coniferous plots (> 75% of the trees were Scots pines or Norway spruces). In the birch-dominated stands, the variation in LAI was too small for reliable regression analysis.

We studied only linear relationships. The strength of the relationship was assessed by the coefficient of determination ($R^2$) and the root mean square error (RMSE).

| Abbr. | Index | Formula | Reference | Bands applied |
|---|---|---|---|---|
| Indices concentrating on the red-edge | | | | |
| SR | Simple Ratio | $SR = R_{ETM+4}/R_{ETM+3}$ | Rouse et al. (1974), Birth & McVey (1968) | ETM+3, ETM+4 |
| NDVI | Normalized Difference Vegetation Index | $NDVI = (R_{ETM+4} - R_{ETM+3})/(R_{ETM+4}+R_{ETM+3})$ | Rouse et al. (1974) | ETM+3, ETM+4 |
| REP | Red Edge Position | $REP = 700 + (((R_{773} + 1,5 {}^*R_{662}) - R_{692}) / (R_{733} - R_{692})) {}^*(740-700)$ | Danson & Plummer (1995) | 773, 662, 692, 733 |
| Indices concentrating on pigment content | | | | |
| PSSRa | Pigment-Specific Simple Ratio – chla | $PSSRa = R_{803}/R_{681}$ | Blackburn (1998) | 681, 803 |
| Water sensitive indices | | | | |
| MSI = ISR | Moisture Stress Index = Infrared Simple Ratio | $ISR = R_{ETM+5}/R_{ETM+4}$ | Rock et al. (1986), Fernandes et al. (2002) | ETM+4, ETM+5 |
| RSR | Reduced Simple Ratio | $RSR = (R_{ETM+4}/R_{ETM+3}) {}^* ((R_{ETM+5\_min} - R_{ETM+5}) / (R_{ETM+5\_max} - R_{ETM+5\_min}))$ | Brown et al. (2000) | ETM+3, ETM+4, ETM+5 |

Table 2. Vegetation indices investigated in this study. The symbol R refers to the HDRF. Subscripts refer to the applied ETM+ band or the central wavelength (in nm) of the Hyperion band

## 2.3 Results

### 2.3.1 General characteristics of forest spectra

Two examples of forest reflectance factors (HDRFs) are presented in Figure 4. To allow relating the vegetation spectra to satellite signals, the sensitivity functions of the corresponding ETM+ bands are shown. Note the correspondence of ETM+2 with the green peak, ETM+3 with the red local minimum and ETM+4 with the plateau in the NIR. The red-edge slope (between ETM+ bands 3 and 4) is not covered by ETM+ bands. ETM+5 and ETM+7 catch the signal in the shortwave infrared region (SWIR-1 (here: 1470-1800 nm) and SWIR-2 (here: 2030-2360 nm) respectively), avoiding the two strong water absorption bands in-between.

The average reflectance of coniferous stands is slightly lower in the green region and decidedly lower in the NIR than the reflectance of birch stands. In SWIR-1 (covered by ETM+5) the reflectances become more comparable, and in SWIR-2 (covered by ETM+7) the signals almost meet.

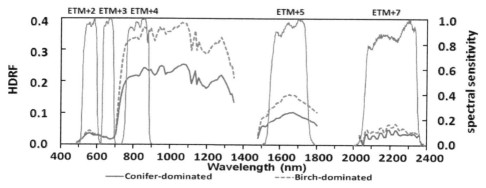

Fig. 4. Average conifer and birch-dominated stand spectra. The grey lines show the spectral sensitivity of the ETM+ bands.

### 2.3.2 Regression analysis for single bands

The different average HDRF for the two forest types (Fig. 4) results in different correlations of the satellite bands to LAI (Fig. 5).

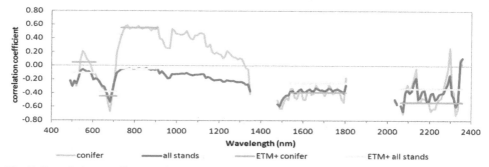

Fig. 5. Correlation coefficient of LAI with ETM+ and Hyperion spectral bands for all sample stands, and separately for conifer sample stands.

The correlation coefficients for all stands varied between -0.6 and -0.038. All correlations were negative, except for the two Hyperion bands centred at 2345 nm and 2355 nm. Two important regions (green and NIR) had almost no correlation with LAI. Only the absorption peak of chlorophyll produced a strong negative correlation at 681nm. The SWIR correlations were also mostly negative.

For conifer stands, correlation coefficients varied between -0.7 and 0.6. The first peak was at 549 nm, in the middle of the green band, followed by a strong negative correlation in the red with a peak at 681 nm. In the NIR a strong positive correlation was observed again. A slight shoulder began at 712 nm, with a plateau at 752 nm. In the SWIR, correlation coefficients were very close to those of all stands.

Fig. 5 also shows the correlation of the ETM+ bands to LAI. The lower spectral resolution averages wider wavelength ranges and therefore shows less variation in correlation coefficients.

### 2.3.3 Correlation of vegetation indices to LAI for all sample plots

The best broadband index analysed here was the Infrared Simple Ratio (ISR, $R^2$ = 0.56), followed by the Reduced Simple Ratio (RSR, $R^2$ = 0.40) (Table 3). The best narrowband combinations (either RI or NDI) showed more potential with $R^2$s exceeding 0.65 (Table 3, Fig. 6). If there were several indices based on neighbouring bands (within 10 nm) we chose the best one to Table 3.

| VI | Bands applied | $R^2$ | RMSE | RMSE Conifer | RMSE Broadleaf |
|---|---|---|---|---|---|
| broadband indices using simulated ETM+ | | | | | |
| ISR | ETM+4, ETM+5 | 0.56 | 0.44 | 0.42 | 0.25 |
| RSR | ETM+3, ETM+4, ETM+5 | 0.40 | 0.52 | 0.59 | 0.31 |
| NDVI | ETM+3, ETM+4 | 0.09 | 0.64 | 0.68 | 0.51 |
| SR | ETM+3, ETM+4 | 0.04 | 0.66 | 0.73 | 0.46 |
| narrowband indices using Hyperion | | | | | |
| RI | 1134, 1790 | 0.71 | 0.36 | 0.34 | 0.38 |
| NDI | 1134, 1790 | 0.68 | 0.38 | 0.36 | 0.39 |
| RI | 732, 1790 | 0.67 | 0.38 | 0.42 | 0.31 |
| RI | 1074, 1790 | 0.67 | 0.38 | 0.40 | 0.34 |
| RI | 885, 1790 | 0.67 | 0.39 | 0.37 | 0.35 |
| RI | 854, 1790 | 0.66 | 0.39 | 0.37 | 0.34 |
| RI | 1003, 1639 | 0.66 | 0.39 | 0.39 | 0.26 |
| RI | 1044, 1790 | 0.66 | 0.39 | 0.39 | 0.37 |
| NDI | 732 1790 | 0.66 | 0.39 | 0.42 | 0.33 |
| NDI | 1084, 1286 | 0.66 | 0.39 | 0.43 | 0.22 |

Table 3. Indices most correlated with LAI for all sample plots. RMSE was also calculated separately for each forest class. Bands for Hyperion refer to the central wavelength (in nm).

The best band combinations for RI and NDI indices were very similar (Fig. 6). A strong correlation with LAI existed for bands combining the region between 730 to 900 nm and 1130 to 1350 nm. Another interesting region was within SWIR-1; especially strong was the correlation around 1780 and 1790 nm. These bands also showed up in the best performing indices for all forest classes combined (Table 3).

The two best narrowband indices for all forest plots were the RI ($R^2$ = 0.71, RMSE = 0.36) and NDI ($R^2$ = 0.68, RMSE = 0.38) based on bands centred at 1134 and 1790 nm (Table 3). This is consistent with the best broadband index (ISR) which also combines NIR and SWIR. The same spectral regions are used by all the other best indices except two cases including a band in the red-edge (732 nm). Examples of the strongest relationships are shown in Fig. 7. However, when looking at the RMSE for conifer and broadleaf stands (Table 3) it became apparent that for some indices (e.g. NDI based on 1084 nm and 1286 nm: RMSE = 0.43 for conifers and RMSE = 0.22 for broadleaf) their LAI was correlated differently to the same VI.

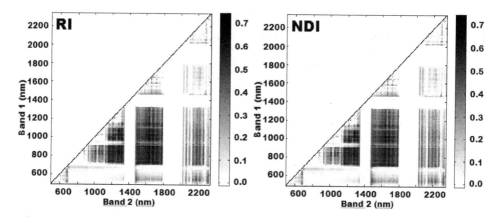

Fig. 6. Matrixes showing the $R^2$ between LAI and simple narrowband indices calculated for all possible combinations of Hyperion bands. The indices are defined as follows: RI=Band1/Band2, and NDI=(Band1-Band2)/(Band1+Band2).

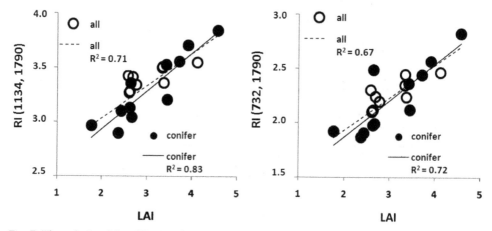

Fig. 7. The relationship of LAI and two best ratio indices (RI).

### 2.3.4 Correlations for coniferous dominated forest plots

The performance of the broadband indices for conifer-dominated stands was much better than over all sample stands. $R^2$ now ranged from 0.60 to 0.79, and NDVI showed the best correlation with LAI, followed by SR.

The best performing narrowband index over coniferous forest was neither RI nor NDI but REP ($R^2 = 0.89$) calculated according to the method of Danson & Plummer (1995) (Table 2). This index combined four bands in the visible and NIR; an area also represented in several of the other indices which best correlated with LAI in coniferous stands.

The matrixes for all band combinations of Hyperion bands over conifer-dominated stands (Fig. 8) showed wider spectral regions of high correlation than for all stands (Fig. 6).

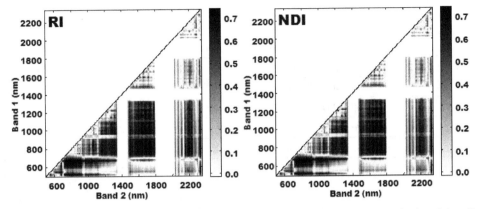

Fig. 8. Matrixes showing the R² between LAI and two narrowband indices calculated for all possible combinations of Hyperion bands for conifer-dominated stands.

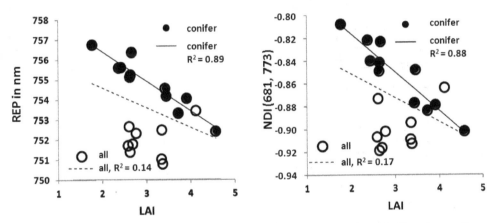

Fig. 9. The relationship of LAI and the two best performing narrowband indices for conifer-dominated stands.

Most of the indices with the highest correlations to LAI in coniferous stands used bands around the red-edge. Almost all of them (e.g. the Pigment-Specific Simple Ratio Index for chlorophyll a, PSSRa) applied the Hyperion band centred at 681nm, the peak of chlorophyll a absorption. Exceptions were the RI and NRI using the bands centred at 1185 and 1790 nm (i.e. combining NIR and SWIR), and RI and NDI using bands centred at 518 and 773 nm (i.e. combining carotene absorption and NIR).

Scatterplots for the two best indices for coniferous stands are shown in Fig. 9. In both cases, coniferous plots differed considerably from the other plots. This was indicated also by the high RMSE for all stands (up to 1.42, Table 4). However, for indices using NIR and SWIR (e.g. RI and NDI based on 1185 and 1790 nm) the differences were less pronounced. The VI showing the lowest RMSE for all stands (0.49) was the RI (1185 and 1790 nm) with an R² for conifer stands of 0.86 and RMSE 0.29.

| VI | Bands applied | R² | RMSE | RMSE All stands |
|---|---|---|---|---|
| broadband indices using simulated ETM+ | | | | |
| NDVI | ETM+3, ETM+4 | 0.79 | 0.36 | 1.20 |
| SR | ETM+3, ETM+4 | 0.78 | 0.36 | 1.56 |
| ISR | ETM+4, ETM+5 | 0.71 | 0.42 | 0.44 |
| RSR | ETM+3, ETM+4, ETM+5 | 0.60 | 0.50 | 0.90 |
| narrowband indices using Hyperion | | | | |
| REP | 671, 702, 742, 783 | 0.89 | 0.26 | 1.29 |
| NDI | 681, 773 | 0.88 | 0.27 | 1.02 |
| RI | 681, 773 | 0.88 | 0.28 | 1.01 |
| RI | 1185, 1790 | 0.86 | 0.29 | 0.49 |
| NDI | 1185, 1790 | 0.86 | 0.30 | 0.50 |
| NDI | 681, 742 | 0.85 | 0.30 | 1.01 |
| NDI | 681, 824 | 0.85 | 0.30 | 0.98 |
| RI | 681, 742 | 0.85 | 0.31 | 0.99 |
| NDI | 518, 773 | 0.85 | 0.31 | 1.42 |
| PSSRa | 803, 681 | 0.85 | 0.31 | 1.30 |
| RI | 518, 773 | 0.85 | 0.31 | 1.39 |

Table 4. Indices most correlated with LAI in conifer-dominated plots. $R^2$ and RMSE for conifer-dominated stands, and RMSE separately for all stands. Bands for Hyperion refer to the central wavelength (in nm).

## 2.4 Discussion

In our case study, the narrowband VIs provided more accurate LAI estimates than the broadband VIs synthesized from the same data in a boreal forest study site. The best narrowband combinations showed relatively strong linear relationships with LAI ($R^2 >$ 0.65), although the Hyperion image was acquired in the middle of the growing season when LAI is the highest. The relationships were even stronger if the analysis was restricted to the conifer stands ($R^2 > 0.85$). The results are promising as common broadband VIs tend to saturate at the highest LAI values. The improvement of estimation accuracy is in agreement with the previous studies, which have emphasized the potential of narrowband VIs for estimating forest canopy LAI (e.g. Lee et al., 2004; Schlerf et al., 2005; Brantley et al., 2011; Wu et al., 2010).

Most of the narrowband VIs showing the strongest relationships with LAI were based on reflectances in the far red and at the red edge (680−740 nm), NIR (e.g. 885 and 1134 nm) and SWIR (e.g. 1639 nm and 1790 nm) wavelength regions (Figure 10). Many of the most important spectral regions are not covered by the ETM+ spectral bands, and the spectral regions are very narrow in comparison to the ETM+ bands.

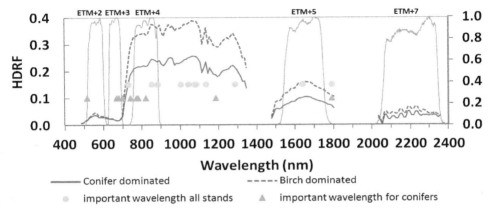

Fig. 10. Spectral regions used by the indices showing the strongest relationships with LAI over all sample stands and conifer stands.

The NIR and SWIR spectral bands were particularly important when all sample plots were analyzed together. This is in agreement with the best broadband indices, ISR and RSR. The importance of NIR and SWIR bands has been emphasized also in previous studies testing narrowband VIs for estimating forest LAI (e.g. Lee et al., 2004; Schlerf et al., 2005). The leaf (needle) reflectance at those wavelengths is mainly controlled by water absorption, although leaf biochemical components such as proteins, cellulose and lignin also contribute to absorption in the infrared (e.g. Curran, 1989). The amount of water at the canopy level is directly related to LAI, which explains strong correlations. The bands centered at 1134 nm and 1790 nm are among the Hyperion bands, which are closest to the water absorption regions centered at approximately 1200 nm and 1940 nm. The spectral bands close to the water absorption regions at 970 nm and 1400 nm are also employed in some of the best indices. The spectral bands of the broadband sensors are usually placed in the middle of the atmospheric windows to avoid atmospheric absorption. However, it seems that narrow spectral bands close to the water absorption regions are particularly interesting for estimating LAI. In these wavelength regions, the reflectance seems to be relatively insensitive to tree species or composition of the understory vegetation, as suggested earlier by the studies using broadband indices (e.g. Brown et al., 2000).

When pure coniferous stands were studied separately, the relationships became stronger and the far red and red edge spectral bands were included in several of the best VIs. However, the improvement in accuracy relative to the best VI based on NIR and SWIR reflectance (RI based on bands centered at 1185 nm and 1790 nm) was rather modest. The best broadband indices were NDVI and SR, which are based on ETM+ red and NIR bands. Usually, NDVI has shown relatively weak relationships with LAI in conifer dominated boreal forest (e.g. Stenberg et al., 2004).

The strongest relationship with LAI was provided by the red edge position (REP) calculated by the method proposed by Danson and Plummer (1995). In general, the REP is considered to be sensitive to leaf and canopy chlorophyll content, so that increasing the amount of chlorophyll, or LAI, is related to the longer REP wavelength because of the widening of the chlorophyll absorption region at approximately 680 nm (Danson & Plummer, 1995; Dawson and Curran, 1998; Sims & Gamon, 2002; Pu et al., 2003). In comparison to SWIR spectral bands, the far red and red edge spectral region is sensitive to species composition, shown as poor relationships over mixed vegetation. However, sometimes poor relationships between the REP and LAI have been reported even for pure coniferous stands (Blackburn, 2002). However, although the REP calculated in this study showed strong correlation with coniferous LAI, the estimated wavelengths do not correspond to the Red Edge Inflection Point (REIP), i.e. the steepest slope of the red-edge. The wavelengths are considerably longer. Therefore, the unusual inverse relationship between REP and LAI in this study is explained by the calculation method (Danson and Plummer, 1995). Alternative calculation methods for REP are summarized, for example, by Pu et al. (2003).

Although many studies testing narrowband VIs for LAI estimation have stressed the potential of the red edge and SWIR spectral regions, the specific spectral bands providing the strongest relationships with LAI vary between the studies. Also in our case study, the optimal band combinations provided stronger relationships with LAI than VIs collected from the literature. This is somewhat expected, as the number of spectral bands and their possible combinations is so large that empirically determined optimal band combinations are likely to depend heavily on the local environmental conditions and type of satellite image data. For example, approximately 150 useful spectral bands of Hyperion make more than 20,000 two-band combinations. Because of this, the optimal indices cannot necessarily be generalized very well. Furthermore, a large number of spectral bands combined with a small number of sample plots increase the risk that the regression models are overfitted. However, this should be mostly a problem with multivariate approaches (e.g. Lee et al., 2004). Moreover, when comparing broadband and narrowband indices, it should be noted that we used only synthesized ETM+ data and the results could differ to some extent if true ETM+ data would have been used instead (Lee et al., 2004). This is because the synthetic broadband data is affected by the lower signal-to-noise ratio of the narrow spectral bands, even if data are averaged.

## 3. Future perspectives

Wider use of imaging spectroscopy data is hampered by the availability of the data. Today, mostly airborne instruments are used to produce remote sensing data with high spectral resolution. Airborne measurements are associated with relatively small spatial coverage and high operating costs falling directly to data users. The Hyperion sensor used in this case study is a rare exception: it is the only true imaging spectrometer in orbit today, providing wide spectral coverage with uniform spectral resolution and contiguous bands. The scene, however, is about to change. At the end of the decade (i.e., around 2020), NASA is planning to launch the HyspIRI mission, providing narrowband data with routine global coverage (Samiappan et al., 2010). Before HyspIRI, several national space programs are striving to launch satellites with capability to produce narrowband data (e.g. the EnMAP instrument, Segl et al., 2010). Therefore, the need for developing algorithms that would make use of the advanced properties of narrowband data, compared to the more traditional multispectral data, is evident.

In this case study, we used narrowband VIs to relate forest LAI to remotely sensed reflectance signals. Historically, vegetation indices have been among the very first tools in interpreting multispectral remote sensing data from vegetated areas. Later, physically-based reflectance modelling has taken over the role of the preferred method in large-scale retrievals of vegetation biophysical variables. Similar developments may take place in the interpretation of narrowband imaging spectroscopy data. However, let us first take a closer look at narrowband indices as they are used in the current study.

As discussed above (section 1.3), VIs are usually treated as empirical (or, at least semi-empirical) tools in remote sensing. However, it has been known for a long time that the reflectance indices convey also some information on the physical processes related to the interaction of light with plant elements. Indeed, Myneni et al. (1995) showed that the common indices are actually derivatives of canopy reflectance and are physically related to abundances of absorbing pigments. For this reason, indices commonly make use of two spectral regions: one inside the spectral region where the absorption of a pigment is strong, and one outside the absorption band. The use of red and near-infrared wavelengths thus corresponds to measuring the abundance of one of the most vital plant pigments, chlorophyll.

Can such an interpretation be extended to narrowband indices? From the point-of-view of the physics of radiative transfer, there is no fundamental difference between broad- and narrowband indices. However, for calculating a spectral derivative, there is little use of well-tuned and potentially much noisier narrow spectral bands. For detecting pigments whose absorption spectra span tens, if not hundreds of nanometers, broadband indices seem a much more robust tool. Further, vegetation indices, especially early ones like the NDVI, have been shown both empirically and on the basis of theoretical studies, to be sensitive to factors others than those of interest, such as soil brightness changes and atmospheric effects. Most narrowband indices can be viewed as finely tuned versions of their older broadband counterparts. Site-specific selection of wavelengths leads to a better explanatory power of narrowband VIs as we also demonstrated in this case study. Unfortunately, the fine tuning for eliminating environmental effects makes narrowband indices potentially even more site-specific than broadband ones.

The comparison of narrowband and broadband VIs presented above did not concern indices capturing truly narrowband effects, e.g. the photochemical reflectance index PRI (Gamon et al., 1992) or various red edge parameters. Intrinsically narrowband VIs are based on effects that cannot be detected from broadband data. These indices are not more site-specific than broadband indices and do indeed, due to a finer spectral resolution, provide additional information on vegetation cover on all scales. Similarly, the red edge parameters calculated above make use of the high spectral resolution of narrowband data in a manner which is not site-specific. Therefore, it is not surprising that they provide a good fit for estimating forest stand variables regardless of dominating species.

An alternative to using narrowband indices would be to invert a full canopy reflectance model: the goals of both methods are to retrieve information on some biophysical variable of interest (Rautiainen et al., 2010). As discussed in this chapter, the theoretical foundations of the two approaches are somewhat similar. However, obvious limitations of index-based inversions lie in that it is not possible to define a spectral index sensitive to only one process, nor is it possible to design a universal spectral index which would be optimal for all applications everywhere and all the time (Verstraete & Pinty, 1996). Further, since vegetation indices carry only part of the information available in the original channel reflectances, they assume that the information of interest is contained exclusively in the observed spectral variations. VIs also often neglect the effects of surface anisotropy associated with the specific geometry of illumination and observation at the time of the measurements (Govaerts et al., 1999). Last, but not least, a fundamental shortcoming of the index-based approach lies in its potentially wide application area. A user not directly working in the field of remote sensing science may be distracted by a statistically strong dependence between a variable of interest (e.g. an ecological parameter describing diversity) and a vegetation index. However, canopy reflectance signals can carry information only on what are known as state variables of radiative transfer (abundances of optically active substances, canopy amount and structure, etc.). Other variables may be correlated with one or more of the state variables, but before drawing conclusions based on such correlations, the nature and application range of the correlation should be clarified.

Naturally, physical canopy reflectance models are immune to the problems listed above. When working in the forward mode, a modern reflectance model can reliably predict the spectral reflectance signal of a vegetation canopy given the required inputs (e.g. Widlowski et al., 2007). When run in inverse mode, the models should be able to produce an estimate of the state variables of radiative transfer based on measured spectral reflectance values. Unfortunately, due to the large number of the state variables and the mathematical nature of the inverse problem, a robust result is difficult to achieve (Baret & Buis, 2008). Despite the present-day problems with inverting canopy reflectance models, it is clear that physical models hold a clear advantage over index-based biophysical parameter estimation, especially when using imaging spectroscopy data. Physical models account for changes in environmental conditions and estimate all state variables simultaneously. They also have the advantage of failing if unphysical data is fed to them (e.g. due to sensor failure or preprocessing error) instead of producing unrealistic results. The problem with the large number of state variables can be solved by the larger information content of imaging spectroscopy data (compared with that produced by multispectral sensors) and development of novel physically based parameterizations allowing a more efficient description of canopy structure. However, until the full potential of imaging spectroscopy has been utilized by the developers of physical models, narrowband vegetation indices remain valuable tools in exploring the richness of high spectral resolution data.

## 4. Acknowledgment

We thank Anu Akujärvi for assisting in the field measurements. This study was funded by Emil Aaltonen Foundation, University of Helsinki Research and Postdoctoral Funds, and the Academy of Finland. Hyperion EO-1 data was available courtesy of the U.S. Geological Survey.

## 5. References

Asner, G.P. (1998). Biophysical and biochemical sources of variability in canopy reflectance. *Remote Sensing of Environment*, Vol. 64, No. 3, (June 1998), pp. 234-253, ISSN 0034-4257

Baret, F., Buis & S. (2008), Estimating canopy characteristics from remote sensing observations: review of methods and associated problems, In: *Advances in Land Remote Sensing: System, Modeling, Inversion and Application*, S. Liang (ed.), pp. 173-201, Springer-Verlag, ISBN 978-1-4020-6449-4, Heidelberg

Baret, F., Guyot, G. (1991). Potentials and limits of vegetation indices for LAI and APAR assessment. *Remote Sensing of Environment*, Vol. 35, No. 2-3, (February-March 1991), pp. 161-173, ISSN 0034-4257.

Barry, P. (2001). *EO-1/ Hyperion Science Data User's Guide, Level 1_B*, TRW Space, Defense & Information Systems, Redondo Beach, California, USA

Birth, G. S. & McVey, G. R. (1968) Measuring the color of growing turf with a reflectance spectrometer. *Agronomy Journal, Vol. 60 No. 6, (March 1968), pp. 640-643*, ISSN: 0002-1962

Black, T.A., Chen, J.M., Lee, X. & Sagar, R.M.. (1991). Characteristics of shortwave and longwave irradiances under a Douglas fir forest stand. *Canadian Journal of Forest Research* Vol. 21, No. 7, (July 1991), pp. 1020–1028, ISSN 0045-5067

Blackburn, G. A. (1998). Spectral indices for estimating photosynthetic pigment concentrations: a test using senescent tree leaves. *International Journal of Remote Sensing*, Vol. 19, No. 4, (March 1998), pp. 657-675, ISSN 0143-1161

Blackburn, G. A. (2002). Remote sensing of forest pigments using airborne imaging spectrometer and LIDAR imagery. *Remote Sensing of Environment*, Vol. 82, No. 2-3, (October 2002), pp. 311-321, ISSN 0034-4257.

Brantley, S.T., Zinnert, J.C. & Young, D.R. (2011). Application of hyperspectral vegetation indices to detect variations in high leaf area index temperate shrub thicket canopies. *Remote Sensing of Environment* Volume 115, Issue 2, (February 2011), pp. 514-523, ISSN 0034-4257

Broge, N.H. & Leblanc, E. (2000) Comparing prediction power and stability of broadband and hyperspectral vegetation indices for estimation of green leaf area index and canopy chlorophyll density. *Remote Sensing of Environment*, Vol. 76, No. 2, (May 2000), pp. 156-172, ISSN 0034-4257

Brown, L., Chen, J.M., Leblanc, S.G. & Cihlar, J. (2000). A shortwave infrared modification to the simple ratio for LAI retrieval in boreal forests: An image and model analysis. *Remote Sensing Environment*, Vol. 71, No. 1, (January 2000), pp. 16-25, ISSN 0034-4257.

Butera, M.K. (1986). A Correlation and Regression Analysis of Percent Canopy Closure Versus TMS Spectral Response for Selected Forest Sites in the San Juan National Forest, Colorado. *IEEE Transactions on Geoscience and Remote Sensing*, Vol. GE-24, No. 1, (January 1986), pp. 122-129, ISSN 0196-2892.

Carlson, T.N. & Ripley, D.A. (1998). On the Relation between NDVI, Fractional Vegetation Cover, and Leaf Area Index. *Remote Sensing of Environment*, Vol. 62, No. 3, (December 1997), pp. 241-252, ISSN 0034-4257.

Chen, J.M. (1996). Evaluation of vegetation indices and modified simple ratio for Boreal applications. *Canadian Journal of Remote Sensing*, Vol. 22, No. 3, (June 1996), pp. 229-242, ISSN 0703-8992.

Chen, J.M. & Black T.A. (1991). Measuring leaf-area index of plant canopies with branch architecture. *Agricultural and Forest Meteorology* Vol. 57, No. 1-3, (December 1991), pp. 1-12, ISSN 0168-1923

Chen, J.M. & Black, T.A. (1992). Defining leaf area index for non-flat leaves. *Plant Cell & Environment*, Vol. 15, No. 4, (May 1992), pp. 421-429, Online ISSN 1365-3040

Chen, J. & Cihlar, J. (1996). Retrieving leaf area index of boreal conifer forests using Landsat TM images. *Remote Sensing of Environment* Vol. 55, No. 2, (February 1996), pp. 153-162, ISSN 0034-4257

Chen, J., Pavlic, G., Brown, L., Cihlar, J., Leblanc, S., White, P., Hall, R., Peddle, D., King, D., Trofymow, J., Swift, E., Van der Sanden, J., & Pellikka, P. (2002). Derivation and validation of Canada-wide coarse-resolution leaf area index maps using high-resolution satellite imagery and ground measurements. *Remote Sensing of Environment*, Vol. 80, No. 1, (April 2002), pp. 165-184, ISSN 0034-4257.

Curran, P.J. (1989). Remote Sensing of Foliar Chemistry. *Remote Sensing of Environment*, Vol. 30, No. 3, (December 1989), pp. 271-278, ISSN 0034-4257.

Danson, F. M., & Plummer, S. E., 1995, Red edge response to forest leaf area index. *International Journal of Remote Sensing*, 1995, Vol. 16, No. 1, (January 1995), pp. 183-188, ISSN 0143-1161

Dawson, T.P. & Curran, P.J. (1998). A new technique for interpolating the reflectance red edge position. *International Journal of Remote Sensing*, Vol. 19, No. 11, (July 1998), pp. 2133-2139, ISSN 0143-1161

Deblonde, G., Penner, M. & Royer, A. (1994). Measuring Leaf Area Index with LI-COR LAI-2000 in Pine Stands. *Ecology*. Vol. 75, No. 5, (July 1994), pp. 1507-1511, ISSN 00129658

Deng, F., Chen, J., Plummer, S., Chen, M. & Pisek, J. (2006). Algorithm for global leaf area index retrieval using satellite imagery. *IEEE Transactions on Geoscience and Remote Sensing*, Vol. 44, No. 8, (August 2006), pp. 2219-2229, ISSN 0196-2892

Eklundh, L., Harrie, L. & Kuusk, A. (2001). Investigating relationships between Landsat ETM+ sensor data and leaf area index in a boreal conifer forest. *Remote Sensing of Environment*, Vol. 78, No. 3, (December 2001), pp. 239-251, ISSN 0034-4257

Eriksson, H., Eklundh, L., Kuusk, A., & Nilson, T. (2006). Impact of understory vegetation on forest canopy reflectance and remotely sensed LAI estimates. *Remote Sensing of Environment*, Vol. 103, No. 4, (August 2006), pp. 408–418, ISSN 0034-4257

Fernandes, R., Butson, C., Leblanc, S. & Latifovic, R. (2003). Landsat-5 TM and Landsat-7 ETM+ based accuracy assessment of leaf area index products for Canada derived from SPOT-4 VEGETATION data. *Canadian Journal of Remote Sensing*, Vol. 29, No. 2, (April 2003), pp. 241-258, ISSN 0703-8992.

Fernandes, R., Leblanc, S., Butson, C., Latifovic, R. & Pavlic, G. (2002). Derivation and evaluation of coarse resolution LAI estimates over Canada. In: *Proceedings of Geosciences and Remote Sensing Symposium, 2002.* IGARSS 02.2002, IEEE International Publications, Vol. 4, pp. 2097-2099, ISBN 0-7803-7536-X

Gamon, J., Penuelas, J., & Field, C. (1992). A narrow-waveband spectral index that tracks diurnal changes in photosynthetic efficiency. *Remote Sensing of Environment*, Vol. 41, No. 1, (July 1992), pp. 35-44, ISSN 0034-4257

Gao, B-C. (1996). NDWI–A normalized difference water index for remote sensing of vegetation liquid water from space. *Remote Sensing of Environment,* Vol. 58, No. 3, (December 1996), pp. 257-266, ISSN 0034-4257.

Gitelson, A. A., Kaufmann, Y. J. & Merzlyak, M. N. (1996) Use of a green channel in remote sensing of global vegetation from EOS-MODIS. *Remote Sensing of Environment,* Vol. 58, No. 3 (December 1996), pp. 289-298, ISSN 0034-4257

Gong, P., Pu, R., Biging, G.S. & Larrieu, M.R. (2003). Estimation of forest leaf area index using vegetation indices derived from Hyperion hyperspectral data. *IEEE Transactions on Geoscience and Remote Sensing* Vol. 41, No. 6, (June 2003), pp. 1355-1362, ISSN 0196-2892

Goodenough, D.G., Dyk, A., Niemann, K.O., Pearlman, J.S., Chen, H., Han, T. Murdoch, M. & West, C. (2003). Processing Hyperion and ALI for Forest Classification. *IEEE Transactions on Geoscience and Remote Sensing,* Vol. 41, No. 6, (June 2003), pp. 1321-1331, ISSN 0196-2892

Govaerts, Y. M., Verstraete, M. M., Pinty, B. & Gobron, N. (1999). Designing optimal spectral indices: a feasibility and proof of concept study. *International Journal of Remote Sensing,* Vol. 20, No. 9, (June 1999), pp. 1853-1873, ISSN 0143-1161

Häme, T. (1991). Spectral interpretation of changes in forest using satellite scanner images. *Acta Forestalia Fennica,* Vol. 222, The Society of Forestry in Finland & The Finnish Forest Research Institute, ISBN 951-651-092-2, Helsinki

Heiskanen, J., Rautiainen, M., Korhonen, L., Mõttus, M., Stenberg, P. (2011). Retrieval of boreal forest LAI using a forest reflectance model and empirical regressions. *International Journal of Applied Earth Observation and Geoinformation,* Vol. 13, No. 4, (August 2011), pp. 595-606, ISSN 0303-2434.

Horler, D.N.H. & Ahern, F.J. (1986). Forestry information content of Thematic Mapper data. *International Journal of Remote Sensing,* Vol. 7, No. 3, (January 1986), pp. 405-428, ISSN 0143-1161.

Huete, A.R. (1988) A soil-adjusted vegetation index (SAVI). *Remote Sensing of Environment,* Volume 25, Issue 3, (August 1988) , Pages 295-309, ISSN 0034-4257

Huete, A., Didan, K., Miura, T., Rodrigues, E.P., Gao, X. & Ferreira, L.G. (2002). Overview of the radiometric and biophysical performance of the MODIS vegetation indices. *Remote Sensing of Environment,* Vol. 83, No. 1-2, (November 2002), pp. 195-213, ISSN 0034-4257.

Jupp, D.L.B., Datt, B., Lovell, J., Campbell, S., King, E., et al. (2002). *Discussions around Hyperion data: background notes for the Hyperion data users workshop.* CSIRO Earth Observation Centre, Canberra. (accessed: 10.10.2011), Available from ftp://ftp.eoc.csiro.au/pub/djupp/Hyperion/Workshop/Minimal_Set/Documents/Hyp_Wsn.pdf

Jonckheere, I., Fleck, S., Nackaerts, K., Muys, B., Coppin, P., Weiss, M. & Baret, F. (2004). Reviews of methods for in situ leaf area index determination. Part I. Theories, sensors, and hemispherical photography. *Agricultural and Forest Meteorology,* Vol. 121, No. 1-2, (January 2004), pp. 19-35, ISSN 0168-1923

Kokaly, R. F., Asner, G.P., Ollinger, S.V., Martin, M.E. & Wessman, C.A. (2009). Characterizing canopy biochemistry from imaging spectroscopy and its application to ecosystem studies. *Remote Sensing of Environment*, Vol. 113, Supplement 1, (September 2009), pp. S78-S91, ISSN 0034-4257

Lang, A.R.G. (1986). Leaf area and average leaf angle from transmission of direct sunlight. *Australian Journal of Botany*, Vol. 34, No. 3, (May 1986), pp. 349 - 355. ISSN 0067-1924

Lee, K.-S., Cohen, W. B., Kennedy, R. E., Maiersperger, T. K. & Gower, S. T. (2004) Hyperspectral versus multispectral data for estimating leaf area index in four different biomes. *Remote Sensing of Environment*, Vol. 91, No. 3-4, (June 2004), pp. 508-520, ISSN 0034-4257.

LI-COR (1992). LAI-2000 *Plant Canopy Analyzer: Instruction Manual*. Lincoln, Nebraska, LI-COR, Inc.

le Maire, G., François, C., Soudani, K., Berveiller, D., Pontailler, D., Bréda, N., Genet, H.,Davi, H. & Dufrêne, E. (2008). Calibration and validation of hyperspectral indices for the estimation of broadleaved forest leaf chlorophyll content, leaf mass per area, leaf area index and leaf canopy biomass. *Remote Sensing of Environment*, Vol. 112, No. 10, (October 2008), pp. 3846–3864, ISSN 0034-4257

Matthew, M. W., Adler-Golden, S. M., Berk, A., Richtsmeier, S. C., Levine, R. Y., Bernstein, L. S., Acharya, P. K., Anderson, G. P., Felde, G. W., Hoke M. P., Ratkowski, A., Burke, H.-H., Kaiser, R. D., & Miller, D. P., (April 2000). Status of Atmospheric Correction Using a MODTRAN4-based Algorithm. In: *SPIE Proceedings, Algorithms for Multispectral, Hyperspectral, and Ultraspectral Imagery VI*. Vol. 4049, Accessed (10.10.2011), Available from: http://www.dtic.mil/cgi-bin/ GetTRDoc?AD=ADA447767&Location=U2&doc=GetTRDoc.pdf

Miller, J.B. (1967). A formula for average foliage density. *Australian Journal of Botany*, Vol. 15, No. 1, (February 1967), pp. 141 - 144, ISSN 0067-1924

Monsi, M. & Saeki, T. (1953). Über den Lichtfactor in den Pflanzengesellschaften und seine bedeutung für die Stoff-production. *Japanese Journal of Botany*, Vol. 14, (March 1953) pp. 22-52, ISSN 0075-3424

Monteith, J.L. & Moss, C.J. (1977). Climate and the efficiency of crop production in Britain. *Philosophical Tranactions of the Royal Soiety B*. Vol. 281, No. 980, (November 1977), pp. 277-294. Online ISSN 1471-2970

Morisette, J.T., Baret, F., Privette, J. L., Myneni, R.B., Nickeson, J., Garrigues, S., Shabanov, N., Weiss, M., Fernandes, R., Leblanc, S., Kalacska, M., Sánchez-Azofeifa, G. A., Chubey, M., Rivard, B., Stenberg, P., Rautiainen, M., Voipio, P., Manninen, T., Pilant, A., Lewis, T., Iiames, J., Colombo, R., Meroni, M., Busetto, L., Cohen, W., Turner, D., Warner, E.D., Petersen, G.W., Seufert, G. & Cook, R. (2006). Validation of global moderate resolution LAI Products: a framework proposed within the CEOS Land Product Validation subgroup. *IEEE Transaction on Geosciences and Remote Sensing*, Vol. 44, No. 7, (July 2006), pp. 1804-1817, ISSN 0196-2892

Mutanga, O. & Skidmore, A.K. (2004). Narrow band vegetation indices overcome the saturation problem in biomass estimation. *International Journal of Remote Sensing*, Vol. 25, No. 19, (October 2004), pp. 3999–4014, ISSN 0143-1161

Myneni, R. B., Hall, F. G., Sellers, P. J. & Marshak, A. L. (1995) Interpretation of spectral vegetation indexes. *IEEE Transactions on Geoscience and Remote Sensing*. Vol. 33, No. 2, (March 1995), pp. 481-486, ISSN 0196-2892

Myneni, R.B., Nemani, R.R. & Running, S.W. (1997). Estimation of global leaf area index and absorbed PAR using radiative transfer models. *IEEE Transactions on Geoscience and Remote*, Vol. 35, No. 6, (November 1997), pp. 1380-1393, ISSN 0196-2892.

Nemani, R., Pierce, L., Running, S. & Band, L. (1993). Forest ecosystem processes at the watershed scale: sensitivity to remotely-sensed Leaf Area Index estimates. *International Journal of Remote Sensing*, Vol. 14, No. 13, pp. 2519-2534, ISSN 0143-1161

Nilson, T. (1971). A theoretical analysis of the frequency of gaps in plant stands. *Agricultural Meteorology*, Vol. 8, pp. 25-38, ISSN 0002-1571

Nilson, T. & Peterson, U. (1994). Age dependence of forest reflectance: analysis of main driving factors. *Remote Sensing of Environment*, Vol. 48, No. 3, (June 1994), pp. 319-331, ISSN 0034-4257.

Norman, J. & Jarvis, P. (1975). Photosynthesis in Sitka spruce (Picea sitchensis (Bong.) Carr.). V. Radiation penetration theory and a test case. *Journal of Applied Ecology*, Vol. 12, No. 3. (December 1975), pp. 839-877, ISSN 00218901

Pearlman, J. S., Barry, P.S., Segal, C. C., Shepanski, J., Beiso, D. & Carman, S. L. (2003) Hyperion, a space-based imaging spectrometer. *IEEE Transactions on Geoscience and Remote Sensing*, Vol. 41, No. 6, (June 2003), pp. 1160-1173, ISSN 0196-2892

Pisek, J., Rautiainen, M., Heiskanen,J. & Mõttus, M. (2012). Retrieval of seasonal dynamics of forest understory reflectance in a Northern European boreal forest from MODIS BRDF data. *Remote Sensing of Environment*, Vol. 17, pp. 464-468, ISSN 0034-4257

Pu, R., Gong, P., Biging, G.S. & Larrieu, M.R. (2003). Extraction of Red Edge Optical Parameters From Hyperion Data for Estimation of Forest Leaf Area Index. *IEEE Transaction on Geosciences and Remote Sensing* Vol. 41, No. 4, (April 2003), pp. 916-921, ISSN 0196-2892

Pu, R., Gong, P. & Yu, Q. (2008) Comparative analysis of EO-1 ALI and Hyperion, and Landsat ETM+ data for mapping forest crown closure and leaf area index. *Sensors* 2008, Vol. 8, No. 6, (June 2008), pp. 3744-3766, ISSN 1424-8220

Rautiainen, M., Heiskanen, J., Eklundh, L., Mõttus, M., Lukeš, P. & Stenberg, P. (2010), Ecological applications of physically based remote sensing methods. *Scandinavian Journal of Forest Research* Vol. 25, No. 4, (July 2010), pp. 325–339, ISSN 0282-7581

Rautiainen, M. & Stenberg, P. (2005). Application of photon recollision probability in simulating coniferous canopy reflectance. *Remote Sensing of Environment*, Vol. 96, No. 1, (May 2005), pp. 98-107, ISSN 0034-4257

Rautiainen, M.; Mõttus, M.; Heiskanen, J.; Akujärvi, A.; Majasalmi, T. & Stenberg, P. (2011). Seasonal reflectance dynamics of common understory types in a northern European boreal forest. *Remote Sensing of Environment*, in press, (available online since 27 July 2011), ISSN 0034-4257

Rautiainen, M., Suomalainen, J., Mõttus, M., Stenberg, P., Voipio, P., Peltoniemi, J. & Manninen, T. (2007). Coupling forest canopy and understory reflectance in the Arctic latitudes of Finland. *Remote Sensing of Environment*, Vol. 110, No. 3, (October 2007), pp. 332-343, ISSN 0034-4257

Roberts, D., Ustin, S., Ogunjemiyo, S., Greenberg, J., Dobrowski, S., Chen, J. & Hinckley, T. (2004). Spectral and structural measures of northwest forest vegetation at leaf to landscape scales. *Ecosystems*, Vol. 7, No. 5, (May 2004), pp. 545-562. ISSN 1432-9840

Rock, B. N., Vogelmann, J. E., Williams, D. L., Vogelmann, A. F. & Hoshizaki, T. (1986). Remote Detection of Forest Damage. *Bio Science*, Vol. 36, No. 7, (July-August 1986), pp. 439-445, ISSN 0006-3568

Rouse, J. W., Haas, R. H., Schell, J. A. & Deering, D. W. (1974). Monitoring Vegetation Systems· in the Great Plains with ERTS. In: *Third ERTS Symposium*, pp. 309-317, NASA SP-351 I, United States

Samiappan, S., Prasad, S., Bruce, L.M. & Robles, W. (2010) NASA's upcoming HyspIRI Mission - Precision vegetation mapping with limited ground truth. *Geoscience and Remote Sensing Symposium (IGARSS), 2010 IEEE International*, (July 2010), pp. 3744-3747. ISSN 2153-6996

San, B. T., & M. L. Suzen, (2010), Evaluation of Different Atmospheric Correction Algorithms for EO-1 Hyperion Imagery, *International Archives of the Photogrammetry, Remote Sensing and Spatial Information Science*, Volume XXXVIII, Part 8, (August 2010), pp. 392-397, ISSN 1682-1777

Schlerf, M. et al. (2005). Remote sensing of forest biophysical variables using HyMap imaging spectrometer data. *Remote Sensing of Environment*, Vol. 95, No. 2, (March 2005), pp. 177¬-194, ISSN 0034-4257

Schull, M., Knyazikhin, Y., Xu, L., Samanta, A., Carmona, P., Lepine, L., Jenkins, J., Ganguly, S. & Myneni, R. (2011). Canopy spectral invariants, Part 2: Application to classification of forest types from hyperspectral data. *Journal of Quantitative Spectroscopy and Radiative Transfer*, Vol.112, No.4, (March 2011), pp. 736-750, ISSN 0022-4073

Segl, K., Guanter, L., Kaufmann, H., Schubert, J., Kaiser, S., Sang, B. & Hofer, S. (2010) Simulation of spatial sensor characteristics in the context of the EnMAP hyperspectral mission. *IEEE Transactions on Geoscience and Remote Sensing*, Vol.48, No. 7, (July 2010), pp. 3046-3054, ISSN 0196-2892

Sellers, P.J. (1985). Canopy reflectance, photosynthesis and transpiration. *International Journal of Remote Sensing*, Vol. 6, No. 8, pp. 1335-1372, ISSN 0143-1161

Shinozaki, K., Yoda, K, Hozumi, K. & Kira, T. (1964). A quantitative analysis of plant form - the pipe model theory. I. Basic analyses. *Japanese Journal of Ecology*, Vol. 14, No. 3, (June 1964), pp. 97-105, ISSN 0021-5007

Sims, D.A. & Gamon, J.A. (2002). Relationships between leaf pigment content and spectral reflectance across a wide range of species, leaf structures and developmental stages. *Remote Sensing of Environment*, Vol. 81, No. 2-3, (August 2002), pp. 337-354, ISSN 0034-4257

Smith, N.J., Chen, J.M. & Black, T.A. (1993). Effects of clumping on estimates of stand leaf area index using the LI-COR LAI-2000. *Canadian Journal of Forest Research*, Vol. 23, No. 9, pp. 1940-1943, ISSN 0045-5067

Spanner, M.A., Pierce, L.L., Peterson, D.L. & Running, S.W. (1990). Remote sensing of temperate coniferous forest leaf area index: the influence of canopy closure, understory vegetation and background reflectance. *International Journal of Remote Sensing*. Vol. 11, No.1, pp. 95-111, ISSN 0143-1161

Stenberg, P. (2006). A note on the G-function for needle leaf canopies. *Agricultural and Forest Meteorology*, Vol. 136, No. 1-2, (January 2006), pp 76-79, ISSN 0168-1923

Stenberg, P., Linder, S., Smolander, H. & Flower-Ellis, J. (1994). Performance of the LAI-2000 plant canopy analyzer in estimating leaf area index of some Scots pine stands. *Tree Physiology*, Vol. 14, No. 7-8-9, (July 1994), pp. 981-995, Online ISSN 1758-4469

Stenberg, P., Rautiainen, M., Manninen, T., Voipio, P. & Smolander, H. (2004). Reduced Simple Ratio better than NDVI for estimating LAI in Finnish pine and spruce stands. *Silva Fennica*, Vol. 38, No. 1, (March 2004), pp. 3-14, ISSN 0037-5330.

Sun, L., Neville, R., Staenz, K., & White, H.P. (2008). Automatic destriping of Hyperion imagery based on spectral moment matching. *Canadian Journal for Remote Sensing*. Vol. 34, Supplement 1, (May 2008), pp. 68-81, ISSN 0703-8992

Tucker, C. J. (1979) ETM+3 and photographic infraETM+3 linear combinations for monitoring vegetation. *Remote Sensing of Environment*, 8, 127-150, ISSN 0034-4257

Ustin, S. L., Gitelson, A.A., Jacquemoud, S., Schaepman, M., Asner, G.P., Gamon, J.A., & Zarco-Tejada, P. (2009). Retrieval of foliar information about plant pigment systems from high resolution spectroscopy. *Remote Sensing of Environment*, Vol. 113, Supplement 1, (September 2009), pp. S67-S77, ISSN 0034-4257.

Validation of Land European Remote Sensing Instruments (VALERI) Network. Cited: 16.8.2011. Available from: http://www.avignon.inra.fr/valeri/

Verstraete, M. M. & Pinty, B. (1996) Designing optimal spectral indexes for remote sensing applications. *IEEE Transactions on Geoscience and Remote Sensing*, Vol. 34, No. 5, (September 1996), pp. 1254-1265, ISSN 0196-2892

Waring, R.H., Schroeder, P.E. & Oren, R. (1982). Application of the pipe model theory to predict canopy leaf area. *Canadian Journal of Forest Research*. Vol. 12, No. 3. (September 1982), pp. 556-560. ISSN 0045-5067

Wilson, J.W. (1965). Stand structure and light penetration. I. Analysis by point quadrats. *Journal of Applied Ecology*, Vol. 2, No. 2, (November 1965), pp. 383-390, ISSN: 00218901

Watson, D. (1947). Comparative physiological studies in the growth of field crops. I. Variation in net assimilation rate and leaf area between species and varieties, and within and between years. *Annales of Botany*, Vol. 11, No. 1, (January 1947), pp. 41-76, ISSN 0305-7364

Welles, J.M. & Cohen, S. (1996). Canopy structure measurement by gap fraction analysis using commercial instrumentation. *Journal of Experimental Botany*, Vol. 47, No. 9, (September 1996), pp. 1335-1342, ISSN 0022-0957

Widlowski, J. L., Taberner, M., Pinty, B., Bruniquel-Pinel, V., Disney, M., Fernandes, R., Gastellu-Etchegorry, J. P., Gobron, N., Kuusk, A., Lavergne, T., Leblanc, S., Lewis, P. E., Martin, E., Mottus, M., North, P. R. J., Qin, W., Robustelli, M., Rochdi, N., Ruiloba, R., Soler, C., Thompson, R., Verhoef, W., Verstraete, M. M. & Xie, D. (2007). Third Radiation Transfer Model Intercomparison (RAMI) exercise: Documenting progress in canopy reflectance models. *Journal of Geophysical Research-Atmospheres* 112, D09111, (May 2007), 28 pages, ISSN 0148-0227

Williams, D. (1991). A comparison of spectral reflectance properties at the needle, branch, and canopy level for selected conifer species. *Remote Sensing of Environment* Vol. 35, No. 2-3, (February-March 1991), pp. 79-93, ISSN 0034-4257

Wu, C., Han, X., Niu, Z. & Dong, J. (2010). An evaluation of EO-1 hyperspectral Hyperion data for chlorophyll content and leaf area index estimation. *International Journal of Remote Sensing*, Vol. 31, No. 4, (February 2010), pp. 1079-1086, ISSN 0143-1161

Zarco-Tejada, P.J., Rueda, C.A. & Ustin, S.L. (2003). Water content estimation in vegetation with MODIS reflectance data and model inversion methods. *Remote Sensing of Environment*, Vol. 85, No. 1, (April 2003), pp. 109-124, ISSN 0034-4257.

# 2

# Seasonal Variability of Vegetation and Its Relationship to Rainfall and Fire in the Brazilian Tropical Savanna

Jorge Alberto Bustamante, Regina Alvalá and Celso von Randow
*Brazilian National Institute for Space Researches*
*Brazil*

## 1. Introduction

The Brazilian savanna, named locally Cerrado, is the second largest Brazilian biome, covering approximately two million km², especially in the Central Highlands (Ratter *et al.*, 1997). This biome is composed predominantly of tropical savanna vegetation and is considered as one of the world's biodiversity hotspots, a priority area for biodiversity conservation in the world (Myers et al., 2000). The Cerrado region is considered the last agricultural frontier in the world (Borlaug, 2002), which has been converted in the last 50 years especially for agriculture and pasture purposes, where natural and mainly anthropogenic annual burning is a common practice. Currently, around 50% of natural vegetation in the Cerrado region has been converted to pastures and crops (PROBIO-MMA, 2007).This conversion has impacted the biological diversity, the hydrological cycle, the energy balance, the climate and the carbon dynamics at local and regional scales due to habitat fragmentation, invasive alien species, soil erosion, pollution of aquifers, degradation of ecosystems and changes in fire regimes (Klink & Machado, 2005; Aquino & Miranda, 2008). The knowledge of spatial distribution, temporal dynamics and biophysical characteristics of the vegetation types, are important elements to improve the understanding of what is the interaction like between vegetation, precipitation and fire.

The objective of this study is to determine the relationship of environmental variables, such as precipitation and fire, with spatial and temporal distribution patterns of main vegetation type of the Brazilian tropical savanna. Thus, we seek to answer the question: how environmental variables, like rain and fire, influence the main vegetation types, like herbaceous, shrubs, deciduous trees and evergreen trees, in the Cerrado biome taking in account the seasonal patterns of the variables involved?

In this study, the potential of multi-temporal satellite data, like TRMM data for precipitation, MODIS vegetation indices products for land cover mapping, and others sensors like GOES and MODIS for fire detection is explored by the use of remote sensing and geographic information systems (GIS) techniques.

### 1.1 Seasonality of Cerrado vegetation

Phenological parameters of vegetation, such as start and end of the growing season, are strongly influenced by atmospheric conditions (like precipitation, temperature and humidity) at different time scales (intrannual, inter-annual, interdecadal, and so on). Atmospheric conditions at intrannual scale influence the main phenological events that the plant experiences during the annual cycle of growth (Reed et al. 1994). At greater time scales, climate influence on the spatial and temporal distribution of vegetation (Schwartz, 1994). On the other hand, the vegetation influence atmosphere while maintaining or modifying the flows of matter and energy, albedo, roughness, $CO_2$, which in turn affect the regional and/or global climate.

Savanna ecosystems that cover approximately 20% of the global land surface have mechanisms that control the flow of matter and energy in tropical savannas. These ecosystems are not well understood, which has hindered the inclusion of this biome in studies of regional and global modeling (Law et al., 2006).

## 1.2 Climate and precipitation regime

Climate patterns from intra-seasonal to decadal and century scales directly influence the timing, magnitude (productivity), and spatial patterns of vegetation growth cycles, or phenology (Reed et al., 1994; Schwartz, 1994).

The Savanna biome has a wet/dry climate. Its Köppen climate group is **Aw**. The *A* stands for a tropical climate, and the *w* for a dry season in the winter and the rainy season in the summer. During the dry season of a savanna, most of the plants shrivel up and die. Some rivers and streams dry up (Parker, 2000; Ritter, 2006). In the wet season all of the plants are lush and the rivers flow freely. The temperature of the savanna climate ranges from 20° to 30° C. In the winter, it is usually about 20° to 25° C. In summer the temperature ranges from 25° to 30° C. The savanna temperature does not change a lot, although when it does, it is very gradual and not drastic.

Because of its latitudinal position, the Brazilian savanna region is characterized by the transition between the warm climates of low latitudes and mesothermal climates of middle latitudes (Nimer, 1989). This region is considered almost homogeneous on the length and location of the dry and rainy periods (Rao & Hada, 1990). However, Castro et al. (1994) show that this region has a certain degree of heterogeneity due to the variation of length in the dry and rainy periods. This heterogeneity is determined by the interaction of atmospheric circulation systems in the lower and upper troposphere over the region. Some of these systems are: The South Atlantic anticyclone also known as South Atlantic Convergence Zone (SACZ), Polar anticyclone and Chaco low. SACZ is one of the main phenomena that determine the rainfall across the region (Satyamurty et al., 1998). In general, rainfall in the region ranges from 1000 to 1500 mm.

The climate of the Cerrado is tropical warm and semi-humid, with just two seasons, a dry one from May to September and a rainy one from October to April. Monthly rainfall in dry season (that include fall and winter) reduces considerably, reaching zero, resulting in a dry period that varies from three to five months duration (Coutinho, 2000). The rainy season (spring and summer) sometimes has short dry periods named locally "*veranicos*". The mean annual temperatures vary between 22 and 27°C and the mean annual precipitations between 600 and 2.200 mm.

## 1.3 Fire regime and detection

Fire is one of the most important drivers that influence vegetation function and structure. Fire incidence, in a given area or ecosystem, is part of a fire regime which has specific patterns of fire occurrences, frequency, size, severity, and sometimes vegetation and fire effects as well. For example, savanna fires are often of low intensity and high frequency (often annual), while forest fires are often of low frequency (once every few centuries) and very high intensity (Bowman & Murphy, 2010). Most of the wildland fires occur by the combination of edaphic, climatic and human activities (Roy, 2004). Natural fires are generally started by lightning, with a very small percentage started by spontaneous combustion of dry fuel such as sawdust and leaves. This kind of fire is insignificants in comparison to number of fires started by humans (Roy, 2004). Most tropical fires are set intentionally by humans (Bartlett 1955, 1957, 1961) and are related to several main causative agents (Goldammer, 1988): deforestation activities (conversion of natural vegetation to other land uses, e.g. agricultural lands pastures, exploitation of other natural resources); traditional, but expanding slash-and-burn agriculture; grazing land management (fires set by graziers, mainly in savannas and open forests with distinct grass strata); use of non-wood forest products (use of fire to facilitate harvest or improve yield of plants, fruits, and other forest products, predominantly in deciduous and semi-deciduous forests); wildland/residential interface fires (fires from settlements, e.g. from cooking, torches, camp fires etc.); other traditional fire uses (in the wake of religious, ethnic and folk traditions; tribal warfare) and socio-economic and political conflicts over questions of land property and land use rights.

Satellite-borne sensors can detect fires in the visible, thermal and mid-infrared bands. These sensors have been used most extensively for detecting and monitoring fire activity from landscape to global scales (Justice et al., 2003; Diaz-Delgado et al., 2004; Allan et al., 2003; Brandis & Jacobson, 2003; Miller et al. 2003; Rollins et al., 2004; Bowman et al., 2003). Justice et al. (2003) analyzed global remote sensing data and showed that occurrence of landscape fire is not random across the world, which is strongly influenced by climatic variables, like moisture deficit, wind speed, relative humidity and air temperature.

## 2. Methodology

### 2.1 Study area

The study area represents almost all (more than 90%) of the Brazilian savanna (Cerrado) biome, excluding only the southern region, which is characterized by few small isolated patches of savannas with intense anthropic activities like agriculture and ranching. The Cerrado vegetation exhibits a wide range of physiognomies. Following the "forest-ecotone-grassland" concept (Coutinho, 1978), the Cerrado ranges from *campo limpo*, a grassland, to *cerradão*, a tall woodland. The intermediate physiognomies (*campo sujo* - a shrub savanna, *campo Cerrado* - a savanna woodland, and *Cerrado sensu stricto* - a woodland) are considered ecotones of the two extremes.

The soil surface dries out during the dry season, leading the herbaceous and sub shrub plants suffering water stress. Thus, leaves dry out and die, while the underground plant structures are kept alive. The presence of dead leaves by water stress and also by frost greatly increases the litterfall and, consequently, the risk of fire (Nimer, 1977; Coutinho, 2000).

### 2.2 Methodology

The methodology involves the use of two spatial approaches, regional and local, to analyze the spatio-temporal relationships between environmental variables (precipitation and fire) and vegetation (NDVI).

The analysis unit at the local approach is the point, a specific pixel, which is obtained from the grid of points that were selected using a stratified random sampling. This grid contains separately the following types of vegetation: herbaceous, shrubs, deciduous trees, and evergreen trees of the Brazilian savanna in our study area.

At the regional approach, the entire region is considered another analysis unit, which means the Cerrado vegetation was not classified into four vegetation types. In this case, we calculated a NDVI mean value, keeping together all vegetation types (from grassland to forest) to each 16-days composite of the NDVI time series data.

The procedure applied to the vegetation data is also applied to the precipitation and fire data. The results are seasonal profiles to each variable along the annual cycle which were related using correlation and regression techniques. These seasonal profiles allow calculating a gradient of vegetation seasonality, which is defined by the difference of highest and lowest values of NDVI, precipitation, or fire. In the case of vegetation, the degree of seasonality is directly related to the degree of deciduousness, that is, the degree of leaf biomass loss during the dry season, when most plants suffer some degree of water stress.

The spatial and temporal resolutions of the data are: 250m and 16-day, 1km and 1-day, ~20km and 3 hours, for MODIS NDVI, fire hotspot and precipitation, respectively. These data are arranged to standardize them in the same 16-day temporal scale. Data from 2002, 2005 and 2008 were collected since they are considered as years under normal climatic condition, without the influence of El Niño-Southern Oscillation events.

### 2.2.1 Vegetation seasonality

The Normalised Difference Vegetation Index (NDVI), normalised ratio between near infrared reflectance (NIR) and red reflectance (red), has been widely used in satellite-based vegetation monitoring and modelling. NDVI is computed as:

$$NDVI = (NIR - red)/(NIR + red) \tag{1}$$

Index values can range from -1.0 to 1.0, but vegetation values typically range between 0.1 and 0.7. Higher index values are associated with higher levels of healthy vegetation cover, whereas clouds and snow will cause index values near zero, making it appear that the vegetation is less green.

Six Moderate Resolution Imaging Spectroradiometer (MODIS) Normalized Difference Vegetation Index (NDVI) tiles (h12v09, h12v10, h12v11, h13v09, h13v10 e h13v11) were joined to create a mosaic for the entire study area. Three annual time-series were prepared for the following years: 2002, 2005 and 2008. Each annual dataset consists of 23 MODIS NDVI data, at 16-day composite intervals, and 250 m spatial resolution. These data were used to classify and analyze seasonal and phenology profiles of the Brazilian Cerrado vegetation.

At this point, the methodology consists first in the classification of vegetation, from the annual NDVI time-series data and using the tree decision technique, into four types: grasses and herbs, shrubs, deciduous trees and evergreen trees. This classification uses the phenological parameter named end of vegetation growing season, which corresponds to the following ranges of NDVI for each vegetation type: grasses and herbs (E1) from 100 to174, shrubs (E2) from 175 to 199, deciduous trees (E3) from 200 to 219 and evergreen trees (E4) from 220 to 255. Ground truth data was used to validate this classification.

The second part consists of selecting representative spatial points of vegetation types (Figure 1), which are obtained from the vegetation classification image. Each point corresponds to a pixel on the image and is defined as our unit of analysis. A stratified random sampling technique was used for the selection of points in the classification image. The number of points to each vegetation types was proportional to its spatial coverage in the study area. So, herbaceous (E1) represents 52% of the points, shrubs (E2) 24%, deciduous trees (E3) 15% and evergreen trees (4) 9%.

The total number of points identified in the study area was N = 639, which are distributed as follows: 251 points of herbaceous, 318 of shrubs, 59 of deciduous trees and 11 of evergreen trees (Figure 1).

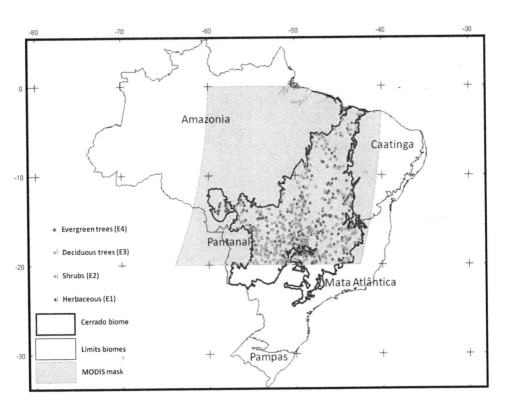

Fig. 1. Location of Brazilian biomes highlighting the savanna (Cerrado) biome. The shaded area is a mosaic of 4 MODIS-13Q1 tiles. Dots of different colors correspond to stratified random sampling of the following vegetation types: herbaceous (E1), shrubs (E2), deciduous trees (E3), and evergreen trees (E4).

## 2.2.2 Fire seasonality

First, daily data of fire hot spot obtained as latitude/longitude coordinates, in ASCII format, are converted in XYZ vector format data using a geographical information system (GIS) tool. These daily vectors were used to create a new vector data set of 16-days composites accumulating these daily data. Each composite were used to create a raster data set of fire hotspot density by the use of the Kernel density estimator, according to the following equation (Silverman, 1986):

$$\widehat{f}(x;H) = n^{-1}\sum_{i=1}^{n}K_H(x - X_i) \tag{2}$$

Where:

- $X_1, X_2,...,X_n$ is sample of $n$ data points (fire hot spot)
- H is bandwith matrix
- $K_H x - X_i$ is normal probability density function (pdf) with mean $X_i$ and variance H

Kernel Density calculates the density of point features around each output raster cell. The kernel function is based on the quadratic kernel function as described in Silverman (1986). Conceptually, a smoothly curved surface is fitted over each point. The surface value is highest at the location of the point and diminishes with increasing distance from the point, reaching zero at the Search radius distance from the point. Only a circular neighborhood is possible. The volume under the surface equals the Population field value for the point, or 1 if NONE is specified. The density at each output raster cell is calculated by adding the values of all the kernel surfaces where they overlay the raster cell center.

## 2.2.3 Precipitation

We used two kinds of data for precipitation in the study area for the years 2002, 2005 and 2008. First, Tropical Rainfall Measuring Mission (TRMM) multisatellite rainfall data (3B42 product), which has 0.25 degree spatial resolutions and 3-hours temporal resolution. Second, meteorological station rainfall data scattered throughout the study area, which has 1-hour temporal resolution.

These two datasets (TRMM and observed data) are combined following the approach of Vila et al. (2009), which use the Barnes objective analysis (Barnes, 1973; Koch et al., 1983) for data interpolation. This analysis allows the incorporation of observed data in a grid of estimated data and also improves its spatial resolution. As result, the new precipitation data has 0.2-degree spatial resolution and 1-day temporal resolution.

## 3. Results and discussion

### 3.1 Regional analysis

Figure 2 shows seasonal profiles of vegetation and precipitation in the Cerrado region for the three years analyzed (2002, 2005 and 2008). These results show that the Cerrado vegetation seasonality is well defined, which in turn has a direct relationship to the seasonality of precipitation. However, there is a time lag ranging from 1 (16 days) to 3 (48

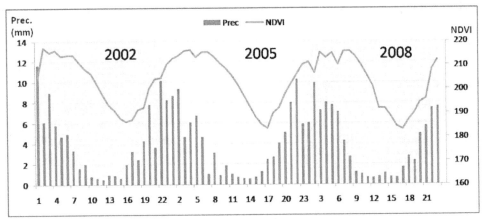

Fig. 2. Annual seasonality of vegetation and precipitation in the years 2002, 2005 and 2008 for the Cerrado biome. Each year consists of 23 16-days composite periods. Precipitation is the daily mean rainfall values for a 16-days composite period in mm (first y axis) and vegetation the mean NDVI values for the same period (second y axis).

days) periods between the beginning of the rainy season and the beginning of the vegetation growing season.

Figure 3 shows seasonal profiles of vegetation and fire in the Cerrado region for the three years analyzed (2002, 2005 and 2008). These results show, as in Figure 2, that the fire occurrence in the Cerrado has well-defined seasonality, which in turn has a direct negative relationship to the seasonality of vegetation. That means, the highest fire occurrence during the growing cycle of fire is related to the greatest loss of plant cover during the dry season, with a time lag ranging from 0 to 3 periods (0 to 48 days).

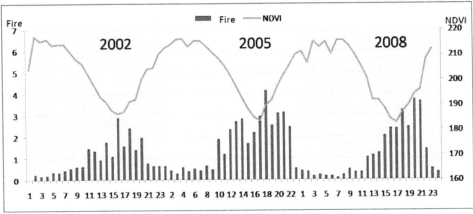

Fig. 3. Annual seasonality of vegetation and fire in the years 2002, 2005 and 2008 for the Cerrado biome. Each year consists of 23 16-days composite periods. Fire is the daily mean value of the density of hotspot within a 10km radius for a 16-days composite period and vegetation the mean NDVI values for the same period (second y axis).

## 3.2 Local analysis

The results presented show the seasonality of vegetation, rainfall and fire in places (points) defined by the grid points representing the four vegetation types analyzed in the study.

### 3.2.1 Seasonality of vegetation

Figure 4 shows the seasonal profile of four vegetation types over the three years analyzed. These results show a clear difference, regarding the degree of vegetation seasonality, among the four types of vegetation analyzed, according to the following gradient: herbaceous (E1), with strong seasonality, shrubs (E2), deciduous trees (E3), and evergreen trees (E4), with weak seasonality.

The vegetation phenology metrics are shown in Figure 5. Figure 5a shows annual maximum and minimum NDVI values, indicating the highest and lowest vegetation productivity respectively, for each type of vegetation in the three years analyzed. Figure 5b shows the difference between the maximum and minimum NDVI as a percentage, indicating the degree of seasonality. Also Figure 5a shows a slight difference between the maximum NDVI values, high plant productivity in the four vegetation types, while the difference between the minimum NDVI values, lower productivity, in the four vegetation types is significant. In general, the degree of seasonality of the vegetation (Figure 5b) was consistently detected in the four vegetation types. That is, small plants with low canopy (shrubs and herbaceous) have higher degree of seasonality than tall one with high canopy, which in turn have lower degree of seasonality.

### 3.2.2 Seasonality of precipitation

Figure 6 shows the seasonal profile of rainfall recorded in the same sampling points of the four vegetation types over the three years analyzed. These results show, in the beginning of the year during the rainy season, lower rainfall at sites where herbaceous and shrubs were registered than at sites where deciduous and evergreen trees are predominant. This result is a first indicator that shows a relationship between rainfall gradient and vegetation cover gradient. These gradients range from sites with higher precipitation, associated with high canopy plants (evergreen trees), to those with less precipitation, associated with a lower canopy plants (herbaceous).

### 3.2.3 Seasonality of fire

Figure 7 shows the pattern of the fire season recorded in the same sampling points of the four vegetation types over the three years analyzed. The results of fire occurrence throughout the Cerrado region show that there is a pronounced seasonality in all vegetation types analyzed with a peak in the months of greatest drought in the dry season.

The results show a well-defined gradient of fires in the four types of vegetation. This gradient varies from lower fire density in evergreen trees (E4), with shorter periods of time (12 to 22) throughout the annual cycle, to higher fire density in herbaceous plants (E1), with more periods of time (1 to 23, except 2), as seen in Figure 7. Most of the fire occurrences in the four types of vegetation were recorded in 2005 and 2008 indicating the occurrence of an inter-annual variability of fire. The higher fires were recorded between the periods from 15 to 21 taking into account the four vegetation types and the three years analyzed.

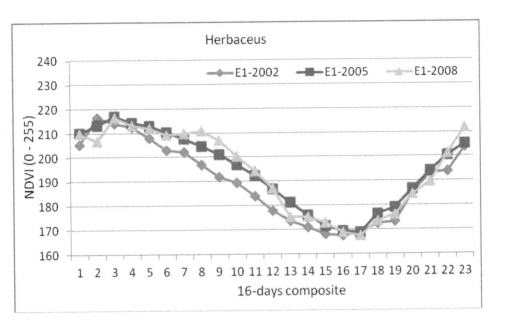

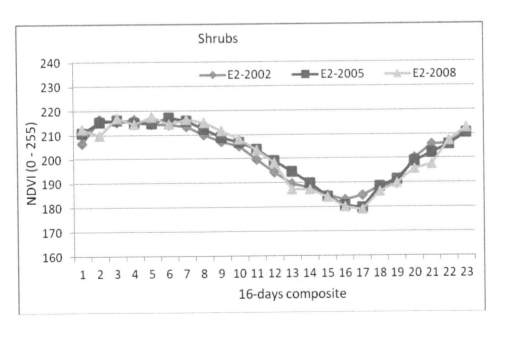

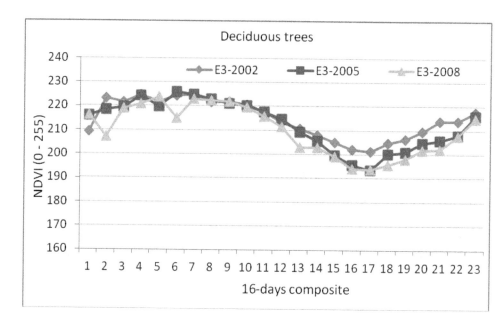

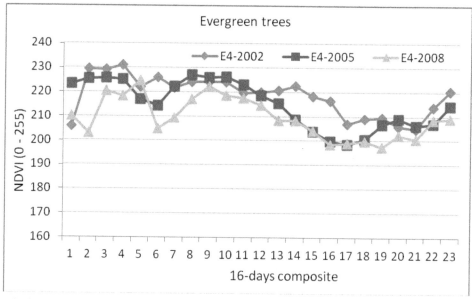

E1: herbaceous; E2: shrubs; E3: deciduous trees; and E4: evergreen trees.

Fig. 4. Annual seasonality of vegetation derived from NDVI data for the years 2002, 2005 and 2008 in the four vegetation type analyzed. Each year consists of 23 16-days composite periods.

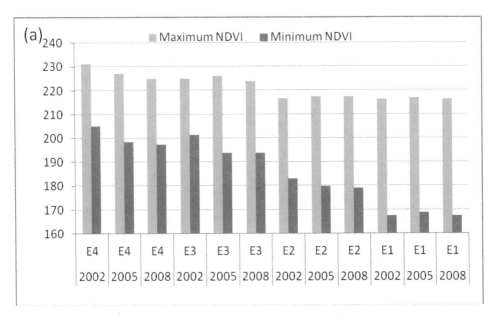

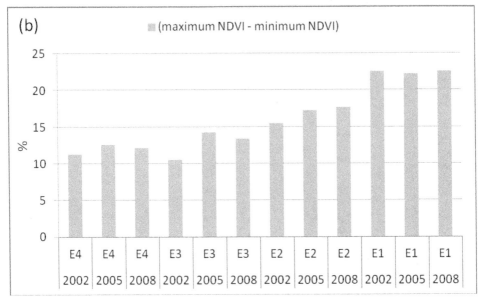

E1: herbaceous; E2: shrubs; E3: deciduous trees; and E4: evergreen trees

Fig. 5. Metrics of vegetation phenology derived from NDVI data used in Fig. 4. Maximum and minimum NDVI values indicate periods of higher and lower plant productivity (left) respectively, and the difference of both, as a percentage, indicates the degree of seasonality of each vegetational type in the three years analyzed.

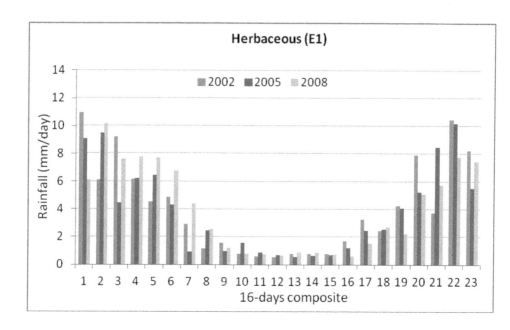

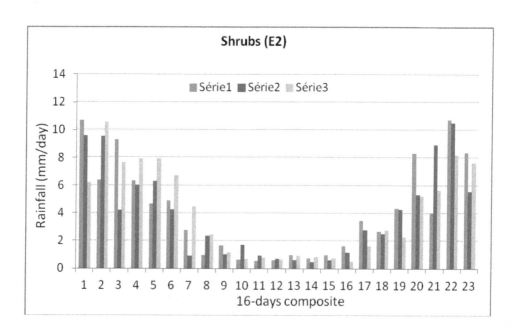

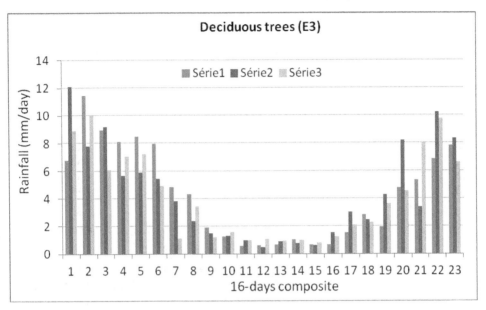

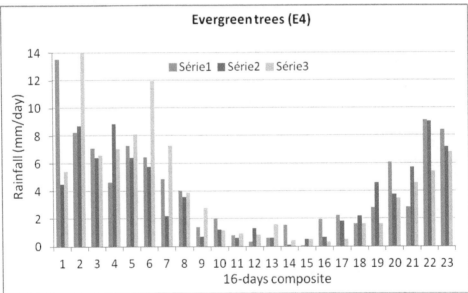

E1: herbaceous; E2: shrubs; E3: deciduous trees; and E4: evergreen trees.

Fig. 6. Annual seasonality of precipitation for the years 2002, 2005 and 2008, in places where we sampled the four vegetation types analyzed. Each year consists of 23 16-days composite periods. Precipitation in mm is the daily mean rainfall values for a 16-days composite period.

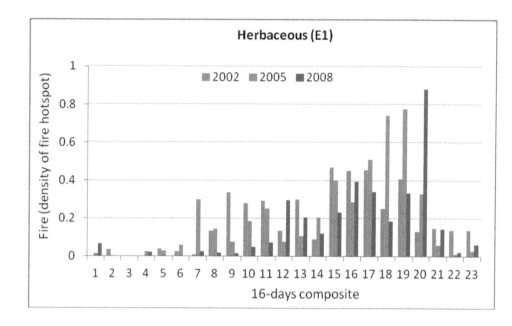

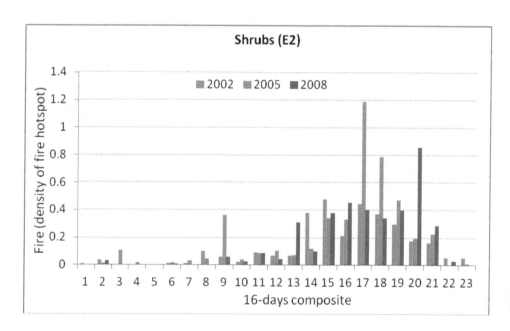

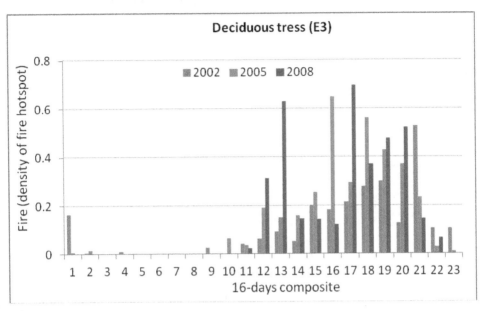

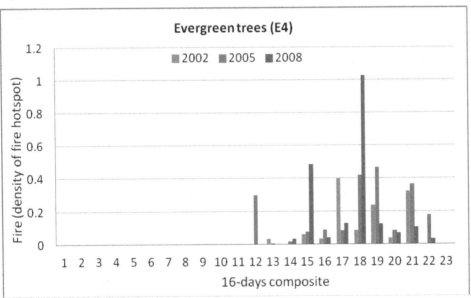

E1: herbaceous; E2: shrubs; E3: deciduous trees; and E4: evergreen trees.

Fig. 7. Annual seasonality of fire for the years 2002, 2005 and 2008, in places where we sampled the four vegetation types analyzed. Each year consists of 23 16-days composite periods. Fire is the daily mean value of the density of hotspot within a 10km radius for a 16-days composite period.

### 3.2.4 Relationship between vegetation (NDVI) and environmental variables (precipitation and fire)

The results showed in Table 1 indicate significant positive correlation between NDVI and precipitation in herbaceous, shrubs and deciduous trees, and negative correlation between NDVI and fire in the same three vegetation types. In the case of evergreen trees, the correlation between NDVI and precipitation is positive but not significant, and between NDVI and fire is negative, but also not significant. These results are corroborated in subsequent analysis.

|          | E1 Prec | E2 Prec | E3 Prec | E4 Prec | E1 Fire | E2 Fire | E3 Fire | E4 Fire |
|----------|---------|---------|---------|---------|---------|---------|---------|---------|
| E1-NDVI  | **0.60** | 0.58 | 0.67 | 0.70 | **-0.69** | -0.68 | -0.66 | -0.41 |
| E2-NDVI  | 0.52 | **0.51** | 0.60 | 0.65 | -0.68 | **-0.72** | -0.70 | -0.43 |
| E3-NDVI  | 0.20 | 0.19 | **0.31** | 0.36 | -0.61 | -0.73 | **-0.75** | -0.51 |
| E4-NDVI  | 0.00 | -0.01 | 0.16 | **0.09** | -0.43 | -0.57 | -0.66 | **-0.47** |
| E1-Fire  | **-0.40** | -0.39 | -0.43 | -0.49 | 1.00 | 0.77 | 0.65 | 0.35 |
| E2-Fire  | -0.31 | **-0.30** | -0.34 | -0.42 | 0.77 | 1.00 | 0.65 | 0.41 |
| E3-Fire  | -0.25 | -0.24 | **-0.31** | -0.38 | 0.65 | 0.65 | 1.00 | 0.47 |
| E4-Fire  | -0.09 | -0.08 | -0.19 | **-0.20** | 0.35 | 0.41 | 0.47 | 1.00 |

Table 1. Correlation matrix of vegetation, rainfall and fire variables, highlighting the significant correlations between the following couple of variables: NDVI and rainfall, NDVI and fire, and rainfall and fire; which taking into account the four types of vegetation analyzed (E1: herbaceous; E2: shrubs; E3: deciduous trees; and E4: evergreen trees).

Figure 8 shows the result of the linear regression analysis between vegetation and precipitation for each vegetation type. Each line in this figure with a specific color shows the degree of fit between the points distributed for both variables by type of vegetation. Although this degree of fit between both variables is low, the results indicate that there is a gradient of fit between precipitation and vegetation, here named as precipitation gradient, which ranges from high to low coefficient of correlation (R2) following the sequence: herbaceous-E1 (high R2), shrubs-E2, deciduous trees-E3 and evergreen trees-E4 (low R2).

Thus, as the R2 value increases the influence of precipitation on vegetation increases, so herbaceous is more dependent on rainfall, in the annual cycle, than the other types of vegetation analyzed. That means, herbaceous are strongly dependent on rainfall in order to increase its vegetation cover. In the dry season, these kinds of species lose their leaves or even die.

At the opposite end of the precipitation gradient, where the evergreen trees-E4 are positioned, precipitation has weak influence on the vegetation cover, which means that in

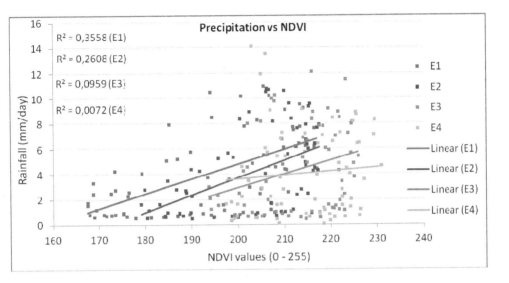

E1: herbaceous; E2: shrubs; E3: deciduous trees; and E4: evergreen trees.

Fig. 8. Regression of precipitation (independent variable) and NDVI (dependent) for each vegetation type analyzed. NDVI values range from 0 to 255 (x-axis). Precipitation in mm is the daily mean rainfall values for a 16-days composite period (y-axis). N = 69.

the dry season, evergreen trees are able to capture water from the vicinity of river courses, as occurs in gallery forests, or from deep soil, where the length of tree roots reach deep and moist soil layers, allowing these trees to replace their leaves throughout the year, which gives them their evergreen nature.

An analyses of variance (ANOVA) performed to evaluate these regressions is shown in table 2. Results indicate that, except for the regression between NDVI and Precipitation for the evergreen trees (E4) class, all regressions are significant at the 0.99 confidence level. Moreover, the relationships between NDVI and Fire were significant for all classes.

| | NDVI x Prec | | | NDVI x Fire | | |
|---|---|---|---|---|---|---|
| | $R^2$ | F | p | $R^2$ | F | p |
| E1 | 0.3558 | 37.01 | <0.01 | 0.4795 | 61.72 | <0.01 |
| E2 | 0.2608 | 23.64 | <0.01 | 0.5172 | 71.77 | <0.01 |
| E3 | 0.0959 | 7.11 | <0.01 | 0.5575 | 84.40 | <0.01 |
| E4 | 0.0072 | **0.49** | **0.49** | 0.2218 | 19.10 | <0.01 |

Table 2. Analysis of Variance (ANOVA) of linear regression NDVI x Precipitation (Prec) and NDVI x Fire. Bold values indicate the case where regression was not significant.

Figure 9 shows the result of the linear regression analysis between vegetation and fire for each vegetation type. Each line in this figure with a specific color shows the degree of fit between the points distributed for both variables by type of vegetation. The results indicate that there is a gradient of fit between fire and vegetation, here named as fire gradient, which ranges from high to low coefficient of correlation (R2) following the sequence: deciduous trees-E3 (high R2), shrubs-E2, herbaceous-E1, and evergreen trees-E4 (low R2).

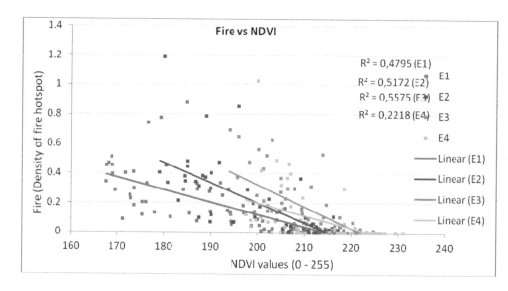

E1: herbaceous; E2: shrubs; E3: deciduous trees; and E4: evergreen trees.

Fig. 9. Regression of fire (independent variable) and NDVI (dependent) for each vegetation type analyzed. NDVI values range from 0 to 255 (x-axis). Fire is the daily mean value of the density of hotspot within a 10km radius for a 16-days composite period. N = 69.

The fire gradient identified above indicates that there is direct relationship between NDVI of the main vegetation types (herbaceous, shrubs and deciduous trees), which make up the Cerrado vegetation, and fire, indicating the role of fire in the maintenance of these vegetation types.

Fire occurs with greater intensity at the end of dry season. First of all, fire consumes part of the burk and organic matter of the plant, after the first rains, in the beginning of the rainy season, these partially burned plant sprouts new shoots with greater vigor.

At the opposite end of the fire gradient, where the evergreen trees-E4 are positioned, the fire occurs in lower proportion in these trees, however, unlike what happens with other types of vegetation, the effect of fire is pernicious, it can damage or even eliminate some species in this vegetation type according to the intensity level.

The multiple regression analysis indicates that there is a direct relationship between precipitation and fire, and vegetation index (NDVI) in the four vegetation types of the savanna vegetation. The multiple coefficients of determinations ($R^2$) show that the environmental variables as a whole (precipitation an fire) follow a gradient of high influence in vegetation types with low vegetation cover (herbaceous $R^2=0.67$ and shrubs $R^2=0.65$) to low influence in that with high vegetation cover (deciduous trees $R^2= 0.55$ and evergreen trees $R^2=0.27$). Results from the ANOVA of the multiple regression presented in Table 3 indicate that, when the analysis is performed considering both independent variables, the multiple regression gives statistically significant parameters, for all classes of vegetation. However, an univariate test of significance performed for each independent variable show that precipitation alone is not significant correlated to the vegetation index for both tree classes (E3 and E4).

|    | Whole model R | | | Univariate test of significance | | | |
|----|------|------|------|--------|--------|--------|--------|
|    | $R^2$ | F | p | F_prec | F_fire | p_prec | p_fire |
| E1 | 0.6001 | 49.52 | <0.01 | 19.90 | 40.31 | <0.01 | <0.01 |
| E2 | 0.6141 | 52.52 | <0.01 | 16.59 | 60.43 | <0.01 | <0.01 |
| E3 | 0.5648 | 42.83 | <0.01 | **1.12** | 71.11 | **0.29** | <0.01 |
| E4 | 0.2220 | 9.42 | <0.01 | **0.01** | 18.22 | **0.92** | <0.01 |

Table 3. Analysis of Variance (ANOVA) of the multiple regression between NDVI (dependent variable) and precipitation and fire (independent variables). Bold values indicate the cases where regression was not significant.

## 4. Conclusions

The response of vegetation NDVI is more related to the variation of fire than to variations in precipitation in Cerrado region. Vegetation NDVI responds to variation of precipitation with a time lag ranging from 16 to 48 days, while vegetation NDVI responds to variation of fire with a time lag ranging from 0 to 48 days.

The relationship between vegetation types, derived from NDVI, and precipitation, derived from TRMM, shows a gradient of positive correlations in vegetation types with low vegetation cover, herbaceous (r= 0.60) and shrubs (r= 0.51), to very little or none with high vegetation cover, deciduous trees (r= 0.31) and evergreen trees (r= 0.09). On the other hand, the relationship between vegetation and fire hotspot shows a gradient of negative correlation, which is stronger in herbaceous (r= 0.72), shrubs (r= 0.74) and deciduous trees (r= -0.73) than in evergreen trees (r= -0.52).

Our analyses show that vegetation cover increases are related to increases in precipitation and decreased in density of fire hotspots. We also found high density of fire hotspot in the dry season in deciduous trees, shrubs and herbaceous which suggesting the high removal of $CO_2$ (greenhouse gas) of the land cover to the atmosphere somehow influencing the dynamic equilibrium of this (atmosphere) in the region of the Brazilian tropical savanna.

# 5. References

Allan, G.; Johnson, A.; Cridland, S. & Fitzgerald, N. (2003). Application of NDVI for predicting fuel curing at landscape scales in northern Australia: can remotely sensed data help schedule fire management operations? *International Journal of Wildland Fire* 12, 299–308.

Aquino, F.G.; Miranda, G.H.B. (2008). Consequencias ambientais da fragmentação de habitats no cerrado. *In* Sano, S.M.; Almeida, S.P.; Ribeiro, J.F. (Eds.). *Cerrado: ecologia e flora*. Brasília: Editora Embrapa, p.385-398.

Barnes, S. L. (1973). Mesoscale objective analysis using weighted time-series observations, NOAA Tech. Memo. ERL NSSL-62 National Severe Storms Laboratory, Norman, OK 73069, 60 pp. [NTIS COM-73-10781.], 1973. 2359

Bartlett, H.H. (1955, 1957, 1961). Fire in relation to primitive agriculture and grazing in the tropics: annotated bibliography, Vol. 1-3. Mimeo. Publ. Univ. Michigan Bot. Gardens, Ann Arbor, USA.

Borlaug, N.E. (2002). Feeding a world of 10 billion people: the miracle ahead. In: R. Bailey (ed.). *Global warming and other eco-myths*. pp. 29-60. Competitive Enterprise Institute, Roseville, EUA.

Bowman, D.M.J.S. & Murphy, B.P. (2010). Fire and Biodiversity. In Sodhi, N.S.; Ehrlich, PR., eds 2010. *Conservation Biology for all* , pp. 163-180.

Bowman, D.M.J.S., Zhang, Y., Walsh, A. & Williams, R.J. (2003). Experimental comparison of four remote sensing techniques to map tropical savanna fire-scars using Landsat-TM imagery. *International Journal of Wildland Fire* 12, 341–48.

Brandis, K. & Jacobson, C. (2003). Estimation of vegetative fuel loads using LandsatTM imagery in New South Wales, Australia. *International Journal of Wildland Fire,* Vol. 12, 185-94.

Castro, L.H.R.; Moreira, A.M. & Assad, E.D. (1994). Definição e regionalização dos padrões pluviométricos dos Cerrados brasileiros. In: ASSAD, E.D. *Chuvas nos Cerrados: análise e espacialização*. Brasília, Embrapa-CPAC/Embrapa-SPI, 1994. 423p.

Coutinho, L.M. (1978). O conceito de cerrado. *Revista Brasileira de Botânica* 1(1):17-23.

Coutinho, L. M. (2000). O bioma do cerrado, in: Klein, A. L. Eugen Warming e o cerrado brasileiro: um século depois. UNESP, 156 p.

Diaz-Delgado, R.; Lloret, F. & Pons, X. (2004). Statistical analysis of fire frequency models for Catalonia (NE Spain, 1975–1998) based on fire scar maps from Landsat MSS data. *International Journal of Wildland Fire* 13, 89–99.

Goldammer, J.G. (1988). Rural land use and fires in the tropics. *Agroforestry Systems*, Vol.6: 235-253

Justice, C.O.; Smith, R.; Gill, A.M. & Csiszar, I. (2003). A review of current space-based fire monitoring in Australia and the GOFC/GOLD program for international coordination. *International Journal of Wildland Fire*, 12, 247–258.

Klink, C.A. & Machado, R.B. (2005). A conservação do cerrado brasileiro. *Megadiversidade*, Vol.1: 147-155.

Koch S. E.; desJardins, M. & Kocin, P. J. (1983). An interactive Barnes objective map analysis scheme for use with satellite and conventional data. *J. Climate Appl. Meteor.*, Vol.22, 1487–1503.

Law, R. M.; Kowlaczyk, E. A. & Wang, Y.P. (2006). Using atmospheric CO2 data to assess a simplified carbon-climate simulation for the 20th century. *Tellus*, Vol.58B, 427-437.

Miller, J.D.; Danzer, S.R.; Watts, J.M.; Stone, S. & Yool, S.R. (2003). Cluster analysis of structural stage classes to map wildland fuels in a Madrean ecosystem. *Journal of Environmental Management*, Vol. 68, 239–52.

Myers, N. et al. 2000. Biodiversity hotspots for conservation priorities. *Nature* 403: 853-858.

Nimer, E. (1989). *Climatologia do Brasil*. Secretaria de Planejamento e Coordenacao da Presidencia da Republica e IBGE, Rio de Janeiro, 421p.

Nimer, E. 1977. Clima. Pp 47-48 in: IBGE. *Geografia do Brasil: região Nordeste*. Rio de Janeiro.

Parker, A. (2000). Wet-Dry Tropical Climate (Aw), In: Climate Savanna, 01.06.2000, Available from
http://www.blueplanetbiomes.org/savanna_climate_page.htm

PROBIO-MMA (Projeto de Conservação e Utilização Sustentável da Diversidade Biológica Brasileira) do Ministério do Meio Ambiente. (2007). *Mapeamento de Cobertura Vegetal do Bioma Cerrado*. Technical Report. Brasilia, Brasil.

Rao, V.B. & Hada, K. (1990). Characteristics of rainfall over Brazil: annual variations and connections with the southern oscillations. *Theoretical and Applied Climatology*, Vol.2, p.81-91.

Ratter, J.A.; Ribeiro, J.F. & Bridgewater, S. (1997). The brazilian Cerrado vegetation and threats to its biodiversity. *Annals of Botany*, Vol.*80*: 223-230.

Reed, B.C.; Brown, J.F.; Vanderzee, D.; Loveland, T.R.; Merchant, J.W. & Ohlen, D.O. (1994). Measuring phenological variability from satellite imagery. *J. Veg. Sci.* Vol.5, 703-714.

Ritter, M.E. (2006). *The Physical Environment: an Introduction to Physical Geography*. Available from
http://www.uwsp.edu/geo/faculty/ritter/geog101/textbook/ title_page. html

Rollins, M.G.; Keane, R.E. & Parsons, R.A. (2004). Mapping fuels and fire regimes using remote sensing, ecosystem simulation, and gradient modeling. *Ecological Applications*, Vol.14, 75–95.

Roy, P.S. (2004). Forest fire and degradation assessment using satellite remote sensing and geographic information system. In: Sivakumar, M.V.K.; Roy, P.S.; Harmesen, K.; Saha, S.K., eds, 2010. *Satellite Remote Sensing and GIS Applications in Agricultural Meteorology* pp. 361-400.

Satyamurty, P.; Nobre, C. A. & Silva Dias, P. L. (1998). South America. *Meteorological Monographs*, Vol.27, n. 49 (Southern Hemisphere Meteorology, cap. 3C), p. 119-139.

Silverman, B.W. (1986). Density estimation for statistics and data analysis. London; New York: Chapman & Hall, 1986. 175 p.

Schwartz, M.D. (1994). Monitoring global change with phenology: the case of the spring green wave. *Int. J. Biometeorol.* Vol.38, 18-22.

Vila, D.A.; de Goncalves, L.G.G.; Toll, D.L. & Rozante, J.R. (2009). Statistical Evaluation of Combined Daily Gauge Observations and Rainfall Satellite Estimates over Continental South America. *J. Hydrometeor.* Vol.10, 533–543.

# 3

# Crop Disease and Pest Monitoring by Remote Sensing

Wenjiang Huang, Juhua Luo, Jingcheng Zhang,
Jinling Zhao, Chunjiang Zhao, Jihua Wang,
Guijun Yang, Muyi Huang, Linsheng Huang and Shizhou Du
*Beijing Research Center for Information Technology in Agriculture, Beijing*
*China*

## 1. Introduction

Plant diseases and pests can affect a wide range of commercial crops, and result in a significant yield loss. It is reported that at least 10% of global food production is lost due to plant diseases (Christou and Twyman, 2004; Strange and Scott, 2005). Excessive pesticides are used for protecting crops from diseases and pests. This not only increases the cost of production, but also raises the danger of toxic residue in agricultural products. Disease and pest control could be more efficient if disease and pest patches within fields can be identified timely and treated locally. This requires obtaining the information of disease infected boundaries in the field as early and accurately as possible. The most common and conventional method is manual field survey. The traditional ground-based survey method requires high labor cost and produces low efficiency. Thus, it is unfeasible for large area. Fortunately, remote sensing technology can provide spatial distribution information of diseases and pests over a large area with relatively low cost. The presence of diseases or insect feedings on plants or canopy surface causes changes in pigment, chemical concentrations, cell structure, nutrient, water uptake, and gas exchange. These changes result in differences in color and temperature of the canopy, and affect canopy reflectance characteristics, which can be detectable by remote sensing (Raikes and Burpee 1998). Therefore, remote sensing provides a harmless, rapid, and cost-effective means of identifying and quantifying crop stress from differences in the spectral characteristics of canopy surfaces affected by biotic and abiotic stress agents.

This chapter introduces some successful studies about detecting and discriminating yellow rust and aphid (economically important disease and pest in winter wheat in China) using field, airborne and satellite remote sensing.

## 2. Detecting yellow rust of winter wheat by remote sensing

Yellow rust *(Biotroph Puccinia striiformis)*, also known as stripe rust, is a fungal disease of winter wheat *(Triticum aestivum L.)*. It produces leaf lesions (pustules), which are yellow in color and tend to be grouped in patches. Yellow rust often occurs in narrow stripes, 2–3 mm wide that run parallel to the leaf veins. Yellow rust is responsible for approximately 73–85%

of recorded yield losses, and grain quality is also significantly reduced (Li et al. 1989). Consequently, effective monitoring of the incidence and severity of yellow rust in susceptible regions is of great importance to guide the spray of pesticides and to provide data for the local agricultural insurance services. Fortunately, remote sensing technology provides a possible way to detect the incidence and severity of the disease rapidly.

The interaction of electromagnetic radiation with plants varies with the wavelength of the radiation. The same plant leaves may exhibit significant different reflectance depending on the level of health and or vigor (Wooley 1971, West et al. 2003, Luo et al., 2010). Healthy and vigorously growing plant leaves will generally have

1. Low reflectance at visible wavelengths owing to strong absorption by photoactive pigments (chlorophylls, anthocyanins, carotenoids).
2. High reflectance in the near infrared because of multiple scattering at the air-cell interfaces in the leaf's internal tissue.
3. Low reflectance in wide wavebands in the short-wave infrared because of absorption by water, proteins, and other carbon constituents.

The incidence and severity of yellow rust can be monitored according to the differences of spectral characteristics between healthy and disease plants. In this chapter, we will report several successful studies on the detection and identification of yellow rust in winter wheat by remote sensing.

## 2.1 Detecting and discriminating yellow rust at canopy level

Hyperspectral remote sensing is one of the advanced and effective techniques in disease monitoring and mapping. However, the difficulty in discriminating a disease from common nutrient stresses largely hampers the practical use of this technique. This is because some common nutrient stresses such as the shortage or overuse of nitrogen or water could have similar variations of biochemical properties and plant morphology, and therefore result in similar spectral responses. However, for the remedial procedures for stressed crops, there is a significant difference between disease and nutrient stresses. For example, applying fungicide to water-stressed crops would lead to a disastrous outcome. Therefore, to discriminate yellow rust from common nutrient stresses is of practical importance to crop growers or landowners.

The specific objectives of this study are to: (1) systematically test the sensitivity and consistency of several commonly used spectral features to yellow rust disease during major growth stages; (2) for those spectral features that are consistently sensitive to yellow rust disease, we will further examine their sensitivity to nutrient stresses to determine whether there are specifically sensitive to yellow rust disease, but insensitive to water and nitrogen stresses.

### 2.1.1 Materials and methods

### 2.1.1.1 Experimental design and field conditions

The experiments were conducted at Beijing Xiaotangshan Precision Agriculture Experimental Base, in Changping district, Beijing (40°10.6'N, 116°26.3'E) for the growing seasons of 2001-2002 and 2002-2003. Table 1 summarizes the soil properties including organic matter, total nitrogen, alkali-hydrolysis nitrogen, available phosphorus and available potassium for both growing seasons. Three cultivars of winter wheat used in 2001-2002 experiment (2002 Exp) were Jingdong8, Jing9428 and Zhongyou9507, while the cultivars used in 2002-2003 (2003 Exp) were Xuezao, 98-100 and Jing411. All the cultivars applied in both growing seasons included erective, middle and loose with respect to the canopy morphology.

| Items | | Disease inoculation experiment | Nutrient stress experiment |
|---|---|---|---|
| Growth period | | Sep 2002-Jun 2003 | Sep 2001-Jun 2002 |
| Top soil nutrient status (0-0.3m depth) | Organic matter | 1.42%-1.48% | 1.21%-1.32% |
| | Total nitrogen | 0.08%-0.10% | 0.092%-0.124% |
| | Alkali-hydrolysis nitrogen | 58.6-68.0 mg kg$^{-1}$ | 68.8-74.0 mg kg$^{-1}$ |
| | Available phosphorus | 20.1-55.4 mg kg$^{-1}$ | 25.2-48.3 mg kg$^{-1}$ |
| | Rapidly available potassium | 117.6-129.1 mg kg$^{-1}$ | 96.6-128.8 mg kg$^{-1}$ |
| Cultivars | | Xuezao, 98-100, Jing411 | Jingdong8, Jing9428, Zhongyou9507 |
| Treatments | | Normal; YR1: 3mg 100$^{-1}$ ml spores solution; YR2: 9mg 100$^{-1}$ ml spores solution; YR3: 12mg 100$^{-1}$ ml spores solution (all treatments applied 200 kg ha$^{-1}$ nitrogen and 450 m$^3$ ha$^{-1}$ water) | Normal: 200 kg ha$^{-1}$ nitrogen, 450 m$^3$ ha$^{-1}$ water; W-SD: 200 kg ha$^{-1}$ nitrogen, 225 m$^3$ ha$^{-1}$ water; W-SED: 200 kg ha$^{-1}$ nitrogen, 0 m$^3$ ha$^{-1}$ water; N-E: 350 kg ha$^{-1}$ nitrogen, 450 m$^3$ ha$^{-1}$ water; N-D: 0 kg ha$^{-1}$ nitrogen, 450 m$^3$ ha$^{-1}$ water; W-SED+N-E: 350 kg ha$^{-1}$ nitrogen, 0 m$^3$ ha$^{-1}$ water; W-SED+N-D: 0 kg ha$^{-1}$ nitrogen, 0 m$^3$ ha$^{-1}$ water; |
| Spectral reflectance measurements (on day after sowing) | | 207, 216, 225, 230, 233 | 196, 214, 225, 232, 239 |

Table 1. Basic information of disease inoculation experiment and nutrient stress experiment

For 2002 Exp, six stress treatments of water and nitrogen were applied, and the treatments were based on local conditions, which usually suffered from yellow rust in the northern part of China. Each treatment was applied on 0.3 ha area, and the treatments were 200 kg ha$^{-1}$ nitrogen and 225 m$^3$ ha$^{-1}$ water (slightly deficient water, W-SD),200 kg ha$^{-1}$ nitrogen and no irrigation (seriously deficient water, W-SED), 350 kg ha$^{-1}$ nitrogen and 450 m$^3$ ha$^{-1}$ water (excessive nitrogen, N-E), no fertilization and 450 m3 ha-1 water (deficient nitrogen, N-D), 350 kg ha$^{-1}$ nitrogen and no irrigation (seriously deficient water and excessive nitrogen, W-SED+N-E), and no fertilization and no irrigation (seriously deficient water and deficient nitrogen, W-SED+N-D). A 0.3 ha reference area (Normal) was applied with the recommended rate which received 200 kg ha$^{-1}$ nitrogen and 450 m$^3$ ha$^{-1}$ water. Three cultivars were evenly distributed in each treatment plot.

For 2003 Exp, according to the National Plant Protection Standard (Li et al. 1989), three levels of concentration of summer spores of yellow rust were applied, and they were 3 mg 100$^{-1}$ ml$^{-1}$ (Yellow rust 1, YR1), 9 mg 100$^{-1}$ ml$^{-1}$ (Yellow rust 2, YR2) and 12 mg 100$^{-1}$ ml$^{-1}$ (Yellow rust 3, YR3), with a dosage of 5 ml spores solution per square meter. The reference area (Normal) that was not inoculated yet was applied with the recommended amount of fungicide to prevent the occasional infection. Each treatment involved 1.2 ha area, with even constitution of three cultivars. All plots in 2003 Exp received the recommended rates of nitrogen (200 kg ha$^{-1}$) and water (450 m$^3$ ha$^{-1}$).

### 2.1.1.2 Canopy spectral measurements

A high spectral resolution spectrometer, ASD FieldSpec Pro spectrometer (Analytical Spectral Devices, Boulder, CO, USA) fitted with a 25 field of view fore-optic, was used for in-situ measurement of canopy spectral reflectance for both 2002 Exp and 2003 Exp. All canopy spectral measurements were taken from a height of 1.3m above ground (the height of the wheat is 90±3 cm at maturity). Spectra were acquired in the 350-2,500 nm spectral range at a spectral resolution of 3 nm between 350 nm and 1,050 nm, and 10 nm between 1,050 nm and 2,500 nm. A 40 cm × 40 cm BaSO4 calibration panel was used for calculation of reflectance. All irradiance measurements were recorded as an average of 20 scans at an optimized integration time. Prior to subsequent preprocessing, all spectral curves were resampled with 1 nm interval. All measurements were made under clear blue sky conditions between 10:00 and 14:00 (Beijing Local Time).

The spectral measurements were taken 5 times from 196 days after sowing (DAS) to 239 DAS for 2002 Exp, which covered the growth stages of stem elongation, booting, anthesis and milk development. For 2003 Exp, the spectral measurements were taken 5 times from 207 DAS to 233 DAS, which covered the growth stages of booting, anthesis and milk development. The detailed measurement dates for both experiments were given in Table 1. The stem elongation and anthesis stages are essential for the control of yellow rust development, whereas the milk development stage is important for yield loss assessment.

### 2.1.1.3 Selection of spectral features

The spectral features that we adopted were related to several commonly used vegetation indices (VIs), which were proved to be sensitive to variations of pigments and stresses. Furthermore, in order to conduct a thorough investigation of various types of spectral features, we also included a number of spectral features that were based on derivative transformation and continuum removal transformation (Gong et al. 2002; Pu et al. 2003;2004). Therefore, the total 38 spectral features are shown in Table 2.

| Variable | Definition | Description | Literatures |
|---|---|---|---|
| **Derivative transformed spectral variables** | | | |
| $D_b$ | Maximum value of 1st derivative within blue edge | Blue edge covers 490-530nm. $D_b$ is a maximum value of 1st order derivatives within the blue edge of 35 bands | Gong et al., 2002 |
| $\lambda_b$ | Wavelength at $D_b$ | $\lambda_b$ is wavelength position at $D_b$ | Gong et al., 2002 |
| $SD_b$ | Sum of 1st derivative values within blue edge | Defined by sum of 1st order derivative values of 35 bands within the blue edge | Gong et al., 2002 |
| $D_y$ | Maximum value of 1st derivative within yellow edge | Yellow edge covers 550-582nm. $D_y$ is a maximum value of 1st order derivatives within the yellow edge of 28 bands | Gong et al., 2002 |
| $\lambda_y$ | Wavelength at $D_y$ | $\lambda_y$ is wavelength position at $D_y$ | Gong et al., 2002 |
| $SD_y$ | Sum of 1st derivative values within yellow edge | Defined by sum of 1st order derivative values of 28 bands within the yellow edge | Gong et al., 2002 |
| $D_r$ | Maximum value of 1st derivative within red edge | Red edge covers 670-737nm. $D_r$ is a maximum value of 1st order derivatives within the red edge of 61 bands | Gong et al., 2002 |
| $\lambda_r$ | Wavelength at $D_r$ | $\lambda r$ is wavelength position at $D_r$ | Gong et al., 2002 |
| $SD_r$ | Sum of 1st derivative values within red edge | Defined by sum of 1st order derivative values of 61 bands within the red edge | Gong et al., 2002 |
| **Continuous removal transformed spectral features** | | | |
| DEP550-750 | The depth of the feature minimum relative to the hull | In the range of 550nm-750nm | Pu et al., 2003;2004 |
| DEP920-1120 | | In the range of 920nm-1120nm | |
| DEP1070-1320 | | In the range of 1070nm-1320nm | |
| WID550-750 | The full wavelength width at half DEP (nm) | In the range of 550nm-750nm | Pu et al., 2003;2004 |
| WID920-1120 | | In the range of 920nm-1120nm | |
| WID1070-1320 | | In the range of 1070nm-1320nm | |
| AREA550-750 | The area of the absorption feature that is the product of DEP and WID | In the range of 550nm-750nm | Pu et al., 2003;2004 |
| AREA920-1120 | | In the range of 920nm-1120nm | |
| AREA1070-1320 | | In the range of 1070nm-1320nm | |

| Variable | Definition | Description | Literatures |
|---|---|---|---|
| **VI-based variables** | | | |
| GI | Greenness Index | $R_{554}/R_{677}$ | Zarco-Tejada et al., 2005 |
| MSR | Modified Simple Ratio | $(R_{800}/R_{670}-1)/(R_{800}/R_{670}+1)^{1/2}$ | Chen, 1996; Haboudane et al., 2004 |
| NDVI | Normalized Difference Vegetation Index | $(R_{NIR}-R_R)/(R_{NIR}+R_R)$, where $R_{NIR}$ indicates 775-825nm, $R_R$ indicates 650nm-700nm, that include most key pigments | Rouse et al., 1973 |
| NBNDVI | Narrow-band normalised difference vegetation index | $(R_{850}-R_{680})/(R_{850}+R_{680})$ | Thenkabail et al., 2000 |
| NRI | Nitrogen reflectance index | $(R_{570}-R_{670})/(R_{570}+R_{670})$ | Filella et al., 1995 |
| PRI | Photochemical Physiological Reflectance Index | $(R_{531}-R_{570})/(R_{531}+R_{570})$ | Gamon et al., 1992 |
| TCARI | The transformed chlorophyll Absorption and Reflectance Index | $3*[(R_{700}-R_{670})-0.2*(R_{700}-R_{550})*(R_{700}/R_{670})]$ | Haboudane et al., 2002 |
| SIPI | Structural Independent Pigment Index | $(R_{800}-R_{445})/(R_{800}-R_{680})$ | Peñuelas et al., 1995 |
| PSRI | Plant Senescence Reflectance Index | $(R_{680}-R_{500})/R_{750}$ | Merzlyak et al., 1999 |
| PhRI | The Physiological reflectance index | $(R_{550}-R_{531})/(R_{550}+R_{531})$ | Gamon et al., 1992 |
| NPCI | Normalized Pigment Chlorophyll ratio Index | $(R_{680}-R_{430})/(R_{680}+R_{430})$ | Peñuelas et al., 1994 |
| ARI | Anthocyanin Reflectance Index | $ARI=(R_{550})^{-1}-(R_{700})^{-1}$ | Gitelson et al., 2001 |
| TVI | Triangular Vegetation Index | $0.5[120(R_{750}-R_{550})-200(R_{670}-R_{550})]$ | Broge and Leblanc, 2000; Haboudane et al., 2004 |
| CARI | Chlorophyll Absorption Ratio Index | $(|(a670+R_{670}+b)|/(a^2+1)^{1/2})\times(R_{700}/R_{670})$ $a=(R_{700}-R_{550})/150, b=R_{550}-(a \times 550)$ | Kim et al., 1994 |

| Variable | Definition | Description | Literatures |
|---|---|---|---|
| DSWI | Disease Water Stress Index | $(R_{802}+R_{547})/(R_{1657}+R_{682})$ | Galvão et al., 2005 |
| MSI | Moisture Stress Index | $R_{1600}/R_{819}$ | Hunt and rock, 1989; Ceccato et al., 2001 |
| SIWSI | Shortwave Infrared Water Stress Index | $(R_{860}-R_{1640})/(R_{860}+R_{1640})$ | Fensholt and Sandholt, 2003 |
| RVSI | Red-Edge Vegetation Stress Index | $[(R_{712}+R_{752})/2]-R_{732}$ | Merton and Huntington, 1999 |
| MCARI | Modified Chlorophyll Absorption in Reflectance Index | $(R_{701}-R_{671})-0.2(R_{701}-R_{549})]/(R_{701}/R_{671})$ | Daughry et al., 2000 |
| WI | Water Index | $R_{900}/R_{970}$ | Peñuelas et al., 1997 |

Table 2. Definitions of spectral features used in this study

### 2.1.1.4 Preprocessing and normalization of spectral reflectance data

### Aggregating spectral reflectance data

**As the first step,** all spectra were processed with the following transformation to suppress possible difference in illumination. The spectral regions with wavelength of 1330-1450 nm, 1770-2000 nm and 2400-2500 nm were removed due to strong absorption by water vapor. We then normalized the spectral curves by dividing the mean band reflectance of the curve (Yu et al., 1999). The normalized reflectance for the band$_i$ is given as:

$$Ref'_i = \frac{Ref_i}{\frac{1}{n}(\sum_{i=1}^{n} Ref_i)}$$

where $Ref'_i$ is the normalized reflectance for band$_i$; $Ref_i$ is the original reflectance of the band; n is the total number of bands. Fig. 1(a) shows a plot of unnormalized $Ref_i$ versus band wavelength for six observations (three YR3 curves and three Normal curves) on 233 DAS. Fig. 1(b) shows the corresponding curves in Fig.1(a) after normalization. The normalization clearly separated the diseased spectra from the normal spectra especially over the near infrared region (approximately from 770 nm to 1300 nm). The benefit of eliminating spectral difference caused by the change of illumination conditions was also mentioned by Yu et al. (1999).

### Normalization of the difference in measuring dates

As shown in Table 1, although both experiments conducted in five growth stages in 2002 and 2003, most measurement dates were not consistent, except for 255 DAS. Hence, to improve the comparability of two datasets, we adapted the 2002 Exp data to match the dates

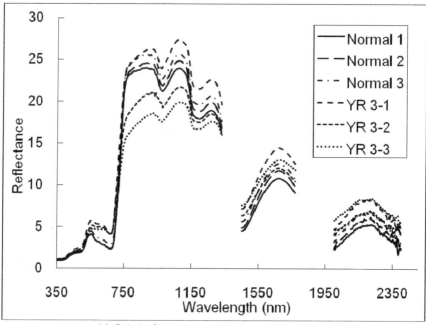

(a) Original spectra on 233 days after sowing

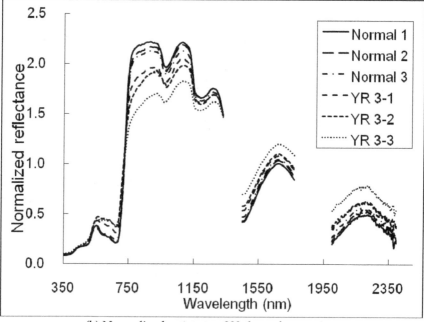

(b) Normalized spectra on 233 days after sowing

Fig. 1. Comparison between original spectra and normalized ones

of 2003 Exp, by using a linear interpolation method. The reflectance curve of a certain date could be obtained based on the spectra from the adjacent data before and after the measurement date (using days after sowing as a time scale). Each band of the spectra should be processed as:

$$Ref_{current} = Ref_{before} - \frac{DAS_{current} - DAS_{before}}{DAS_{after} - DAS_{before}}(Ref_{before} - Ref_{after})$$

where $Ref_{current}$ represents the reflectance transformed from the date corresponding to an ideal date in 2003 Exp; $Ref_{before}$ and $Ref_{after}$ represent reflectances, respectively, from $DAS_{before}$ and $DAS_{after}$; $DAS_{current}$ indicates an ideal date in 2003 Exp while $DAS_{before}$ and $DAS_{after}$ are the adjacent dates in 2002 Exp before and after the ideal date in 2003 Exp.

Fig. 2 provides an example of the progress of the normalization of measurement dates. The averaged reflectance at central wavelengths of green band (560 nm) and near-infrared band (860 nm) of Landsat-5 TM for normal samples were plotted against the measured dates in both 2002 Exp and 2003 Exp. The date normalized reflectance values were marked as triangle symbol in the graph. Through this step, the datasets collected in these two years could be considered as acquired in the same dates, which thereby facilitated the subsequent comparisons and analysis.

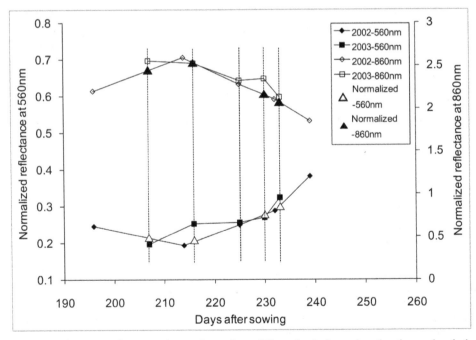

Adaptation of average reflectance of normal samples at 560 nm (central wavelengths of green band of Landsat-5 TM) and 860 nm (central wavelengths of near-infrared band of Landsat-5 TM) to match the dates of 2003 Exp, by using a linear interpolation method

Fig. 2. An example for normalization of measuring dates

**Normalization of the difference from cultivars and soil backgrounds**

The canopy spectra of winter wheat were not only supposed to respond to stresses, but are also determined and influenced by several other aspects such as cultivars and soil properties. Although the both 2002 Exp and 2003 Exp were conducted in the same fields that had approximately identical climate and environmental conditions, the difference in cultivars and soil properties between 2002 Exp and 2003 Exp should not be ignored (Table 1). To minimize this discrepancy, we calculated a ratio spectral curve for each of measured dates (after the normalization of the measuring dates) by the averaged spectral curve from normal samples in 2002 Exp divided by the averaged spectral curve from normal samples in 2003 Exp, resulting in a total of five ratio curves corresponding to each growth stage (Fig. 3). After that, all the spectral data measured at different growth stages were multiplied by the corresponding ratio curves to yield a set of normalized spectra. It should be pointed out that the present normalization processing to raw spectral measurements will only enhance the comparability between the 2002 Exp and 2003 Exp with little change in internal relations among different treatments because all the spectral data at one growth stage were processed with the same ratio curve. The ultimate goal of all these preprocessing and normalization steps above is to mitigate effects of the variation of illumination conditions, measurement dates, cultivars and soil properties between the 2002 Exp and 2003 Exp on target spectra.

**2.1.1.5 Spectral features calculation and statistical analysis**

With the spectra normalized using the methods above, we calculated 38 spectral features. An analysis of variance (ANOVA) was employed to investigate the spectral differences between the normal samples and all forms of stressed samples. Firstly, on different measured dates, both the yellow rust disease data and nutrient stressed data were compared with the normal data by ANOVA. For those spectral features that were consistently sensitive to yellow rust disease, we not only tested their differences between the normal treatment and different forms of stresses, but also tested the differences between various kinds of nutrient stresses and varying levels of disease stresses with ANOVA. Statistical analyses were conducted using SPSS 13.0 procedure.

**2.1.2 Results**

**2.1.2.1 Spectra after normalizations**

The spectral ratio curves in Fig 3 reflect the deviations between 2002 Exp and 2003 Exp's reflectance datasets at different wavelength positions. The ratio value close to 1.0 indicates no difference in reflectance exists between the two years. Generally, the ratio values ranged from 0.7 to 1.3, with an uneven distribution along the wavelength axis (Fig 3). The ratio tended to deviate from 1.0 in the regions of 350 - 730 nm, 1450 - 1570 and 2000 - 2400 nm, but stayed around 1.0 in the regions of 730 - 1330 nm and 1570 - 1770 nm. To assess the improvement in comparability, we examined the difference of normalized datasets of normal samples between 2002 Exp and 2003 Exp through an ANOVA with all 38 spectral features. The result showed that the differences of all spectral features were insignificant at all growth stages ($p$-value>0.05), with an average $p$-value (for all measuring dates) of 0.94, indicating a relatively high level of similarity between two datasets. Therefore, we confirmed that such normalization processes minimized the spectral difference originated

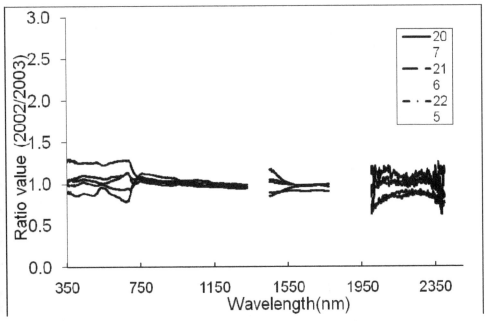

Fig. 3. Ratios of spectra for normalization with different years and varieties

from variation of illumination and different measurement dates, etc., and enabled more rational comparisons among different treatments.

### 2.1.2.2 Spectral responses to different forms of stresses

The result of ANOVA between normal samples and different forms of stress samples indicated that all spectral features had a response (defined as $p$-value<0.05) to at least one type of stresses at one growth stage, except for the WID1070-1320, which had no response to any form of stresses at all growth stages. Total 37 spectral features responded to water associated stresses (W-SD, W-SED, W-SED+N-E, W-SED+N-D) at least at one growth stage, followed by 35 spectral features to yellow rust disease, whereas only15 spectral features had a response to solely nitrogen stress (N-E, N-D). As summarized in Table 3, most spectral features were sensitive to yellow rust infection at least at one growth stage, except for $\lambda_b$, $\lambda_r$ and WID1070-1320. In addition, most spectral features tended to be more sensitive at later growth stages than at the early stages. For example, several features such as DEP920-1120, AREA920-1120, $D_y$, GI, NDVI and Triangular Vegetation Index (TVI) only had a response to yellow rust at the last growth stage in our study (233 DAS). However, for the sake of diagnosis, the spectral features with a consistent response to yellow rust during the important growing period would be much more valuable. Therefore, those spectral features that were sensitive to the yellow rust at 4 out of 5 growth stages were selected as candidates for disease diagnosis. This yielded four vegetation indices (VIs): PRI, PhRI, NPCI and ARI.

### 2.1.2.3 One way ANOVA of four disease sensitive spectral features

Particularly for the four identified VIs that closely associated with yellow rust disease, a throughout one way ANOVA was conducted to compare their differences between the

| Spectral features | Days after sowing | | | | |
|---|---|---|---|---|---|
| | 207 | 216 | 225 | 230 | 233 |
| DEP550-770 | √ | | | √ | √ |
| AREA550-770 | √ | | | √ | √ |
| WID550-770 | | | √ | √ | √ |
| DEP920-1120 | | | | | √ |
| AREA920-1120 | | | | | √ |
| WID920-1120 | | | | | √ |
| DEP1070-1320 | | | | | √ |
| AREA1070-1320 | | | | | √ |
| Db | | | √ | √ | |
| SDb | | | √ | √ | √ |
| Dy | | | | | √ |
| λy | | | | | √ |
| SDy | | | | | √ |
| Dr | | | | √ | |
| SDr | | | | √ | √ |
| GI | | | | | √ |
| MSR | | | | √ | √ |
| NDVI | | | | | √ |
| NBNDVI | | | | √ | √ |
| NRI | | | | | √ |
| PRI | | √ | √ | √ | √ |
| TCARI | | | √ | √ | |
| SIPI | | | | | √ |
| PSRI | √ | | | √ | √ |
| PhRI | | √ | √ | √ | √ |
| NPCI | √ | | √ | √ | √ |
| ARI | √ | | √ | √ | √ |
| TVI | | | | | √ |
| CARI | | | √ | √ | √ |
| DSWI | | | | | √ |
| MSI | | | | | √ |
| SIWSI | | | | | √ |
| RVSI | | | √ | √ | |
| MCARI | | | √ | √ | √ |
| WI | | | | | √ |

Table 3. Responses of spectral features to yellow rust

normal sample and various kinds of stressed samples. Moreover, their differences among each pairs of stress forms were also compared. We conducted this ANOVA based on the data on 207 DAS, 225 DAS and 233 DAS respectively, which were essential growth stages for carrying out fungicide spraying and yield loss assessing procedures. In addition to the $p$-value of ANOVA, we also provided the change direction of spectral features. Positive sign indicates the average spectral feature value of diseased or nutrient stressed samples is greater than that of normal samples, and negative sign indicates the opposite cases to the positive sign. As shown in Table 4, it was observed that for the treatments of N-E and N-D, all four VIs failed to show any response at all growth stages. For the results of other treatments, the responses of four VIs behaved in a varied pattern at three growth stages.

For the results on 207 DAS (Table 4a), compared to the normal samples, the NPCI and ARI had responses to all three levels of yellow rust treatments (YR 1, YR 2, YR 3), and appeared to be more sensitive than PRI and PhRI. For nutrient stresses, the PRI, NPCI and ARI were sensitive to W-SED and W-SED+N-E treatments. Among them, NPCI and ARI showed stronger responses ($p$-value<0.01) to W-SD, W-SED, W-SED+N-E and W-SED+N-D treatments than the other two VIs. For the comparisons between diseased samples and nutrient stressed samples, significant differences between W-SED and W-SED+N-E treatments and YR2 and YR3 treatments were identified for PRI, NPCI and ARI. Moreover, the change directions of the three VIs for diseased and nutrient stressed samples were identical. At this 207 DAS growth stage, PhRI did not show a significant response to any of three levels of disease treatments, but responded to W-SD, W-SED and W-SED+N-E treatments. It is interesting that the change direction of diseased samples of PhRI was contrary to that of the nutrient stressed samples, suggesting a discriminating potential of the index.

For the results on 225 DAS (Table 4b), compared to the normal samples, all four VIs revealed a clear response to level 2 and level 3 of yellow rust treatments (YR2, YR3). For nutrient stresses, PRI, NPCI and ARI also appeared to be sensitive to W-SD, W-SED, W-SED+N-E and W-SED+N-D treatments. However, PhRI was insensitive to all nutrient stresses. In addition, when we looked at the difference of those VIs between diseased samples and nutrient stressed samples, only PhRI showed clear differences between YR2 and YR3 treatments and W-SD, W-SED, W-SED+N-E, and W-SED+N-D treatments. Although a significant difference between YR3 treatment and W-SED treatment also existed for ARI and NPCI, the change directions of both treatments were identical. However, for PhRI, the change directions of all levels of disease treatments were different from those of the nutrient stress treatments.

For the results on 233 DAS (Table 4c), with further development of disease symptoms, compared to the normal samples, all four indices showed responses to all three levels of disease treatments. Comparing to YR1 treatment, the four VIs had shown a stronger significant level ($p$-value<0.01) for YR2, YR3 treatments. For nutrient stresses, PRI, NPCI and ARI exhibited clear responses to W-SED, W-SED+N-E and W-SED+N-D treatments as well. For comparisons between diseased and nutrient stressed samples, PRI and NPCI appeared to be significantly different between YR2 and YR3 treatments and W-SD treatment. However, the change directions of both treatments were identical. Unlike the other three VIs, PhRI remained insensitive to the nutrient stresses, but was significantly different among all levels of disease treatments (YR1, YR2, and YR3) and all forms of nutrient stresses. More

| Treatments | YR 1 | | | | YR 2 | | | | YR 3 | | | | Normal | | | |
|---|---|---|---|---|---|---|---|---|---|---|---|---|---|---|---|---|
| | PRI | PhRI | NPCI | ARI | PRI | PhRI | NPCI | ARI | PRI | PhRI | NPCI | ARI | PRI | PhRI | NPCI | ARI |
| Normal | (-) | (-) | (+)* | (-)* | (-)* | (+) | (+)* | (-)* | (-) | (+) | (+)** | (-)*** | | | | |
| W-SD | (-) | (-) | (+) | (+) | (-) | (-)* | (+) | (+) | (-) | (-)* | (+) | (-) | (-) | (-)* | (+)** | (-)** |
| W-SED | (+) | (-) | (+)* | (+)* | (+) | (-)* | (+)* | (+)* | (+)* | (-)* | (+)* | (+) | (-)*** | (-)* | (+)*** | (-)*** |
| N-E | (-) | (+) | (-) | (-) | (-)* | (-) | (-) | (-) | (-) | (-) | (-)* | (-)** | (+) | (-) | (-) | (+) |
| N-D | (-) | (+) | (-) | (-) | (-) | (+) | (-) | (-) | (-) | (-) | (-)* | (-)** | (-) | (+) | (-) | (+) |
| W-SED+N-E | (+) | (-) | (+)* | (+)* | (+) | (-)* | (+)* | (+)* | (+)* | (-)* | (+) | (+) | (-)*** | (-)* | (+)*** | (-)*** |
| W-SED+N-D | (+) | (-) | (+) | (+) | (+) | (-) | (+) | (+) | (+) | (-)* | (+) | (+) | (-)* | (-) | (+)** | (-)*** |

(a) 207 DAS

| Treatments | YR 1 | | | | YR 2 | | | | YR 3 | | | | Normal | | | |
|---|---|---|---|---|---|---|---|---|---|---|---|---|---|---|---|---|
| | PRI | PhRI | NPCI | ARI | PRI | PhRI | NPCI | ARI | PRI | PhRI | NPCI | ARI | PRI | PhRI | NPCI | ARI |
| Normal | (+) | (+) | (+) | (+) | (+)** | (+)** | (+)* | (+)* | (+)*** | (+)*** | (+)** | (+)** | | | | |
| W-SD | (+)** | (-)* | (+)** | (+)** | (-) | (-)** | (+) | (+) | (-) | (-)** | (+) | (+) | (+)*** | (-) | (+)*** | (+)** |
| W-SED | (+)** | (-) | (+)*** | (+)** | (-) | (-)** | (+) | (+) | (+) | (-)* | (+)** | (+)** | (+)*** | (-) | (+)*** | (+)*** |
| N-E | (-)* | (-) | (-) | (-) | (-)* | (-)* | (-) | (-) | (-)* | (-)* | (-)* | (-)** | (-) | (-) | (-) | (-) |
| N-D | (-) | (+) | (-) | (-) | (-) | (-)* | (-) | (-) | (-)* | (-)* | (-)* | (-)** | (+) | (+) | (+) | (-) |
| W-SED+N-E | (+)* | (-) | (+)* | (+)* | (-) | (-)** | (+) | (+) | (-) | (-)* | (+) | (+) | (+)*** | (-) | (+)*** | (+)** |
| W-SED+N-D | (+)* | (-) | (+)* | (+)* | (-) | (-)** | (+) | (+) | (-) | (-)* | (+) | (+) | (+)*** | (-) | (+)** | (+)** |

(b) 225 DAS

| Treatments | YR 1 | | | | YR 2 | | | | YR 3 | | | | Normal | | | |
|---|---|---|---|---|---|---|---|---|---|---|---|---|---|---|---|---|
| | PRI | PhRI | NPCI | ARI | PRI | PhRI | NPCI | ARI | PRI | PhRI | NPCI | ARI | PRI | PhRI | NPCI | ARI |
| Normal | (+)** | (+)* | (+)* | (+)* | (+)*** | (+)** | (+)*** | (+)*** | (+)*** | (+)** | (+)*** | (+)*** | | | | |
| W-SD | (-) | (-)** | (-) | (+) | (-)*** | (-)** | (-)* | (-) | (-)** | (-)** | (-)** | (-)* | (+) | (-) | (+)* | (+)* |
| W-SED | (+) | (-)** | (+) | (+) | (-)* | (-)** | (+) | (-) | (-)* | (-)** | (-) | (-) | (+)*** | (-) | (+)*** | (+)*** |
| N-E | (-)* | (-)** | (-) | (-) | (-)** | (-)** | (-)** | (-)* | (-)*** | (-)** | (-)*** | (-)** | (-) | (-) | (-) | (+) |
| N-D | (-)* | (-)* | (-) | (-) | (-)*** | (-)** | (-)** | (-)* | (-)*** | (-)** | (-)** | (-)* | (-) | (+) | (+) | (+) |
| W-SED+N-E | (+) | (-)** | (+) | (+) | (-)** | (-)** | (+) | (+) | (-)* | (-)** | (-) | (+) | (+)*** | (-) | (+)*** | (+)*** |
| W-SED+N-D | (+) | (-)** | (+) | (+) | (-)*** | (-)** | (+) | (-) | (-)* | (-)** | (-) | (-) | (+)*** | (-) | (+)*** | (+)*** |

(c) 233 DAS

*mean difference is significant at 0.950 confidence level; **mean difference is significant at 0.990 confidence level;*** mean difference is significant at 0.999 confidence level. (+) means the average spectral feature value of diseased or nutrient stressed samples greater than that of normal samples; or means the average spectral feature value of nutrient stressed samples greater than that of diseased samples; (-) means the opposite cases to the case of (+). The definitions of treatments are as follows: "Normal" represents normal samples; "W-SD" represents samples treated with slightly deficient water; "W-SED" represents samples treated with seriously deficient water; "N-E" represents samples treated with excessive nitrogen; "N-D" represents samples treated with deficient nitrogen; "W-SED+N-E" represents samples treated with seriously deficient water and excessive nitrogen; "W-SED+N-D" represents samples treated with seriously deficient water and deficient nitrogen

Table 4. ANOVA for four VIs separately on 207 DAS, 225 DAS and 233 DAS

importantly for the PhRI, the change directions of diseased samples were opposite to those of nutrient stressed samples throughout the entire analysis.

In summary, all four VIs showed a significant sensitivity to yellow rust disease on 207 DAS, 225 DAS and 233 DAS. However, most of them also appeared to be sensitive to water associated stresses to a varing extent, except for PhRI, which was only sensitive to disease yet insensitive to any forms of nutrient stresses on 225 DAS and 233 DAS. More importantly, the change directions of PhRI to disease treatments were always opposite to those to the nutrient stress treatments at all relevant growth stages. This further confirmed the discriminating characteristic of PhRI.

## 2.1.3 Conclusion

Combining with a dataset of yellow rust disease inoculation and a dataset of various forms of nutrient stress treatments, we examined the responses of 38 commonly used spectral features at five important growth stages from booting stage to milk development stage using a one-way analysis of variance (ANOVA). There were 37 spectral features sensitive to water associated stresses, 35 spectral features sensitive to yellow rust disease and only 15 spectral features sensitive to sole nitrogen stresses in at least one growth stage. It was observed that more spectral features appeared to have a response to yellow rust disease at later growth stages. A throughout ANOVA was conducted particularly on PRI, PhRI, NPCI and ARI, which showed a consistent response to yellow rust disease at 4 out of 5 growth stages. However, PRI, NPCI and ARI were also responsible for water associated stresses, suggesting a risk of confusion in detecting yellow rust disease. Only PhRI was sensitive to yellow rust disease, but insensitive to different forms of nutrient stresses. The discriminative response of PhRI could provide a means of identifying and detecting yellow rust disease under complicated farmland circumstances. This finding can serve the basis of remote sensing system for detecting yellow rust disease.

## 2.2 Detecting yellow rust using field and airborne hyperspectral data

The aim of this study was to evaluate the accuracy of the spectro-optical, photochemical reflectance index (PRI) for quantifying the disease index (DI) of yellow rust in wheat using in-situ spectral reflectance measurements, and its applicability in the detection of the disease using hyperspectral imagery.

### 2.2.1 Materials and methods

#### 2.2.1.1 Experimental design and field conditions

Experimental design and field conditions was same as 1.1.1. Experimental data from 2002 Exp were used to establish the statistical models, and the data for 2003 Exp were used to validate the models developed.

#### 2.2.1.2 Inspection of disease severity

To quantify the severity of the disease of yellow rust, the leaves of plants were grouped into one of 9 classifications of disease incidence (x): 0,1, 10, 20, 30, 45, 60, 80 and 100% covered by rust. 0% represented no incidence of yellow rust, and 100% was the greatest incidence. The disease index (DI) was then calculated using (Li et al. 1989):

$$DI(\%) = \frac{\sum(x \times f)}{n \times \sum f} \times 100$$

where $f$ is the total number of leaves of each degree of disease severity and $n$ is the degree of disease severity observed (in this work, n ranged from 0 to 8). In each plot, 20 individuals were randomly selected for check.

#### 2.2.1.3 Canopy spectral measurements

The method of canopy spectral measurements and data was same as the part 1.1.1.2 above.

### 2.2.1.4 Airborne hyperspectral imaging

Airborne hyperspectral images of the trial field were acquired in 2003 using the Pushbroom Hyperspectral Imager (PHI) designed by the Chinese Academy of Science (CAS) and flown onboard a Yun-5 aircraft (Shijiazhuang Aircraft Manufacturing Company, China). The PHI comprises a solid state, area array, and silicon CCD device of 780 × 244 elements. It has a field of view of 21º, and is capable of acquiring images of 1 m × 1 m spatial resolution at an altitude of 1000 m above ground. The wavelength range is 400–850 nm with a spectral resolution of 5 nm. Images of the target field were acquired in 2003 at the phenological growth stages of stem elongation (April 18, 2003, Zadoks stage 3), anthesis (May 17, 2003, Zadoks stage 5) and milky maturity (May 31, 2003, Zadoks stage 8). The inoculated wheat was adequately infected by rust on April 18, obviously infected by May 17, and seriously infected by May 31. Measurements of DI were made and in situ canopy reflectance spectra were also acquired on the same dates. All images were geometrically and radiometrically corrected using an array of georeferenced light and dark targets (5 m x 5 m) located at the extremes of the field site. The aforementioned field spectrometer was used to calibrate these targets relative to BaSO4. The location of each target, as well as field measurements of DI were recorded using a differential global positioning system (Trimble Sunnyvale California, USA).

### 2.2.1.5 Photochemical reflectance index (PRI)

Because yellow rust epiphyte reduced foliar physiological activity by destroying foliar pigments, the photochemical reflectance index (PRI) was selected as the spectrophotometric method of estimating the disease index. PRI was calculated by the formula in Table 2.

## 2.2.2 Results

### 2.2.2.1 PRI versus DI

Fig. 4 shows a plot of the measured DI as a function of PRI for all varieties. The data points associated with the variety Xuezao dominate in the top-left region of the scatter plot (relatively high range of DI), while those associated with the variety 98-100 are located in the mid region (mid-range DI) and those associated with Jing 411 dominate the lower right region. This distribution trend is consistent with the relative susceptibility of these varieties to rust; Xuezao is the least resistant and Jing 411 has the greatest resistance. The regression equation of DI using PRI in 2002 Exp was obtained as following (n = 64):

$$DI(\%) = -721.22(PRI) + 2.40 \qquad \left(-0.14 \leq PRI \leq 0.02; r^2 = 0.91\right)$$

An important feature in, the associated regression equation (Fig. 4) was that the spectrally-derived PRI explained 91% of the variance observed in the disease index. This explanation also encompassed the three varieties of wheat as well as the four stages of crop development for each variety. In the subsequent validation of the PRI-DI regression equation with the 2003 Exp data (Fig. 5), the coefficient of determination ($R^2$) between the estimated and measured values was 0.97 (n = 80).

In Fig. 5, the locations of data points associated with individual varieties wew consistent with the levels of resistance to rust. Xuezao dominated the top right-hand region of the scatter plot (relatively high range of DI), the variety 98-100 had points scattered all along the

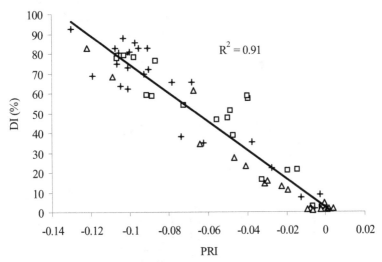

Fig. 4. Plot of measured disease index (DI) as a function of measured photochemical reflectance index(PRI) for all varieties combined in 2002 Exp. Δ: Jing 411; +: Xuezao; □: 98–100

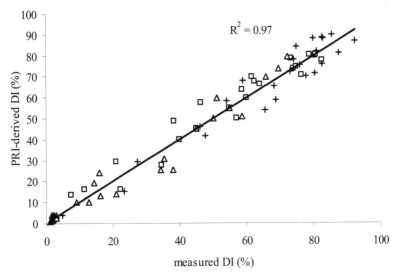

Fig. 5. Comparison of measured DI and PRI-estimated DI for 2003 Exp; 'Δ' = Jing 411; '+' = Xuezao; '□' = 98–100

regression line (predominantly mid-range DI), and Jing 411 was concentrated in the central lower-left region (lower range DI).

### 2.2.2.2 Application of multi-temporal PHI images for DI estimation

The DI was estimated on a pixel-by-pixel basis in each of the acquired PHI images using the regression equation. To map the degree of yellow rust infection in the trial field, the DI was binned into the following classes; very Serious (DI > 80%), serious (45% < DI ≤ 80%), moderate (10% < DI ≤ 45%), slight (1% < DI ≤ 10%) and none (0 < DI ≤ 1%) (Fig. 6).

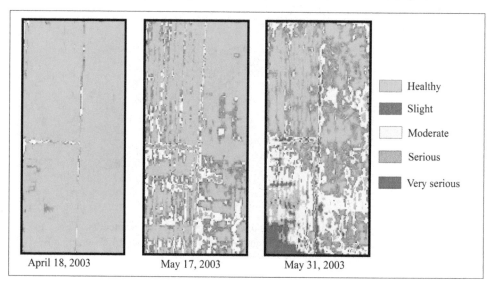

| April 18, 2003 | May 17, 2003 | May 31, 2003 |

Healthy
Slight
Moderate
Serious
Very serious

Fig. 6. Classified DI images derived from PHI airborne images of the trial site in 2003 Exp

Fig. 7 shows the relationship between the DI calculated from the multi-temporal PHI images and the actual measured DI from the 120 sample sites located within the field (R²=0.91).

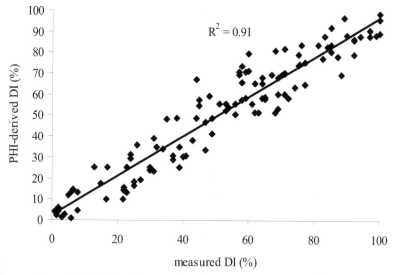

Fig. 7. Comparison of PHI-derived estimates of DI and actual DI values for 2002 Exp. Data were extracted from all three imaging times, although the DI values were< 20% for the April 18 image

### 2.2.3 Conclusion

The results of this work confirm PRI is a potential candidate for monitoring of yellow rust, and could form the basis of an on-the-go sensor and variable-rate spray applicator or remote detection and mapping process.

## 2.3 Detecting yellow rust in winter wheat by spectral knowledge base

In most cases, statistical models for monitoring the disease severity of yellow rust are based on hyperspectral information. The high cost and limited cover of airborne hyperspectral data make it impossible to apply such data for large scale monitoring. Furthermore, the established models of disease detection cannot be used for most satellite images because of the wide range of wavelengths in multispectral images (Zhang et al., 2011).

To resolve this dilemma, the study presents a novel approach by constructing a spectral knowledge base (SKB) of winter wheat diseases, which takes the airborne images as a medium and links the disease severity with band reflectance from moderate resolution remotely sensed data, such as environment and disaster reduction small satellite images (HJ-CCD) accordingly. To achieve this goal, several algorithms and techniques for data conversion and matching are adopted in the proposed system, including minimum noise fraction (MNF) transformation and pixel purity index (PPI) function. The performance of SKB is evaluated with both simulated data and field measured data.

### 2.3.1 materials and methods

Experimental design and field conditions was same as the part of 1.1.1.1

#### 2.3.1.1 Inspection of disease severity

Please refer to the part of 1.2.1.2 above.

#### 2.3.1.2 Airborne hyperspectral imaging

Please refer to the part of 1.2.1.4 above about airborne hyperspectral imaging and image processing.

#### 2.3.1.3 Acquisition of moderate resolution satellite images

In this study, the SKB is designed to fit the charge coupled device (CCD) sensor, which is on the environment and disaster reduction small satellites (HJ-1A/B). The basic parameters of the CCD sensor (using 'HJ-CCD' in the following) are given in Table.5. The four bands of

| Properties of HJ-CCD | | | | |
|---|---|---|---|---|
| Band | Wavelength range (nm) | Spatial resolution (m) | Swath (km) | Revisit time (day) |
| Blue | 0.430–0.520 | | | |
| Green | 0.520–0.600 | 30 | 360 | 2 |
| Red | 0.630–0.690 | | | |
| Near-infrared | 0.760–0.900 | | | |

Table 5. Properties of the environment and disaster reduction small satellites (HJ-CCD)

HJ-CCD covered the visible and near infrared spectral regions. The HJ-CCD sensor has spectral and spatial characteristics that are similar to those of Landsat-5 TM, but the HJ-1A/B satellites have more frequent revisit capability (2 days) than the Landsat-5 satellite (16 days), which is of great importance for agricultural monitoring.

### 2.3.1.4 Construction of the spectral knowledge base

The SKB in this study can be interpreted as a pool of relationships between spectral characteristics and prior knowledge. Here, prior knowledge stands for the degree of severity of yellow rust, and the spectral characteristics are the reflectance of the initial four bands of the HJ-CCD image. Hence, there are two major steps involved in constructing the SKB. First, the relationship between hyperspectral information and severity is obtained with a stable empirical reversion model. Then, through the RSR function of the HJ-CCD sensor, the hyperspectal data can be transferred to the wide-band reflectance. In this way, a one-to-one correspondence between the disease severity of yellow rust and reflectances from the HJ-CCD sensor is established at the pixel level. The SKB can represent disease severity in two ways: the DI (%) value and the class of disease severity. The following sections describe each step for establishing the SKB, including data selection, the reversion model, simulation of the wide-band reflectance and estimating the degree of severity. A technical flow diagram of SKB construction is summarized in Fig. 8.

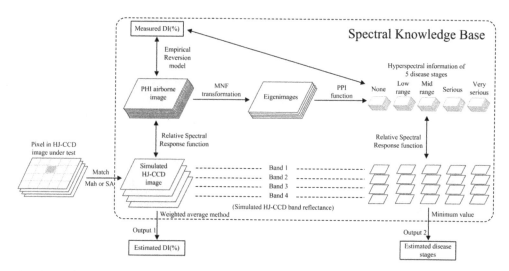

Fig. 8. The flow chart for monitoring of DI(%) of winter wheat stripe rust, b1-b4 represented the reflectance of the four bands of HJ-CCD images

As noted above, the SKB in this study comprised PHI pixels. The predicted accuracy obtained by the SKB was determined primarily by the amount of prior knowledge, which indicated the heterogeneity of disease severity. The design of the yellow rust fungus inoculation ensured a considerable variation in disease severity within the experimental field, from healthy plants to very diseased plants. In addition, to avoid using pixels on or near the ridge in the field that are considered as mixed signals, we chose three rectangular shaped areas that were within the field and comprised 7918 'crop-only' pixels for constructing the SKB.

### 2.3.1.5 Reversion model

The reversion model construction was the first step of establishing the SKB. Based on the conclusion of the part above, PRI was a suitable vegetation index for monitoring the severity of yellow rust disease in winter wheat. Therefore, in this study, PRI was used to establish the linkage between the disease severity and the hyperspectral data. Specifically, the yellow rust infection would be apparent at anthesis stage, and this should be closely related with the subsequent yield loss. Therefore, we chose the PHI image at this stage to form the SKB. To obtain a better fitting model, we reanalyzed the PHI-PRI and corresponding DI (%) data at the anthesis stage specifically, and obtained a linear regression model. It should be noted that the data range of DI must be between 0 and 100%. Any predicted DI results that were>100% or <0% were redefined as DI = 100% and DI = 0% to represent very infected plants and healthy plants, respectively.

### 2.3.1.6 Simulation of the wide band reflectance

The second step of constructing the SKB is to transform the hyperspectral reflectance of PHI-pixels to wide band reflectance of HJ-pixels. To achieve this goal, the best approach is the inherent relative spectral response (RSR) function of the HJ-CCD sensor. By integrating the hyperspectral reflectance of PHI-pixels on the RSR function, the band reflectance of HJ-CCD sensor was thus obtained. Besides, although the wavelength range of the fourth band of HJ-CCD sensor (760 nm-900 nm) was slightly exceeded the maximum wavelength of PHI sensor (850 nm), for most ground measured spectra, the reflectance basically kept on steady from 760 nm to 900 nm. Hence, the simulating results generated using the incomplete range of wavelength (760nm-850nm) should approach to the true value. The integration can be shown as follows:

$$R_{TM} = \int_{b_{start}}^{b_{end}} f(x)dx$$

where $R_{TM}$ is the simulated reflectance of a certain band; $b_{start}$ and $b_{end}$ indicate the beginning and the end wavelength of this band respectively; $f(x)$ indicates the RSR function, which is obtained from CRESDA.

### 2.3.1.7 Spectral characteristics of different degrees of disease severity

Another way to define the disease severity of an undefined pixel, apart from the DI (%) value, is to quantify disease severity by severity classes. The criterion of severity class provided by Huang et al. (2007) was adopted, which corresponded to the major physiological alteration of diseased plants. The DI (%) thresholds for each severity class were: DI<1% indicated not infected (NI), 1% <DI<10% indicated a low degree of infection (LI), 10% <DI<45% indicated mid-range infection (MI), 45% <DI<80% indicated seriously infected (SI) and DI (%)>80% indicated very seriously infected (VI). The MNF transformation and PPI function, which are used for noise reduction and end-member identification, were applied here to select the most representative pixels from the PHI image, and to form the typical spectrum for each severity class.

## 2.3.1.8 Spectral matching algorithms

The basic idea of spectral matching is to identify a set of pixels in the SKB that are the closest to the undefined pixel in terms of spectral characteristics. Before matching, each pixel should be standardized to eliminate systematic variation caused by aerosol conditions or other factors as follows:

$$R_{nor} = \frac{R - R_{min}}{R_{max} - R_{min}}$$

where $R_{nor}$ is the standardized reflectance of a certain band, R is the original reflectance, and $R_{min}$ and $R_{max}$ are the minimum and maximum band reflectance values, respectively, of the corresponding pixel.

Mahalanobis distances (Mah) and Spectral angle (SA) were selected as the distance measurement criterion. Both types of distance measurements had been proved to be with high efficiency in reflecting the spectral discrepancy (South et al., 2004; Goovaerts et al., 2005; Becker et al., 2007). The Mah distance can be written as:

$$D_M(x) = \sqrt{(x - x_R)\sum^{-1}(x - x_R)^T} \quad x=(x_1,x_2,x_3,x_4), \, x_R=(x_{R1}, x_{R2}, x_{R3}, x_{R4})$$

where $x_{1-4}$ are the reflectance of the pixel under test in band1 to band4, respectively; $x_{R1-4}$ are the simulated reflectance of a specific pixel in SKB. $\sum$ is the covariance matrix between x and $x_R$. SA can be calculated by the following formula:

$$\theta = \arccos \frac{\sum_{i=1}^{4} x_i x_{Ri}}{\sqrt{\sum_{i=1}^{4} x_i^2}\sqrt{\sum_{i=1}^{4} x_{Ri}^2}} \quad \theta \in \left[0, \frac{\pi}{2}\right]$$

To determine the DI (%) or class of disease severity of an undefined pixel, we have to calculate the Mah and spectral angle from this pixel to each pixel or class in the SKB. A longer distance or larger angle indicates that the pixel deviated from the undefined pixel, whereas a shorter distance or smaller angle indicates that it is similar to the undefined pixel. By selecting the most similar pixel, the severity class of an undefined pixel can be determined. To determine the DI (%) of a certain pixel, the weighted average method was used. According to the distance criteria above, the five most similar pixels were selected from the SKB. For each band of these pixels (here we used the hyperspectral bands extracted from the PHI image), the reflectance was processed according to the following equation:

$$R_E = \frac{\sum_{i=1}^{k} R_i \times \frac{1}{d_i}}{\sum_{i=1}^{k} \frac{1}{d_i}}$$

where $R_E$ is the estimated reflectance of a certain pixel through k-NN estimation; $R_i$ is the reflectance of the $i_{th}$ nearest pixel according to the ranking order of the distance; $d_i$ is the distance between the pixel under test to the $i_{th}$ nearest pixel.

**2.3.1.9 Verification**

To verify the performance of SKB in identifying and monitoring the severity of yellow rust diseases, two datasets were used: the simulated data and the field-measured data with corresponding satellite images.

1.  Verification of SKB using simulated data
    The simulated data comprised 50 randomly selected pixels in the same experimental field, but outside the three regions selected for constructing the SKB. The hyperspectral information of each pixel was used to create the reference DI (%) and severity class with the empirical model and the corresponding threshold for each severity class. To test the performance of SKB in terms of DI (%) value, we estimated the DI value with both distance criteria described above. The samples were split into two: the pixels with a reference DI between 1 and 100%, i.e. the 'diseased' pixels, and those with a reference DI<1%, i.e. 'healthy' pixels. For the diseased pixels, the estimated DIs were compared with the reference DI by Pearson correlation analysis and the normalized root mean square error (NRMSE). For the healthy pixels, we used 'yes or no' to determine whether the estimated value indicated infection or not, which also provided an accuracy ratio. The estimation of severity class was verified by overall accuracy and the kappa coefficient.
2.  Verification of SKB using field surveyed data
    The field surveyed data sets included the ground investigation of disease severity and the corresponding HJ-CCD images. Between June 1-3, 2009, when the winter wheat was at the anthesis stage, we conducted a survey in the southeast of GanSu Province. The climate of the area surveyed is characterized by high humidity and rainfall, and yellow rust disease occurs almost every year. This area has similar environmental conditions and cultivation customs to those where we constructing the SKB in Beijing, and this makes it an appropriate place for model verification. With the aid of the local Department of Plant Protection, 26 plots were randomly selected and surveyed in the area (Fig. 9). To relate the surveyed value to the pixel value of the HJ-CCD image, we defined the plot as a uniformly planted winter wheat region with an area no less than 30 m in radius. The geographical coordinates of each plot were measured by GPS at the centre of the plot. Disease severity was measured as described above. We repeated the measurement in five evenly-distributed sections in each plot, and 20 individual plants were included in each measurement. The HJ-CCD images (ID: 122516, 122518) acquired on June 2, 2009 completely covered the surveyed area. The raw data from the HJ-CCD imagery was calibrated based on the corresponding coefficients provided by CRESDA. The calibrated data were atmospherically corrected with the algorithm provided by Liang et al. (2001), which estimated the spatial distribution of atmospheric aerosols and retrieved surface reflectance under general atmospheric and surface conditions. The images were also geometrically corrected against historical reference images with the same geographical coordinates. The images were rectified with a root mean square error of less than 0.5 pixels. The spectrum of the each plot was extracted from the image according to the GPS records. The estimated accuracy in this step followed the same process as the simulated data.

**2.3.2 Results**

There were 7918 pixels included in the process of constructing the SKB. The linear regression model between DI (%) and PRI at anthesis stage could be illustrated as follows:

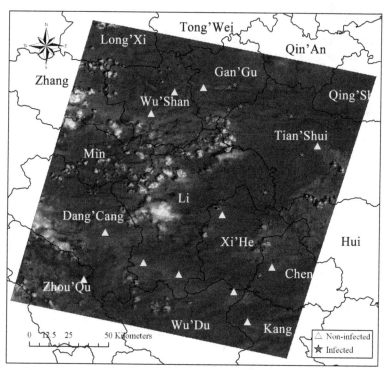

Fig. 9. The field surveyed area in Gansu Province. The base image is the HJ-CCD image
acquired on June 2, 2009

$$DI(\%) = -538.98 \times PRI + 2.0983 \quad (R^2 = 0.88)$$

The pairs of DI (%) and PRI were plotted in Fig.4, which showed a significant correlation ($R^2$
= 0.88). Based on the model, there were 85 pixels with a DI of 100% and 3991 pixels with a
DI between 1% and 100%, indicating 51.5% pixels infected to a varied degree of severity,
whereas the other 48.5% pixels (DI = 0%) were healthy plants. In the experimental field, the
variation in the degree of severity of yellow rust from totally healthy plants to very infected
plants provided the essential diversity or heterogeneity of infection, which then enabled
establishment of the SKB. The MNF transformation resulted in 9 leading eigenvectors with
eigenvalues greater than 4.0 (Fig. 10), and these were used for further analysis.

### 2.3.2.1 Performance of SKB for simulated data

In the simulated dataset, there were six healthy pixels and 44 diseases affected ones. When
estimating DI (%), one pixel with no infection was estimated as infected by the Mah distance
criterion, whereas with the SA criterion two were mislabeled. Fig.11 shows the scatter of the
disease affected pixels plotted in relation to reference DI and estimated DI; the average
reference DI is 36%. The reference DIs and estimated DIs were strongly and linearly
correlated for both the Mah distance ($R^2$ = 0.90) and SA ($R^2$ = 0.84) criteria. Further, the
NRMSE of Mah distance and SA were 0.20 and 0.24, respectively, indicating that the SKB
can estimate DIs accurately from the simulated multi-band reflectance.

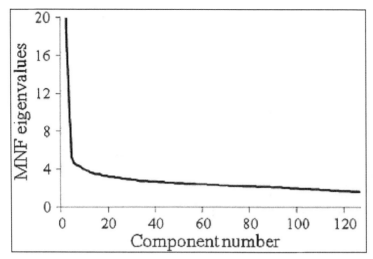

Fig. 10. MNF eigenvalues variation trend

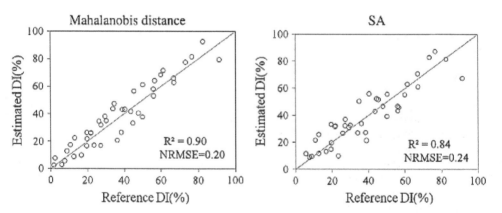

Fig. 11. Estimated DI(%) using simulated data

Table 6 gives the reference class of disease severity and the estimated class in the form of an error matrix. The overall accuracy with Mah distance and the SA criterion were 0.80 and 0.76, respectively, whereas the kappa coefficients were 0.71 and 0.65, respectively. However, we noticed that all the misclassified pixels were assigned to no more than one class adjacent to the reference class. Therefore, for simulated data, the classification accuracy was satisfactory in determining the severity class of yellow rust by SKB.

### 2.3.2.2 Performance of SKB for field surveyed data

Apart from the verification against simulated data, more importantly, the field surveyed data can be also used to assess the performance of the SKB. The field investigation showed that eight out of 26 plots were infected with DI ranged from 4 to 90%, whereas the other 18 plots were not affected by yellow rust. The estimation by DI (%) successfully identified the eight infected plots when the Mah distance criterion was used, whereas the SA criterion

| | | Reference | | | | | |
| | | None | Low range | Mid range | Serious | Very serious | Total |
|---|---|---|---|---|---|---|---|
| Estimation (Mah) | None | 6 | 0 | 0 | 0 | 0 | 6 |
| | Low range | 0 | 5 | 2 | 0 | 0 | 7 |
| | Mid range | 0 | 1 | 20 | 2 | 0 | 23 |
| | Serious | 0 | 0 | 1 | 10 | 1 | 12 |
| | Very serious | 0 | 0 | 0 | 1 | 1 | 2 |
| | Total | 6 | 6 | 23 | 13 | 2 | 50 |
| Estimation (SA) | None | 5 | 1 | 0 | 0 | 0 | 6 |
| | Low range | 1 | 4 | 1 | 0 | 0 | 6 |
| | Mid range | 0 | 1 | 20 | 2 | 0 | 23 |
| | Serious | 0 | 0 | 2 | 9 | 1 | 12 |
| | Very serious | 0 | 0 | 0 | 2 | 1 | 3 |
| | Total | 6 | 6 | 23 | 13 | 2 | 50 |

Table 6. Error matrix for simulated data

resulted in one misestimated plot. Figure 7 shows the scatter of the eight data plotted in relation to reference DI and estimated DI for both distance criteria. There was a significant linear trend in graphs based on both the Mah distance and SA criteria. The $R^2$ of Mah distance and SA were 0.80 and 0.67, respectively, whereas the NRMSE were as high as 0.46 and 0.55. In real circumstances, approximately 50% error in the estimated disease index is unsatisfactory. On the other hand, however, most of the uninfected plots were correctly identified according to DI (%) estimates (i.e. a DI<1%). For both the Mah distance and SA criteria, 15 out of 18 non-infected plots had been identified correctly, resulting in an accuracy of 77.8%. The results for estimating disease severity by severity class were even more encouraging. The overall accuracy for the Mah distance and SA criteria were 0.77 and 0.73, respectively, whereas the kappa coefficients are 0.58 and 0.49, respectively. Table 3 gives the error matrix for both criteria. The misclassified pixels were also assigned exclusively to the adjacent class.

In general, the above results demonstrate that the proposed SKB scheme has great potential for detecting the incidence and severity of yellow rust through multispectral images. As shown from several previous studies, the image processing method of MNF transformation was efficient in extracting the principle information from the images related to wheat disease infection (Zhang et al. 2003; Franke and Menz 2007). For the present study, we found that coupling MNF transformation with the PPI function was an appropriate way of extracting the principle information on yellow rust disease. To estimate disease severity by DI (%), the proposed SKB has achieved a satisfactory accuracy for simulated data. However, the estimated accuracy for field surveyed data was unsatisfactory, implying that the method tends to underestimate or overestimate the disease severity in practice. Nevertheless, to estimate disease severity through disease severity class has achieved a satisfactory accuracy for both simulated data and field surveyed data. Therefore, the disease severity class seems to be more robust in determining the disease severity. This might be because it is more rough estimation than DI (%). It is understandable that for the same sample, the less precise the criterion, the greater accuracy it would achieve. Moreover, the 5-class disease severity quantification is enough to guide field applications. We suggest that DI (%) should be used for detecting the disease severity of yellow rust by SKB. For the distance criteria used in the process of matching with SKB, the Mah distance criterion might be more appropriate because it performed better than SA in all the analyses conducted in this study (Figs. 11, 12, Tables 6, 7). Some previous studies have already emphasized the potential of hyperspectral imagery (Bravo et al. 2003; Moshou et al. 2004; Huang et al. 2007) and the high-resolution of multispectral imagery (Franke and Menz 2007) for detecting yellow rust disease. The development of SKB in the present study can be viewed as a scaling up method, which has extended the capability of detecting yellow rust disease from hyper- spectral imagery to the moderate resolution of multispectral imagery. However, it should be noted that the task of monitoring the occurrence and degrees of infection of crop diseases is far more complex than the cases described in this study. The spectral characteristics of yellow rust infection might appear similar to other sources of stress. In addition, the impact of phenology, cultivation methods, fragmentation of farmlands and other environmental conditions would also increase the difficulty and uncertainty of the estimation process. Therefore, the SKB developed in this study should correspond to the situation at the anthesis stage exclusively, and is only suitable for those regions with similar environmental characteristics and cultivation methods. For other regions with significantly different environmental characteristics, this purposed SKB may not work well. The possible solution to these problems may include incorporating suitable priors, which would require integration strategies and understanding of the mechanisms underlying some fundamental processes. Further research is required to address the problems mentioned above.

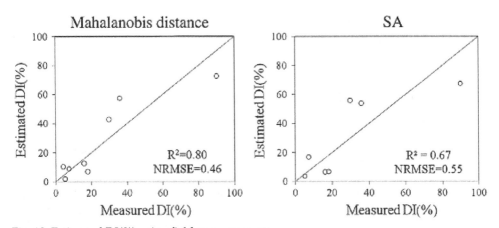

Fig. 12. Estimated DI(%) using field measurements

### 2.3.3 Conclusion

The low spatial resolution and few spectral bands have limited the application of moderate resolution satellite images for monitoring yellow rust disease. The spectral knowledge base developed enabled disease incidence and severity to be detected by moderate resolution satellite images. The SKB supported two ways of estimating disease severity: the disease

| | | Reference | | | | | |
|---|---|---|---|---|---|---|---|
| | | None | Low range | Mid range | Serious | Very serious | Total |
| | None | 16 | 0 | 0 | 0 | 0 | 16 |
| | Low range | 2 | 2 | 1 | 0 | 0 | 5 |
| Estimation | Mid range | 0 | 1 | 3 | 0 | 0 | 4 |
| (Mah) | Serious | 0 | 0 | 0 | 0 | 1 | 1 |
| | Very serious | 0 | 0 | 0 | 0 | 0 | 0 |
| | Total | 18 | 3 | 4 | 0 | 1 | 26 |
| | None | 15 | 0 | 0 | 0 | 0 | 15 |
| | Low range | 3 | 2 | 1 | 0 | 0 | 6 |
| Estimation | Mid range | 0 | 1 | 3 | 0 | 0 | 4 |
| (SA) | Serious | 0 | 0 | 0 | 0 | 1 | 1 |
| | Very serious | 0 | 0 | 0 | 0 | 0 | 0 |
| | Total | 18 | 3 | 4 | 0 | 1 | 26 |

Table 7. Error matrix for ground measured data

index and disease severity class. Both methods of estimation achieved a satisfactory level of accuracy for simulated data. For field surveyed data, estimation by DI (%) resulted in an unsatisfactory level of accuracy, whereas it was satisfactory for severity class. The Mah criterion performed better than spectral angle in all analyses. Therefore, the former should be considered as the more appropriate distance criterion.

Generally, the purposed SKB has a great potential in extending the capability of detecting yellow rust to multispectral remote sensing data, especially when the region of interest has similar environmental conditions to where the SKB was developed. The uncertainties caused by environmental differences should be further investigated in future studies.

### 2.4 Detecting yellow rust of winter wheat using land surface temperature (LST)

The air temperature and humidity are the most direct and important indicators of occurrence of yellow rust fungal. Generally, weather stations can provide the dynamic pattern of meteorological data for site sampled, yet not able to include the information of spatial heterogeneity. Fortunately, remote sensing technology has great potential for providing spatially continuous observations of some variables over large areas (Luo et al., 2010). The aim of the study was to study preliminarily on the relationship between the occurrence of wheat yellow rust and land surface temperature (LST) derived from moderate-resolution imaging spectroradiometer (MODIS) in order to predict and monitor incidence of the yellow rust on large scale.

### 2.4.1 Materials and methods

#### 2.4.1.1 Survey area and field investigations acquisition

Field experiments of winter wheat were conducted during the growing seasons (form April to June) of winter wheat in 2008 and 2009. The investigation locations included Longnan district, Tianshui district, Dingxi district and Pingliang district in GanSu province and Qingyang district in ShanXi province as well as Linxia district in Ningxia Hui Autonomous Region (Fig.1), where the climates are semiarid and subhumid. Survey areas are located between latitude 32°40′N to 35°39′N and longitude 103°10′E to 107°40′E, and the mean altitude is over 2000 meter. The climate condition of surveyed area is characterized by high humidity and rainfall, and yellow rust disease almost occurs every year. It is reported that Longnan district is an important overwintering and oversummering area of yellow rust fungal (Zeng, 2003).

With the aid of the local Department of Plant Protection, 151 plots, including 68 plots from April to June in 2008, and 83 plots from April to June in 2009, were randomly selected and surveyed in the areas. The geographical coordinates of each plot were measured by GPS navigator at the middlemost of the plot. In addition, the disease severity was inspected.

### 2.4.1.2 MODIS land surface temperature (LST) products (MOD11)

**Product description**

MODIS Land Surface Temperature and Emissivity (LST/E) products (named starting with MOD11) provide per-pixel temperature and emissivity values. Temperatures are extracted in Kelvin with a view-angle dependent algorithm applied to direct observations. This method yields the error less than 1 K for materials with known emissivity. The view angle information is included in each LST/E product.

**MOD11 acquisition and processing**

24 MOD11A2 images（MODIS/Terra land surface temperature/emissivity 8-day L3 global 1km SIN grid v005）were acquired for free from Web (http://edc.usgs.gov/#/Find_Data) from April to July in 2008 and 2009, which covered completely the survey area, and 4 scenes images were acquired in every month. The raw data of MOD11A2 imagery were processed and transformed by MRT tool, and LST products were extracted from MODII A2 images. Then the survey area was cut by ENVI from LST images. Followed by that step, 4 scenes 8-day LST images of every month were all averaged, and 6 average LST images, including April, May, June in 2008 and 2009, were obtained. Finally, LST of 151 investigation points were respectively extracted from 6 average LST images.

### 2.4.2 Result

### 2.4.2.1 Determining LST threshold of infected points

The spatial resolution of MODIS temperature products is 1 km, while the DI of every investigation point only stands for the incidence of 30 m in semi diameter plots. Therefore, the scale of MODIS temperature products seemed not satisfied the investigation points for proper relationship between them. However, spatial variability of LST is slim, and the law still exists. A series of results could be found by establishing a two-dimensional spatial coordinate based on DI and LST, in which all investigation points were displayed (Fig 13). Firstly, the DI ranged from 0% to 100%, and most of infected points ranged from 0% to 60%. The LST values were between 292K and 310K with most of infected points distributed in the range from 298K to 306K. In addition, the points in the region of less than 298K were not infected by yellow rust basically; DI were less than 1% expect for one point (296.29K, 16%), which was thought as abnormal point. In addition, the LST values of all investigation points were less than 306K expect for one point (310.09K, 24%), which was abnormal because its LST was far away from LST values of others.

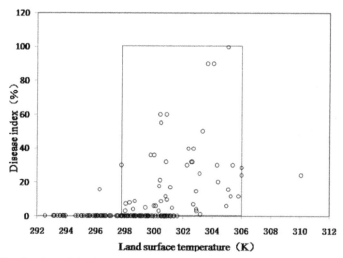

Fig. 13. The distribution of the investigation points

Therefore, without considering other factors, It is concluded that yellow rust can occur when LST is in the region from 298K to 306K.

### 2.4.2.2 Yellow rust incidence analysis based on LST

According to the results illustrated above, the advanced analysis was performed for incidence and possible area of yellow rust. The points in different LST range were done statistical analysis with all points' numbers and the infected points' number, and finally, the incidences were obtained by the number of the infected points dividing the number of all points in the different LST range (Table.8). The result showed that all investigation points in the region of less than 298K were not infected by yellow rust, except for the abnormal point (296.29K, 16%). On the other hand, in the LST region of more than 306K, there was only one point, which was viewed as abnormal point (310.085K, 24%). Thereby, it is quite possible that yellow rust fungus can not survive in the region of more than 306K. The conclusion was consistent with the above result (Fig. 13).

| LST (K) | LST≥2 96 | LST≥2 97 | LST≥ 298 | LST≥ 299 | LST≥ 300 | LST≥ 301 | LST≥ 302 | LST≥ 303 | LST≥ 304 | LST≥ 305 | LST≥ 306 |
|---|---|---|---|---|---|---|---|---|---|---|---|
| Total number | 126 | 112 | 99 | 79 | 61 | 34 | 25 | 16 | 12 | 8 | 1 |
| Number of infected points | 49 | 48 | 47 | 42 | 39 | 27 | 25 | 16 | 12 | 8 | 1 |
| Incidence (%) | 38.89 | 42.86 | 47.47 | 53.16 | 63.93 | 79.41 | 100 | 100 | 100 | 100 | 100 |

Table 8. Statistic analysis in different LST range

Furthermore, there was an increasing trend of incidences with the rising of LST in the region from 296K to 302K. The incidence of yellow rust reached up to 100% when the LST was graeter than 302K (Fig. 14).

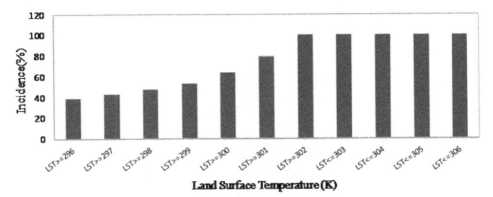

Fig. 14. The incidence of yellow rust in different LST range

### 2.4.2.3 Dividing yellow rust suitable occurrence region based on LST

According to Table 8 and Fig. 14, the survey areas could be divided into yellow rust unsuitable area (NSA), of which LST ranged from 298K to 306K, and yellow rust suitable area (SA), of which the LST was less than 298K and more than 306K. Moreover, the SA was divided into 3 levels according to the infected of yellow rust incidence and LST, and the LST thresholds for each level were: 298K ≤ LST ≤ 299K the low suitable area (LSA), on which the yellow rust occurs with very low possibility (incidence < 60%), 299K ≤ LST ≤ 301K the medium suitable area (MSA), which had moderate possibility for the occurrence of yellow rust (60% <incidence < 100%), and 302K ≤ LST ≤ 306K high suitable area (HSA), of which the environment was highly favorable to yellow rust (incidence=100%).

### 2.4.2.4 Verification

Total 26 points (from May 2008) were applied for the verification the method of estimating the incidence of yellow rust. It should be noted that those points were not used for the defining of the LST thresholds. (Fig. 15). These 26 points were constituted by 18 infected points and 8 non-infected points. Results showed the infected points were all in different suitable areas of wheat yellow rust, while the non-infected points were all in the unsuitable area. Thus the infected situation of yellow rust of these 26 points was consistent with forecast results. Geographically, it seemed that the yellow rust was prone to be prevalent in the northeast of Pingliang, southwest of Qingyang, northeast of Dingxi, the center part of Tianshui, and the west of Longnan, because they all were located in MSA and HAS. This result was consistent with the previous study (Xiao, et al, 2007). To prevent yellow rust from prevalence, more efforts should be placed on the farmlands located in the MSA, HAS and LSA.

### 2.4.3 Conclusions

Plant disease is governed by a number of factors, and the habitat factors play a major role in the development and propagation of fungal pathogens (Sutton et al., 1984; Hélène et al.,

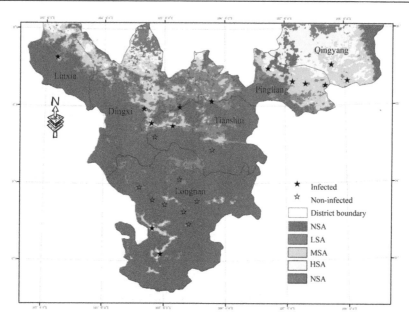

Fig. 15. Forecast map of yellow rust and distribution of measured points in May, 2008 based on LST

2002; Cooke et al., 2006). The yellow rust is no exception. The weather station can only offer points data, and remote sensing, however, can be a promising means for acquiring spatially continuous observations over large area. It has not been reported, if any, that the LST derived from remote sensing data is used to forecast the development of yellow rust.

The study tried to present a method that could forecast the suitable areas of wheat yellow rust by MODIS temperature products in a large scale. And it was proved that LST derived from remote sensing data had potential for predicting the occurrence and development of wheat yellow rust in a large area. From our results, it is clear that preventive measures of yellow rust can been made over large scale area accordingly with different real-time prediction methods based on LST derived from remote sensing data.

## 3. Detecting and discriminating winter wheat aphid by remote sensing

Wheat aphid, Sitobion avenae *(Fabricius)*, is one of the most destructive pests in agricultural systems, especially in temperate climates of the northern and southern hemispheres. Wheat aphid appears annually in the wheat planting area of China, causing great economic damage to plant crops as a result of their direct feeding activities. In high enough densities, wheat aphids can remove plant nutrients, and potentially reduce the number of heads, the number of grains per head, and overall seed weight. The damage is especially high when wheat aphid occurs in the flowering and filling stage of wheat. It is reported that average densities over 20 aphids per plant can cause substantial losses of yield and quality of wheat (Basky & Fónagy, 2003). There are also indirect damages including excretion of honeydew from aphids and as a vector of viruses, most notably two strains of the Luteovirus Barley Yellow Dwarf Virus (BYDV-MAV and BYDV-PAV) (Susan et al, 1992). To prevent the occurrence and prevalence of aphid, large amounts of insecticides are used, causing environment pollution. Therefore, large-scale, real-time prediction and monitoring of wheat aphid incidence and damage degree using remote sensing technology are extremely important.

## 3.1 Detecting winter wheat aphid using hyperspectral data

The study aimed to identify spectral characteristics of wheat leaf and canopy infected by aphid and find the sensitive bands to aphid at canopy level in filling stage of wheat, and to establish an aphid damage hyperspectral index (ADHI) based on those sensitive bands for detecting aphid damage levels in wheat canopy level in filling stage of wheat.

### 3.1.1 Materials and methods

#### 3.1.1.1 Field experiments and field inventory

The field experiment plot was located at Xiaotangshan Precision Agriculture Experiment Base, Changping distract, Beijing (40°10.6′N, 116°26.3′E). The experimental field was about 250 m in length and 80 m in width. The winter wheat was planted in the study area from Oct 3, 2009, and harvested from June 25, 2010. Field inventory was conducted on June 7, 2010 when wheat was in the filling stage. Twenty five ground investigations including different aphid damage levels were selected. Aphid damage level was surveyed according to the investigation rule.

#### 3.1.1.2 Canopy spectral measurements

Please refer to 1.1.1.2 part above.

### 3.1.2 Results

#### 3.1.2.1 Leaf spectral characteristics of wheat infested by aphid

Representative reflectance measured from wheat aphid-infested and uninfested wheat leaves are shown in Fig. 16. It was evident that the spectral response of the wheat leaf was significantly affected by wheat aphid feeding (Fig. 16). The reflectance of wheat leaf infested by aphid was higher in the visible spectrum and short-wave infrared region and lower in near-infrared region than that of uninfested leaf. A significant increase in the reflectance from the wheat aphid-infested leaf in the visible region (400-700 nm) was observed, evidently due to reduction of photosynthetic pigment concentrations in particular chlorophylls caused by wheat aphid feeding (Richardson et al., 2004).

#### 3.1.2.2 Canopy spectral characteristics of wheat infested by aphid

Compared with the canopy spectra of the healthy wheat, the canopy reflectance of aphid-infested wheat was gradually decreased in the range from 350 nm to 1750 nm, especially in the near infrared region (Fig. 17). Previous researches indicated that wheat had higher reflectance at visible wavelengths than the healthy vigorously growing wheat because the photoactive pigments (chlorophylls, anthocyanins, carotenoids) were destroyed. In this study, aphid occurred in the filling stage of wheat and the honeydew excreted by aphid absorbed dust or others from surrounding environment and contaminated (darkened) the leaf surface. As a result, the absorption at light slight wavelengths became stronger instead of weaker.

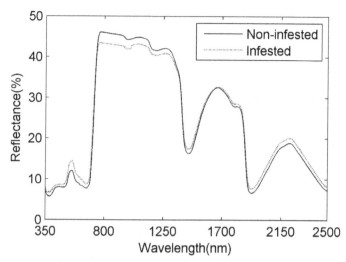

Fig. 16. The spectral reflectance of winter wheat leaf uninfested and infested by aphid

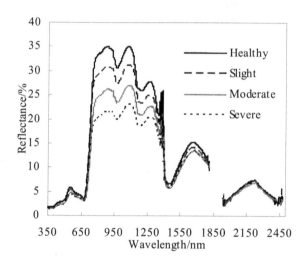

Fig. 17. The spectral reflectance of healthy wheat and wheat infested by various aphid damage levels. (Healthy: the average spectra of healthy wheat samples; Slight: the average spectra of aphid damage level 1and 2; Moderate: he average spectra of aphid damage level 3and 4; Severe: the average spectra of aphid damage level 5 and 6).

### 3.1.2.3 Aphid damage hyperspectral index for detecting aphid damage degree

### Sensitive band selection of aphid infestation based on canopy reflectance

The sensitive bands were selected out by relevance analysis between reflectance and aphid damage levels. The reflectance ranges were from 400 nm to 690 nm, from 700 to 1300 nm and from 1500 to 1800 nm. The most sensitive bands to aphid were 551 nm ($R^2$=0.741) in visible light, 823 nm ($R^2$=0.865) in near infrared (NIR) and 1654 nm in short-wave infrared (SWIR) ($R^2$=0.668), respectively (Fig. 18).

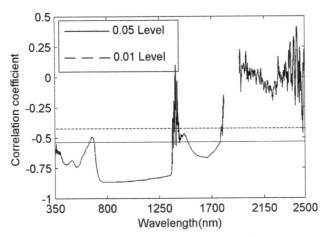

Fig. 18. Correlation coefficient between reflectance and aphid damage levels

Aphid damage hyperspectral index (ADHI) was established based on the most sensitive bands from hyperspectral data in the visible light region, NIR and SWIR and weight coefficient calculated according to rate of change of reflectance between healthy wheat and aphid-infected wheat, respectively.

$$\text{ADHI} = 0.32 \times \frac{R551_{normal} - R551_{infested}}{R551_{normal}} + 0.51 \times \frac{R823_{normal} - R823_{infested}}{R823_{normal}}$$
$$+ 0.17 \times \frac{R1654_{normal} - R1654_{infested}}{R1654_{normal}}$$

where $R551_{normal}$, $R823_{normal}$ and $R1654_{normal}$ are reflectance in 551 nm, 823 nm and 1654 nm of healthy wheat; $R551_{infested}$, $R823_{infested}$, $R1654_{infested}$ are reflectance in 551 nm, 823 nm and 1654 nm of aphid-infected wheat; 0.32, 0.51 and 0.17 are weight coefficients calculated by the contribution to change rates.

Further more, the correlation analysis was conducted between ADHI and aphid damage level from 25 investigation points (Fig. 19). It was concluded that ADHI exhibited high relationship with aphid damage levels ($R^2$=0.839). Therefore, ADHI was an important index to estimate aphid damage level in winter wheat.

### 3.1.3 Conclusions

Hyperspectral remote sensing has gone through rapid development over the past two decades and there is a trend toward the use of hyperspectral image in the application of remote sensing for precision farming. The study analyzed the spectral characteristics of wheat infested by aphid and selected the sensitive bands to aphid damage level. Then, an ADHI was developed using the most sensitive bands in visible light region, NIR and SWIR.

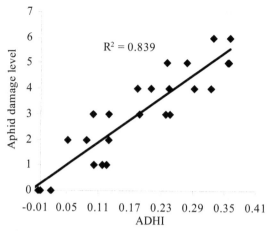

Fig. 19. The correlation between ADHI and aphid damage level

It was concluded that ADHI was a sensitive index to aphid damage levels, and could be used to retrieve aphid damage levels in the filling stage of wheat.

Crop growth is very dynamic processes and monitoring the condition of agricultural corps is a complex issue. It is possible that wheat damage symptoms caused by aphids and its response of canopy reflectance are different in different wheat growth stages. This study revealed that the reflectance of wheat infested by aphid was lower than healthy wheat in filling stage probably because of honeydew excreted by aphid. This was not consistent with previously published results in early detection of aphid infestation. Therefore, whether the ADHI can effectively retrieve aphid damage levels in other wheat growth stages remains as a task of future studies.

## 3.2 Detecting winter wheat aphid incidence using Landsat 5 TM

Wheat aphid occurrence and damage degrees are related to many factors including temperature, humidity, precipitation, field management, enemies, etc.. Most of the present studies on aphid prediction have been conducted based on meteorological data acquired from weather stations, and aphid density was monitored using the spectral characteristics of wheat infested by aphid. However, it is rare to investigate the relationship between environmental parameters, vegetable information derived from satellite images and aphid damage degrees. The aim of the present study is to investigate the relationships of aphid occurrence and damage degree to LST, NDWI, and MNDWI, which are related to vegetation water content derived from multi-temporal Landsat 5 TM. Another goal of the current research is to distinguish the degrees of aphid damage using 2-dimension feature spaces established by LST-NDWI and LST-MNDWI.

### 3.2.1 Materials and methods

#### 3.2.1.1 Study areas

The study areas are selected in Shunyi district (116°28′−116°58′ E，40°00′−40°18′ N) and Tongzhou district (116° 32′−116°56′ E, 39°36′ −40°02′ N,) of Beijing, China (Fig.20-a). The

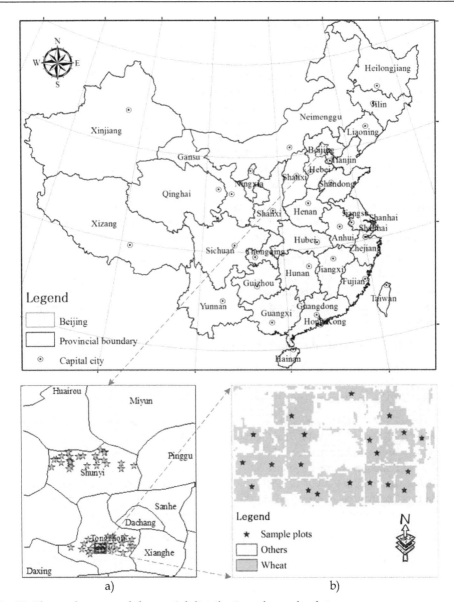

Fig. 20. The study area and the spatial distribution of sample plots

study areas have flat topography, with elevation ranging from 20 m to 40 m. The study areas have semi-humid warm temperate climate with yearly precipitation of 625 mm and mean temperature of 11.5°C in the Shunyi district and yearly precipitation 620 mm and mean temperature of 11.3°C in the Tongzhou district. Both districts are considered main winter wheat planting areas in Beijing, and aphid infestations occur in both areas almost every year.

### 3.2.1.2 Field inventory and data pre-processing

Field inventory was conducted during the growing seasons of winter wheat in 2010. The winter wheat in the study areas were planted between September 25 and October 7, 2009, and harvested between June 19 and June 25, 2010. Based on the combination of representative sampling and random sampling scheme, 70 sample plots with size of 0.09 ha (30 m × 30 m) each were collected as in Fig 1-a. These sample plots had different site conditions, plant densities, and management conditions. Aphid density surveys were carried out respectively on May 4 and May 6 for jointing stage, May 20 and May 21 for the heading stage, and June 3 and June 4 for the filling stage. The geographical coordinates of each plot were measured by global positioning system (GPS) ( GeoExplorer 3000 GPS, with the error within 1m) at the middlemost of the plot.

Each sample covered with an area of 1 m². Then, 10 tillers in each sample plot were randomly selected, and the number of aphids was counted. The aphid densities were then estimated as follows: total aphids /10 tillers.

The survey results were divided into three aphid damage degrees according to the aphid density investigated for facilitating the study. They were S0: non-infested by aphid and no damage to wheat, S1: aphid abundance/per tiller was about 3-10 and damage degree to wheat was slight, and S3: aphid abundance/per tiller was more than 20 and damage degree to wheat was severe.

### 3.2.1.3 Satellite image acquisition and pre-processing

Three Landsat-5 Thematic Mapper (TM) images (path 123/row 32) and three MOD 02 1 KM-Level 1B Calibrated Radiances Production (MOD 02) were acquired on May 4, May 20 and June 5, 2010, respectively. And all images were more than 90% cloud-free.

The Landsat-5 TM images were spectrally corrected to reflectance using the Landsat TM calibration tool and FLAASH (Fast line-of-sight Atmospherics Analysis of Spectral Hypercubes) was used to correct the image for atmospheric effects in ENVI 4.5. The Landsat-5 TM images were geometrically corrected versus a reference IKONOS image (equivalent scale map 1:10000) of the same area, available from a previous study. The resulting root mean square error (RMSE) did not exceed 0.3 pixels, which was adequate for the purposes of the present study.

### 3.2.1.4 Derivation of LST, NDWI and MNDWI from Landsat 5 TM

NDWI and MNDWI are both sensitive to changes in liquid water content of vegetation canopies (Hunt and Rock, 1989). In the current research, both NDWI and MNDWI were used to determine the threshold of aphid occurrence and the aphid damage degree. The indices are of the general form, as shown in the following:

$$NDWI = \frac{R_{NIR} - R_{SWIR}}{R_{NIR} + R_{SWIR}} \quad MNDWI = \frac{R_{GREEN} - R_{SWIR}}{R_{GREEN} + R_{SWIR}}$$

where $R_{GREEN}$, $R_{NIR}$ and $R_{SWIR}$ are the reflectance in the green band, near-infrared band and short wave infrared band, respectively. For Landsat TM/ETM+, $R_{GREEN}$, $R_{NIR}$ and $R_{SWIR}$ correspond to band2, band4 and band5, respectively.

LST is the radioactive skin temperature of the land surface, which plays an important role in farm and ecological environment. The present paper aims to discuss the relationship between LST and aphid occurrence and spread. LST was derived from the thermal infrared band (10.4-12.5μm) data of Landsat-5 TM using generalized single-channel algorithm developed by Jiménez-Muñoz and Sobrino (Jiménez-Muñoz and Sobrino, 2004). Surface emissivity (ε) and atmospheric water vapor content (w) were important parameters in the generalized single-channel algorithm. In the study, w was derived from the reflectance of band2 and band19 of MOD02, (Kaufman and Gao, 1992), and ε was calculated by vegetation coverage (Carlson and Ripley, 1997).

The NDWI, MNDWI and LST of all sample points were calculated and extracted from the Landsat images.

### 3.2.1.5 Subset image selection and wheat extraction

We resized the subset areas with size of 7.2 km$^2$ (3 km × 2.4 km) from the study area image located in Tongzhou district and covered with 20 evenly distributed sample points, and the aphid densities of the sample points were surveyed on May 6, May 20 and June 4, 2010, respectively. The survey results showed that the aphid damage degree of all sample points were S0 on May 6, 18 points for S1 and 2 points for S0 on May 20, and 16 points for S2 and 4 points for S0 on June 4, respectively. The subset areas were small enough and 20 sample points evenly distributed, According to the survey result, the aphid damage degree of the sample plots was basically same. Thus, the change of the aphid damage degree of wheat pixels in the wheat plots was slim or even basically the same as the sample plots. The wheat area of subset image selection area was extracted using classification of decision tree in ENVI 4.5 (Fig 20-b). The LST, NDWI and MNDWI of 2000 wheat pixels were extracted.

### 3.2.1.6 Methods of accuracy assessment

One basic accuracy assessment currently being used is overall accuracy, which is calculated by dividing the correctly classified pixels by the total number of the pixels checked. The Kappa coefficient is a measure of the overall agreement of a matrix introduced to the remote sensing community in early 1983. It has since become a widely used measure for classification accuracy. In contrast to overall accuracy, the Kappa coefficient takes non-diagonal elements into account (Rosenfield and Fitzpatrick-Lins, 1986), and it is calculated by the formula:

$$K = \frac{N\sum_{i=1}^{r} X_{ii} - \sum_{i=1}^{r} X_i + X_{+i}}{N^2 - \sum_{i=1}^{r} X_i + X_{+i}}$$

where r is the number of rows and columns in the error matrix; N is the total number of observations; Xii is the observation in row i and column i; Xi+ is the marginal total of row I; X+i is the marginal total of column i.

### 3.2.2 Results

### 3.2.2.1 2-dimensional feature space based on LST-VI

The minimum value, maximum value, mean values and standard deviations of LST, NDWI and MNDWI with aphid damage degrees of wheat pixels in subset image selection were listed in Table 9 and Table 10. And 2-dimensional feature space coordinates were established with LST as the abscissa and NDWI and MNDWI as the vertical axis, respectively (Figs. 2, 3). LST ranged from 287.5879 to 313.3448, NDWI ranged from 0.0226 to 0.5591 and MNDWI ranged from -0.3402 to -0.1077, respectively.

It is clear that LST was increasing from S0 to S1 to S2. LST was an important driving factor for aphid occurrence and could distinguish wheat non-infected from infested by aphids (Fig. 21 and Table 9). The general trend of NDWI increased firstly and reduced afterward, whereas MNDWI reduced firstly and increased afterward from S0 to S1 to S2.

| Aphid Damage Degree | LST | | NDWI | | MNDWI | |
|---|---|---|---|---|---|---|
| | Minimum value | Maximum value | Minimum value | Maximum value | Minimum value | Maximum value |
| S0 | 287.5879 | 296.2498 | 0.0226 | 0.4405 | -0.3402 | -0.1077 |
| S1 | 297.8084 | 306.0133 | 0.2083 | 0.5591 | -0.6506 | -0.3326 |
| S2 | 300.5391 | 313.3448 | 0.0473 | 0.4542 | -0.4117 | -0.1159 |

Table 9. Minimum and maximum values of LST, NDWI and MNDWI in S0, S1 and S2

| Aphid Damage Degree | LST | | NDWI | | MNDWI | |
|---|---|---|---|---|---|---|
| | Mean value | Standard deviation | Mean value | Standard deviation | Mean value | Standard deviation |
| S0 | 290.8578 | 1.4740 | 0.3029 | 0.0574 | -0.2293 | 0.0296 |
| S1 | 299.9236 | 1.0834 | 0.3998 | 0.0587 | -0.4940 | 0.0362 |
| S2 | 303.9424 | 1.7121 | 0.2979 | 0.0458 | -0.2672 | 0.0402 |

Table 10. Mean value and standard derivation of LST, NDWI and MNDWI in S0, S1 and S2

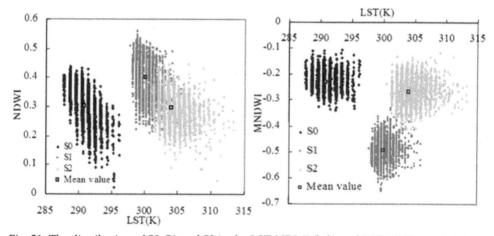

Fig. 21. The distribution of S0, S1 and S2 in the LST-NDWI (left) and LST-MNDWI (right) feature space

### 3.2.2.2 Discriminating aphid damage degrees using LST and MNDWI

In the 2-dimensional feature space coordinate system that was composed by LST and MNDWI, the S0 samples mainly scattered on the left part of the coordinate system, whereas S1 and S2 samples were distributed on the right part. As shown in Fig. 22, when LST was lower than the certain value, aphid did not occur, suggesting that LST served as a key factor of aphid occurrence and the MNDWI was sensitive to aphid damage degree.

Furthermore, $LST_0$ and $MNDWI_0$, which were the cutoff value of threshold values of LST and MNDWI of S0, S1 and S2, were determined by mean values and standard deviations. LST0 and MNDWI0 were calculated by formula as follows:

$$LST_0 = LST\_M1 - 2 \times LST\_SD1$$

$$MNDWI_0 = (M\_M1 + 3 \times M\_SD1) + [(M\_M1 + 3 \times M\_SD1) - (M\_M2 - 3 \times M\_SD2)]/2$$

where LST_M1 and LST_SD1 are the mean value and standard deviation of LST for S1; M_M1and M_SD1 are the mean value and standard deviation of MNDWI for S1; and M_M2 and M_SD2 are the mean value and standard deviation of MNDWI for S2.

According to Table 3, $LST_0$ = 297.7568 and $MNDWI_0$ = -0.3866. Wheat was not infested by aphid when LST< 297.7568, and aphid damage degree was S1 when LST≥297.7568K and -0.6506≤MDNWI ≤-0.3866 and S2 when LST≥297.7568K and -0.3866 ≤MDNWI ≤-0.1077 (Fig. 22).

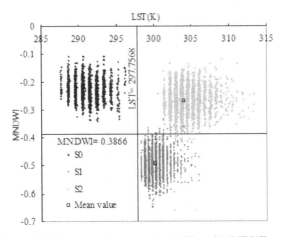

Fig. 22. Discriminating aphid damage degrees using LST and MNDWI

### 3.2.2.3 Verification

All survey samples, except 20 samples in the subset selection image were used to test the aphid prediction accuracy of 2-dimensional feature space based on LST and MNDWI (Fig. 23).

The discrimination accuracy was assessed using overall accuracy and kappa coefficient (Table 11). The results showed that the overall accuracy was 84%, and the Kappa accuracy was 75.67%.

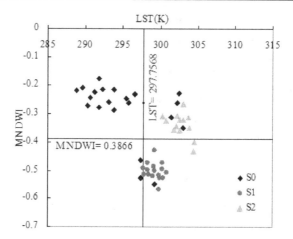

Fig. 23. Distribution of test sample plots in LST-MNDWI feature space

|        | S0 | S1 | S2 | Total |
|--------|----|----|----|-------|
| S0     | 17 | 0  | 0  | 17    |
| S1     | 2  | 14 | 0  | 16    |
| S2     | 4  | 2  | 11 | 19    |
| Total  | 23 | 16 | 11 | 50    |

Kappa coefficient = 0.7567

Table 11. Error matrices of the verification samples

### 3.2.3 Conclusions

This study successfully investigated the relationship between aphid damage degrees and several spectral features, such as NDWI, MNDWI and LST, through 2-dimensional feature space method. The results indicated that LST was the key factor in predicting the occurrence of aphid, and MNDWI was more sensitive to aphid damage degree than NDWI. In the 2-dimension feather space composed by LST and MNDWI, the result showed that S0, S1 and S2 were divided into three regions; S0 was distributed on the left of the space, and S1 and S2 on the right. Further, LST0 and MNDWI0 were calculated according the mean and derivation of S1, S2 as the cutoff value of threshold value to discriminate S0, S1 and S0. Through the verification of discrimination threshold value, it confirmed that the overall accuracy of discrimination was 84% and Kappa coefficient was 0.7567, suggesting that LST and MNDWI were of great potential in discriminating and monitoring the aphid damage degree over a large area, only using thermal infrared band and multi-spectral satellite images.

## 4. References

Basky Z. & Fónagy A. (2003).Glutenin and gliadin contents of flour derived from wheat infested with different aphid species. *Pest Management Science*, 59, 426-430.

Becker, B. L., David, P. L., & Qi, J. G. (2007). A classification-based assessment of the optimal spectral and spatial resolutions for Great Lakes coastal wetland imagery. *Remote Sensing of Environment*, 108, 111–120

Bravo, C., Moshou, D., West, J., McCartney, A., & Ramon, H. (2003). Early disease detection in wheat fields using spectral reflectance. *Biosystems Engineering*, 84, 137–145.

Broge, N. H., & E. Leblanc. 2000. Comparing prediction power and stability of broadband and hyperspectral vegetation indices for estimation of green leaf area index and canopy chlorophyll density. *Remote Sensing of Environment*, 76, 156–72.

Carlson T. N. & Ripley D. A.. On the relation between NDVI, fractional vegetation cover, and leaf area index. *Remote Sensing of Environment*, 1997, 62(3): 241-252.

Ceccato, P., N. Gobron, S. Flasse, B. Pinty, & S. Tarantola. 2002. Designing a spectral index to estimate vegetation water content from remote sensing data: Part 1. Theoretical approach. *Remote Sensing of Environment*, 82, 188–97.

Christou, P., Twyman, R.M., (2004). The potential of genetically enhanced plants to address food insecurity. *Nutrition Research Reviews* 17, 23–42.

Daughtry, C. S., C. L. Walthall, M. S. Kim, E. Brown de Colstoun, & J. E. McMurtrey.(2000). Estimating corn leaf chlorophyll concentration from leaf and canopy reflectance. *Remote Sensing of Environment*, 74, 229-239.

Fensholt, R., & I. Sandholt.(2003). Derivation of a shortwave infrared water stress index from MODIS near-and shortwave infrared data in a semiarid environment. *Remote Sensing of Environment*, 87, 111–21.

Filella, I., Serrano, L., Serra, J., & Penuelas, J. (1995). Evaluating wheat nitrogen status with canopyreflectance indices and discriminant analysis. *Crop Science*, 35, 1400–1405.

Franke, J., & Menz, G. (2007). Multi-temporal wheat disease detection by multi-spectral remote sensing. *Precision Agriculture*, 8, 161–172.

Galvão, L. S., A. Formaggio, R. & Tisot, D. A. (2005). Discrimination of sugarcane varieties in Southeastern Brazil with EO-1 Hyperion data. *Remote Sensing of Environment*, 94, 523–34.

Gamon, J. A., Penuelas, J., & Field, C. B. (1992). A narrow-waveband spectral index that tracks diurnal changes in photosynthetic efficiency. *Remote Sensing of Environment*, 41(1), 35–44.

Gitelson, A. A., Merzlyak, M. N., & Chivkunova, O. B. (2001). Optical properties and nondestructive estimation of anthocyanin content in plant leaves. *Photochemistry and Photobiology*, 74, 38–45.

Goovaerts, P., Jacquez, G. M., & Marcus, A. (2005). Geostatistical and local cluster analysis of high resolution hyperspectral imagery for detection of anomalies. *Remote Sensing of Environment*, 95, 351–367.

Gong, P., Pu R., Heald R.C.(2002). Analysis of in situ hyperspectral data for nutrient estimation of giant sequoia. *International Journal of Remote Sensing*, 23, 1827-1850.

Haboudane, D., Miller, J. R., Tremblay, N., Zarco-Tejada, P. J., & Dextraze, L. (2002). Integrated narrowband vegetation indices for prediction of crop chlorophyll content for application to precision agriculture. *Remote Sensing of Environment*, 81, 416-426.

Haboudane, D., J. R. Miller, E. Pattery, P. J. Zarco-Tejad, & I. B. Strachan. (2004). Hyperspectral vegetation indices and novel algorithms for predicting green LAI of crop canopies: Modeling and validation in the context of precision agriculture. *Remote Sensing of Environment*, 90, 337–52.

Hélène L., Frédéric F., Pierre D., Gaétan B., Isztar Z. (2002),Estimation of the spatial pattern of surface relative humidity using ground based radar measurements and its application to disease risk assessment. *Agricultural and Forest Meteorology*, 111, 223–231.

Huang, W. J., David, W. L., Niu, Z., Zhang, Y. J., Liu, L. Y., & Wang, J. H. (2007). Identification of yellow rust in wheat using in situ spectral reflectance measurements and airborne hyperspectral imaging. *Precision Agriculture*, 8, 187–197.

Hunt E.R., Rock B.N. (1989). Detection of changes in leaf water content using Near- and Middle-Infrared reflectances. *Remote Sensing of Environment*, 30, 43-54.

Jiménez-Muñoz J. C., Sobrino J. A. & Leonardo P. (2004). Land Surface Temperature Retrieval from Landsat TM 5. *Remote Sensing of Environment*, 90, 434-440.

Kaufman Y. J. & Gao B. C. (1992). Remote Sensing of Water Vapor in the Near IR from EOS/MODIS. *IEEE Transactions on Geoscience and Remote Sensing*, 30, 871-884.

Kim, M. S., C. S. T. Daughtry, E. W. Chappelle, & J. E. McMurtrey. 1994. The use of high spectral resolution bands for estimating absorbed photosynthetically active radiation (APAR). *In Proceedings of the 6th International Symposium on Physical Measurements and Signatures in Remote Sensing*, 299–306.

Li, G. B., Zeng, S. M., & Li, Z. Q. (1989). Integrated Management of Wheat Pests (pp. 185–186). Beijing: Press of Agriculture Science and Technology of China (in Chinese).

Luo, J. H., Zhang, J. C., Huang, W. J., Xu, X. G., Jin, N. (2010). Preliminary study on the relationship between land surface temperature and occurrence of yellow rust in winter wheat. *Disaster Advances, 3*, 288-292.

Luo, J. H., Huang, W. J., Zhang, J. C., Xu X. G. & Wang D. C. (2011). The preliminary study on spectral response of wheat under different stresses between field and satellite remote sensing. *Sensor Letters, 9*, 1225-1228.

Merton, R., & J. Huntington.(1999). Early simulation of the ARIES-1 satellite sensor for multi-temporal vegetation research derived from AVIRIS. In *Summaries of the Eight JPL Airborne Earth Science Workshop*, 9–11 February, 299–307. Pasadena, CA: JPL Publication 99-17.

Moshou, D., Bravo, C., West, J., Wahlen, S., McCartney, A., & Ramon, H. (2004). Automatic detection of 'yellow rust' in wheat using reflectance measurements and neural networks. *Computers and Electronics in Agriculture*, 44, 173–188.

Peñuelas, J., Baret, F., & Filella, I.(1995). Semi-empirical indices to assess carotenoids/chlorophyll a ratio from leaf spectral reflectance. *Photosynthetica*, 31, 221–230.

Merzlyak, M. N., Gitelson, A. A., Chivkunova, O. B., & Rakitin, V. Y. (1999). Non-destructive optical detection of pigment changes during leaf senescence and fruit ripening. *Physiologia Plantarum*, 106, 135–141.

Peñuelas, J., Gamon, J. A., Fredeen, A. L., Merino, J., & Field, C. B. (1994). Reflectance indices associated with physiological changes in nitrogen- and water-limited sunflower leaves. *Remote Sensing of Environment*, 48, 135–146.

Peñuelas, J., J. Piñol, R. Ogaya, & I. Filella. (1997). Estimation of plant water concentration by the reflectance water index WI (R900/R970). *International Journal of Remote Sensing*, 18, 2869–75.

Pu, R., Ge S., Kelly N.M., Gong P. (2003). Spectral absorption features as indicators of water status in coast live oak (Quercus agrifolia) leaves. *International Journal of Remote Sensing*, 24, 1799-1810.

Pu, R., Foschi L., Gong P. (2004). Spectral feature analysis for assessment of water status and health level in coast live oak (Quercus agrifolia) leaves. *International Journal of Remote Sensing*, 25, 4267-4286.

Rouse, J. W., R. H. Haas, J. A. Schell, & D. W. Deering.(1973). Monitoring vegetation systems in the Great Plains with ERTS. *Proc 3rd ERTS Symp* 1, 48–62.

Rosenfield G. & Fitzpatrick-Lins K. (1986). A coefficient of agreement as a measure of thematic classification accuracy. *Photogrammetric Engineering and Remote Sensing*, 52, 223-227.

Rules for Resistance Evaluation of Wheat to Diseases and Insect Pests Part 7: Rule for Resistance Evaluation of Wheat to Aphids. NY/T 1443.7-2007

Raikes, C., & L. L. Burpee. (1998). Use of multispectral radiometry for assessment of Rhizoctonia blight in creeping bentgrass. *Phytopathology*, 88, 446-449.

South, S., Qi, J. G., & Lusch, D. P. (2004). Optimal classification methods for mapping agricultural tillage practices. *Remote Sensing of Environment*, 91, 90–97

Susan E. Halbert, June Connelly B., Bishop G. W., et al. (1992).Transmission of barley yellow dwarf virus by field collected aphids (Homoptera: Aphididae) and their relative importance in barley yellow dwarf epidemiology in southwestern Idaho. *Annals of Applied Biology*, 121, 105-121.

Strange, R.N., Scott, P.R., (2005). Plant Disease: A Threat to Global Food Security. *Annual review of Phytopathology*, 40, 83–116.

Sutton, J.C., Gillespie, T.J., Hildebrand, P.D. (1984). Monitoring weather factors in relation to plant disease. *Plant Disease*, 68, 78–84.

Thenkabail, P. S., Smith, R. B., & De Pauw, E. (2000). Hyperspectral vegetation indices and their relationships with agricultural crop characteristics. *Remote Sensing of Environment*, 71, 158–182.

West, J. S., Bravo, C., Oberti, R., Lemaire, D., Moshou, D., & McCartney, H. A. (2003). The potential of optical canopy measurement for targeted control of field crop disease. *Annual Reviews of Phytopathology*, 41, 593–614.

Xiao, Z. Q., Li, Z. M., Fan M., Zhang, Y. Ma, S. j. (2007), Prediction model on stripe rust influence extent of winter wheat in Longnan Mountain area. *Chinese Journal of Agrometeorology*, 28, 350-353.

Yu, B., Ostland, I.M., Gong, P., Pu, R. L. (1999) Penalized discriminant analysis of in situ hyperspectral data for conifer species recognition. *IEEE Transactions on geosciences and remote sensing*, 37, 2569-2577.

Zarco-Tejada, P. J., A. Berjón, R. López-Lozano, J. R. Miller, P. Martín, V. Cachorro, M. R. González, & A. Frutos. (2005). Assessing vineyard condition with hyperspectral indices: Leaf and canopy reflectance simulation in a row-structured discontinuous canopy. *Remote Sensing Environment*, 99:271–87.

Zeng, S. M. (2003). Simulation study on oversummering process of wheat stripe rust caused Puccinia striiformis west. *In China, Acta Phytopathologica Sinica*, 33, 267-278.

Zhang, M. H., Qin, Z. H., Liu, X., & Ustin, S. L. (2003). Detection of stress in tomatoes induced by late blight disease in California, USA, using hyperspectral remote sensing. *International Journal of Applied Earth observation and Geo information*, 4, 295–310.

Zhang, J. C., Huang, W. J., Li, J. Y., Yang, G. J., Luo, J. H., Gu, X. H., & Wang, J. H. (2011). Development, evaluation and application of a spectral knowledge base to detect yellow rust in winter wheat. *Precision Agriculture*, 12, 716-731.

# 4

# Land Cover Change Detection in Southern Brazil Through Orbital Imagery Classification Methods

José Maria Filippini Alba, Victor Faria Schroder
and Mauro Ricardo R. Nóbrega
*Brazilian Agricultural Reasearch Corporation (Embrapa)*
*Brazil*

## 1. Introduction

Remote sensing has been considered a promising technology as support for agriculture since its beginnings, due to its contribution for a climatic perspective or for understanding of processes related to land. However, significant applications occurred only in the late twentieth century, as result of the creation of best orbital systems, with higher spatial resolution, more bands and stereoscopic capture. Several orbital platforms, as AQUA/TERRA, Quickbird and Ikonos are examples in that sense (Moreira, 2005; Embrapa, 2009).

Engineering innovations, new sensors and methods of digital image processing must be performed simultaneously so that the advances in remote sensing will be achieved. Anyway, the incorporation of orbital images on geographic information systems (GIS) and their post-processing appear as significant application since a daily life perspective, specially when classification methods are involved, because of their relation to land use, land cover and easy interpretation.

This chapter considers classification methods applied on orbital imagery in Southern Brazil, in the coastal plain of Rio Grande do Sul state (Fig. 1), where a sequence of lagoons and lakes of different sizes occurs in the context of subtropical to temperate climate with cold winters and hot summers, being organized according to the following four sections:

- About classification methods.
- Evaluation of rice planting area in the vicinity of Caiuba lagoon (1981 – 2009).
- Analysis of land cover evolution in the municipality of Montenegro (1993 – 2008).
- Comparison and evaluation of errors.

All the exposed data are related to research projects of the Embrapa Temperate Climate Research Center, Pelotas, Rio Grande do Sul state, one of the 45 research centers of the Brazilian Agricultural Research Corporation (Embrapa) spread on the national territory.

## 2. About classification methods

Classification methods were created in the statistical context, when a collection of objects or samples could be characterized and separated in different classes (Davis, 1986). The method

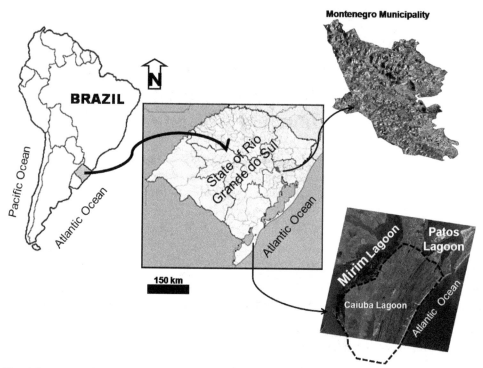

Fig. 1. Location of the areas of study in the context of South America, Brazil and the State of Rio Grande do Sul. Montenegro municipality as gray levels composite of the bands 1, 4 and 5 TM/Landsat 5 (Miranda, 2002, 2005) and Caiuba lagoon in gray levels composite of the bands 3, 2 and 1 ASTER (NASA, 2004) on september 26th, 2000.

was extended for processing of digital images considering the pixels as objects to be classified (Crosta, 1993, Lillesand & Kiefer, 1994, Jensen, 1996). The application of the classification methods on satelite imagery is affected by two main factors:

i.    Intervention of user.
ii.   Criteria of definition for the groups.

The "supervised classification" is included in topic (i), when the user defines the groups through digitalization of uniform spectral answer. Statistics are calculated for each group, so the classification is performed for all the image (pixel by pixel). The unsupervised classification eliminates the intervention of the user; then, the software defines the groups by means of scattergrams "band versus band", where isolines related to distribution density of the pixels are analysed (Crosta, 1993).

Different options for case (ii) are posible, by instance, criteria can use the standard deviation for the parallelepiped method or a minimum distance when an eliptical form is defined for each group or a combination of the later with statistical probability, that is, the maximum likelihood method. Jensen (1996) presented other criteria of classification, as Isodata method and the Fuzzy method.

Spectrometric methods measure the response of target materials in the laboratory or field. Then the spectral patterns are simulated for a specific sensor through a specialized software, so that a sequence of orbital images is classified according to the pattern generated by that software (Lillesand & Kiefer, 1994; Pontara, 1998; Moreira, 2005).

Classification of remote sensing images appear as useful tool in terms of land use, whether in local scale or in regional scale. Filippini-Alba and Siqueira (1999) classified land use in the municipality of Pelotas, Rio Grande do Sul state, Brazil, according to nine classes: agriculture, clay soils, forestry, natural forest, pastures, soil without vegetal cover, urban, water and wetlands. Natural forest and pastures occupied 23% and 30% of the territory respectively, with intense interference between the classes "urban" and "soil without vegetal cover". Similar classes were considered by Bolfe et al. (2009) for land use in Rio Grande do Sul state, but with different results. Agriculture and pastures occurred 32% and 50% of the territory respectively with only 3% for natural forest. The differences between both studies are easy explained in term of scale, because in the two occasions Landsat images were used and a municipality was considered at the former and a state at the latter, with territory difference of 1 to 155 times in size.

Lu et al. (2004) discriminated seven categories of change detection techniques: (i) algebra; (ii) transformation; (iii) classification; (iv) advanced models; (v) Strategies with geographic information systems (GIS); (vi) visual analysis and (vii) miscellanea. The classification methods are detached, with six different modalities. One of them, the "Post-classification comparison", is predominantly used in this chapter. That is, multi-temporal images are classified separately into thematic maps, then the classified images are compared pixel by pixel. The "Post-classification comparison" minimizes the atmospheric impacts, the environmental differences among multi-temporal imagery, as well as differences related to the sensor kind, providing a complete matrix of change information. However, some disadvantages can be appointed, because a great amount of time and expertise is required and, by other side, the final accuracy depends on the quality of the classified image due to the weather condition on that date.

Guild et al. (2004) quantified the areas of deforestation in the Amazonian forest, state of Rondonia, Brazil. The tasselled cup transformation (Crist & Kauth, 1986) was applied with the Landsat imagery from the years 1984, 1986 and 1992. The variables brightness, greenness and wetness were evaluated for each year, then, a file integrated the nine levels of information (three variables by three years). These data were processed through principal components and classification methods with overall accuracy of 79.3, 68.4% and 71.4%, for tasselled cap land cover change classification, tasselled cap with principal components land cover change classification and tasselled cap image differencing, respectivelly. Final classes were a combination between land cover and time, so change detection was quantified.

The two applications present in this chapter consider the Supervised classification method with maximum likelihood as criteria for definition of the classes. The proximity of the study areas and knowledge of the territory justify this option to take advantage of available information. Unsupervised classification is a fast proccess, good for unknown or outlying areas, when truth of field is unavailable and most time-consuming after processing, due to the need of class identification. Maximum likelihood criteira is restricted by software and time-consuming but it represents a improvement in relation simple criteria as the parallelepiped or the minimum distance.

Acording to Lu et al. (2004), methods (iii) and (v) were considered in this chapter. Classification (iii) was applied in both conditions, Caiuba lagoon and Montenegro municipality. The extraction of the poligon corresponding to the "potential area for agriculture" in the vicinity of Caiuba lagoon represents a tipical strategy of GIS (v). Softwares of digital images processing and GIS are very similar. Both can execute multilayer processing, including raster/vector files and logic/mathematical algorithms, but digital images processing is more specific for raster format and GIS for vector format.

## 3. Evaluation of rice planting area in the vicinity of Caiuba lagoon (1981-2009)

The Caiuba lagoon is part of the litoral lacunar complex of Rio Grande do Sul state, southern Brazil and it extends by 3300 hectares, in the municipality of Rio Grande, 15 kilometers to north extreme from the Taim Ecological Reserve and 45 kilometers to south from Patos lagoon (Fig. 1). This significant source of water is used mainly for irrigation of rice, specially when the Merin lagoon is further. Accordingly, the Foundation for Research Support in State of Rio Grande do Sul (FAPERGS) funded a research project leadered by the Federal University of Rio Grande (FURG) attempting to study the sustainability of the productive system, as well as the effects on local biodiversity. The Embrapa Temperate Climate Research Center collaborated to the evaluation of the agricultural area in the period 1973 to 2009 by satellite images. Imagery of Landsat satellite of different years was considered for similar times (Table 1), for the scenes corresponding to orbit 237 points 82 and 83 of the worldwide reference system 1 (MSS sensor) and for the scenes corresponding to orbit 221 points 82 and 83 of the worldwide reference system 2 (TM sensor). Thus, the atmospheric conditions were more or less equivalent, deriving in comparable image quality. Each image was evaluated for the various land uses and the areas occupied for the different classes were calculated in order to study the historical evolution of the process during the above period.

The first satellite of Landsat series was launched in 1972 with the multispectral scanner (MSS), with four bands in visible - near infrared and one in thermal infrared and

| Sensor | Date | Range of wavelengh and IFOV |
|--------|------|------------------------------|
| MSS | Sep. 6th, 1973 | 500 – 600nm, 600 – 700nm and 700 - 800nm, IFOV = 79m |
| MSS | Mar. 13th, 1981 | |
| TM | Jan. 22nd, 1991 | 630 – 690nm, 760 – 900nm and 1550 – 1750nm, IFOV = 30m |
| TM | Dec. 21st, 1996 | |
| TM | Dec. 19th, 2001 | |
| TM | Jan. 20th, 2002 | |
| TM | Jan. 28th, 2005 | |
| TM | Jan. 2nd, 2007 | |
| TM | Jan. 7th, 2009 | |

Table 1. Description of basical parameters of the images of Landsat series used for evaluation of the planting area of rice in the vicinity of tha Caiuba lagoon. IFOV = instantaneous field of view. Source: INPE, 2010b.

instantaneous field of vision (IFOV) of 79 meters and 240 meters respectively. Improvements of the system included more bands (short-medium infrared) and reduction of IFOV to 30 meters and 120 meters respectively, for the thematic mapper (TM) in 1982 (Jensen, 1996). A panchromatic band was developed for the Landsat 7 satellite, with the TM plus sensor, but, the series reached to the end. The Landsat 5 satellite was an engineering success, the platform was launched in 1984 and is still on orbit. Anyway, the Landsat series represents the greatest collection of terrestrial images for environmental applications, specially, since a historical point of view.

Composites of three bands were used, with green band (500 – 600 nm), red band (600 – 700 nm) and near infarred band (700 - 800 nm) for the MSS sensor and the red band (630 – 690 nm), the nearinfarred band (760 – 900 nm) and the shortmedium infrarred band (1550 – 1750 nm) for TM sensor. These games of bands are not equivalent, then similar patterns of colour were adjusted by visual observation.

Digital imagery was registered for the Universal Transverse of Mercator projection(UTM), zone 22 South with the datum WGS84, after that, a mosaic of pairs of scenes was composed, by instance, scene 237/82 and 237/83 for MSS sensor. So the mosaic was cutted evolving the study area and a file with the mentioned three bands was created for each date. Initially, data were processed by the supervised classification according to the maximum likelyhood criteria. Eigth poligons of homogeneous features were digitalized with the software ER-Mapper (1995), deriving in the test areas, then each pixel of the corresponding image was classified according to its similarity with the parameters of each test area (beach/dunes, forestry, rice crops, pastures, sandy fields, soil without vegetal cover, water and wetlands). A second strategy was developed to improve results, so the "potential area for agriculture", that is rice crops, pastures and soil without vegetal cover, was isolated and classified by similar way.

Results of the preliminar process of classification considered a rectangle of 30 km wide and 65 km long for the images of 2001, 2002 and 2005 (Fig. 2). The "potential zone for agriculture" is represented by a "central zone" in direction south - north to the East of Merin Lagoon, where agricultural areas are discriminated. A confusion between rice crop class and wetlands class is observed in the west - north sector of the study area. Sandy fields are long structures related to old movements of the sea (Atlantic Ocean), where a low charge of livestock is a common use and forestry is developed eventually, as observed in the images. The area occupied by water bodies was almost constant, that is 19 - 21% (Table 2), but, the wetlands were reduced in area in 2005, a year of drought probably, then, there was an increment in the area occupied by the class "Soil without vegetal cover" and a reduction of the area occupied by the class "Pastures".

When the "potential zone for agriculture" was isolated, the precison of evaluation of the area occupied by pastures, rice crops and soil without vegetal cover (SWVC) was improved. The kind of sensor, the date of the image and the meteorological conditions induced diferences among the imagery of different dates (Fig. 3). The images of 1973 and 1981 present a different characteristics due to captation with the MSS sensor. The first image corresponds to september, when the culture had not been implanted yet. Some agricultural areas showed different pattern in 2001 and 2002 (same harvest) related to waterlogged soils, probably, due to intense rain in that time. A differencial answer of the vegetation in the agricultural areas was observed since 2005, what suggests a evolution of the vegetal development of the rice

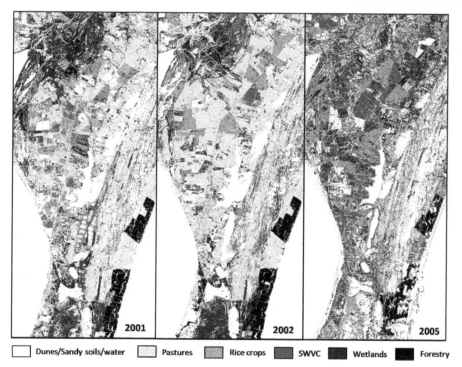

Fig. 2. Preliminary classification in the Caiuba region.

| Image date | Dec. 19th, 2001 | Jan. 20th, 2002 | Jan. 28th, 2005 |
|---|---|---|---|
| Water, % | 19.2 | 21.3 | 20.0 |
| Wetland, % | 17.8 | 18.5 | 11.3 |
| Pastures, % | 20.2 | 12.4 | 8.9 |
| Rice crops, % | 10.2 | 9.1 | 10.4 |
| Sandy fields, % | 17.7 | 13.6 | 15.1 |
| Beach/dunes, % | 6.7 | 6.3 | 10.5 |
| Florestry, % | 4.3 | 3.6 | 4.6 |
| Clouds, % | | 1.5 | |
| SWVC,% | 3.9 | 13.8 | 19.3 |
| Total area, ha | 202,777 | 204,088 | 208,907 |

Table 2. Preliminary areas of land cover calculated by classification methods in the vicinity of Caiubá region through Landsat-TM images in the period 2001 – 2005. SWVC = Soil without vegetal cover.

varieties or, perhaps, the introduction of a new crop. All the images show a intense rotation among pastures, rice and fallow lands, what lets a reduction of inputs, rest of the soil and improvement of productivity.

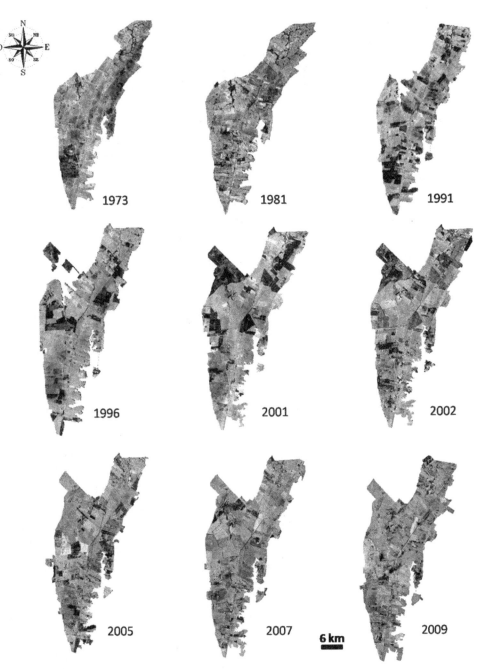

Fig. 3. Images Landsat corresponding to the "Potential zone for agriculture" for different dates.

The poligonal area was classified according to five classes: (1) Undefined; (2) Pastures; (3) Rice crops; (4) Soil without vegetal cover; (5) Water. The class "Undefined" represents rice crops or pastures depending on the year, thus it was incorporated to class "Pastures" in 1973, 1981 e 2001 and to class "Rice crops" in 1991, 1996, 2002, 2005, 2007 e 2009, accordingly the interpretation of the images. So, the classes "Pastures", "Rice crops", "SWVC" and "Water" were evaluated for occupied area (Fig. 4; Table 3).

The occurrence of water is almost insignificant inside the "potencial zone for agriculture", because the irrigation is performed through the water of the lagoons Caiuba and Mirim. The area occupied by the class "Rice crops" seems to depend on the vegetal developping, with restricted values when months previous to January are evolved. This fact was checked with the images of 2001 and 2002, corresponding to the same harvest, Dezember and January respectively.

By this reason, only the data corresponding to the months of january and march, when the vegetal developping of rice is reached, were consider in the graphic of "occupied area" as a function of time (Fig. 5).

| Year | Water | Pastures | Rice crops | SWVC | Total |
|------|-------|----------|-----------|------|-------|
| 06/09/1973 | 173 | 19057 | 12856 | 13565 | 45652 |
| 13/03/1981 | 85 | 28593 | 13915 | 6592 | 49185 |
| 22/01/1991 | 52 | 17110 | 18751 | 15886 | 51798 |
| 21/12/1996 | 825 | 22534 | 12042 | 22534 | 57935 |
| 19/12/2001 | 56 | 30523 | 14404 | 8516 | 53498 |
| 20/01/2002 | 120 | 17579 | 20090 | 16299 | 54087 |
| 28/01/2005 | 49 | 13144 | 21963 | 20246 | 55402 |
| 02/01/2007 | 57 | 25062 | 21054 | 9467 | 55640 |
| 07/01/2009 | 80 | 5302 | 21029 | 31124 | 57535 |

Table 3. Area evaluation of land cover classes for the "potential zone for agriculture". The class "Undefined" was incorporated to the class "Pastures" or the class "Rice crops" according to the year. SWVC = Soils without vegetal cover. Data in hectares.

The area of the "Potential zone for agriculture" was delimited by digitalization, but a soft and constant increment is evident during the period 1981 to 2009. By other side, the area occupied by the class "Rice crops" was evaluated by classification methods; after a period of increment, the class reached a maximum in 2005 with 22 thousand hectares, then there was a stabilization in 2007 - 2009 with about 21 thousand hectares. The classes "Pastures" and "SWVC" showed oscillation in complementary way, because the sum of both classes was almost constant. As classes "Potential zone for agriculture" and "Rice crops" presented linear behavior in the graphic Area against time, thus, linear regression models were adjusted (Table 4).

The parameter $R^2$ is the correlation coefficient between the real variable and the adjusted variable by the model. So, a value near zero indicates bad adjust of the model and a value near one indicates a good adjust of the model. The parameter A indicates the annual growing rate for of the occupation area of the respective class. The area occupied by the

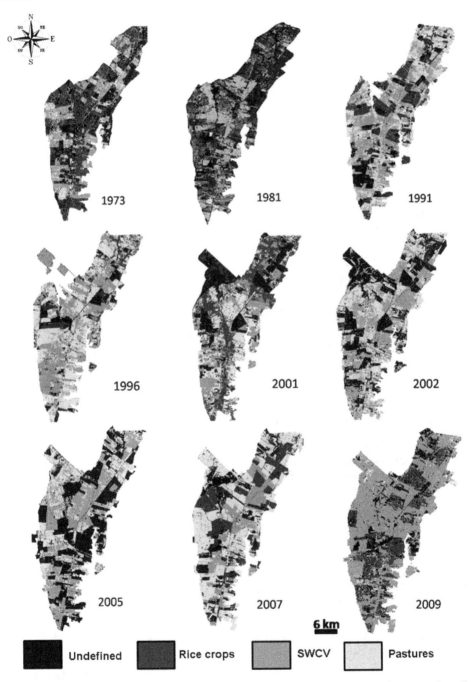

Fig. 4. Images Landsat post-classified corresponding to the "Potential zone for agriculture" for different dates.

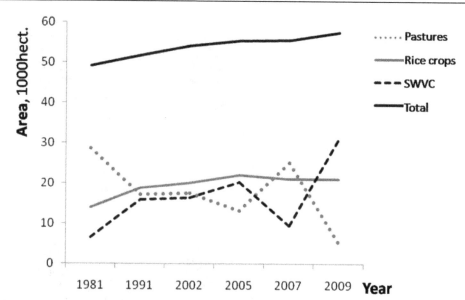

Fig. 5. Areas of land cover as function of the year in the "Potential area for agriculture" (Total), Caiuba region. SWVC = Soils without vegetal cover.

| Class of land cover | A | B | R² | Period |
|---|---|---|---|---|
| Potential zone for agriculture | 271 | 486996 | 0.97 | 1981 - 2009 |
| Rice crops | 302 | 584736 | 0.93 | 1981 - 2005 |

Table 4. Parameters of the linear regression models for the area of classes "Potential zone for agriculture" and "Rice crops" as a function of time (Area = A*year - B, in hectares).

"Potential area for agriculture" grew with a rate of 271 hectares by year, little inferior than the growing rate for the area occupied by the class "Rice crops", that is 302 hectares by year. Parameter B is the value of area in the year zero without real significance in this case.

Data of the municipality of Pelotas (Filippini Alba & Siqueira, 1999) and data for the state of Rio Grande do Sul (Bolfe et al., 2009) were compared to data presented here, after legend conversion. The correlation coefficient of the data discussed here was 0.54 with data of the first paper and 0.77 with data of the second one. Some classes showed significant differences, by instances Bolfe et al. (2009) evaluated 50% of area occupied by "pastures" in the state, but the value was about 30% for the other works. The area occupied by water was 19-20% in the Caiuba region, due to the occurrence of the lagoons. This value was 1% in the municipality context and 3% in the regional one.

## 4. Analisys of land cover evolution in the municipality of Montenegro, Estado do Rio Grande do Sul (1993-2008)

The municipality of Montenegro is located 55 kilometers South from Porto Alegre (state capital), with a territorial area of 420 square kilometers and population about 59,557 inhabitants. Thirty-three municipalities, including Montenegro, integrate the vegetal carbon productive pole. The production of black acacia for the manufacture of tannin is an important activity for the economy of the municipality since 1948, when the first factory of tannin derived from the bark of acacia was installed (TANAC, 2010). Recently, the fruit production is becoming increasingly important in the context of local economy. The intense forest exploitation, the occurrence of new uses of land and a moderate urban occupation oriented the choice of the municipality of Montenegro for this research, focusing on the detection of temporal changes in the territorial organization, during the period 1993 to 2008, in the context of the project "Development and evaluation of products and co-products of the vegetal carbon productive chain in the State of Rio Grande do Sul, aiming for sustentability", with coordination of Embrapa Temperate Climate Reasearch Center.

The topography of the municipality is complex when compared to the previous case, while in the southeastern region occurs a flat terrain changing for slightly wavy; in the north sector occur a basalt plateau with a rugged relief.

Imagery of the Landsat 5 satellite were used, corresponding to the scene of orbit 221, points 80 and 81 for WRS-2 (INPE, 2010b), for three different dates: September 8th, 1993; August 8th, 1999 and October 3th, 2008. The initial data processing was performed with the software Marlin (INPE, 2010a), after that, the software ER-Mapper (1995) was used for classification according to isoclass likelyhood criteria. The images were registered with known ground control points, considering terrestrial features of easy identification, so that, the coordinates systems were uniformized and small errors eliminated. The projection used was the Universal Transector of Mercaptor (UTM), region 22 South, datum WGS 84.

Eigth classes were defined by the supervised classification process according to maximum likelihood criteria. The classes "Annual crops", "Perennial crops" and "Pastures/SWVC" were mapped together in gray tones (Fig. 6). The annual crops reached a maximum area of production in 1999 (Table 5) with poor production in previous and posterior times. By another hand, the perennial crops reached a maximum in 2008, after a significant increment in the previous years, as consequence of an important citrus production. Pastures and SWVC were mapped together due to the dinamic process of changes evolving both classes. A little reduction of the area occupied for both classes was observed.

| Class of land cover | 1993 | | 1999 | | 2008 | |
|---|---|---|---|---|---|---|
| Annual crops | 13.9 | 5835 | 21.2 | 8884 | 12.1 | 5084 |
| Forestry | 19.2 | 8076 | 5.5 | 2299 | 18.3 | 7686 |
| Native forest | 18.9 | 7948 | 23.0 | 9675 | 21.3 | 8957 |
| Pastures/SWVC | 27.0 | 11340 | 26.8 | 11246 | 23.2 | 9733 |
| Perennial crops | 8.9 | 3737 | 18.2 | 7655 | 19.6 | 8226 |
| Unevaluated | 3.3 | 1370 | 1.9 | 810 | 0.4 | 173 |
| Urban | 6.5 | 2729 | 2.1 | 878 | 3.3 | 1377 |
| Water | 2.3 | 968 | 1.3 | 554 | 1.8 | 767 |
| Units | % | hectares | % | hectares | % | hectares |

Table 5. Areas calculated with TM/Landsat 5 imagens for the period 1993 – 2008 through classification methods for Montenegro municipality (Schroder & Filippini-Alba, 2010a).

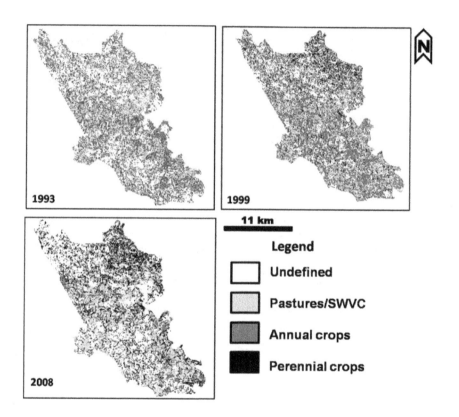

Fig. 6. Evolution of land cover related to agriculture and pastures/SWVC in the municipality of Montenegro based on Landsat 5 imagery (1993 – 2008).

The class "Forestry" showed a minimum of planting area in 1999, what is evident in the map of spatial distribution (Fig. 7.), but the class "Native forest" showed a maximum that year (Fig. 8). The class "Urban" includes other features besides the urban regions, by instance outcrops, which explains its high value in 1993. The density of the central spot in the image of 2008 suggest a real increment of urban population that year.

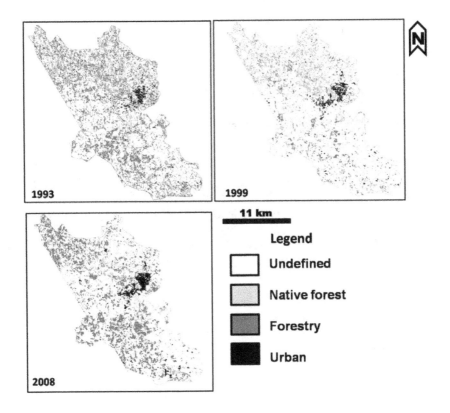

Fig. 7. Evolution of land cover related to the classes "Native forest", "Forestry" and "Urban" in the municipality of Montenegro based on Landsat 5 imagery (1993 – 2008).

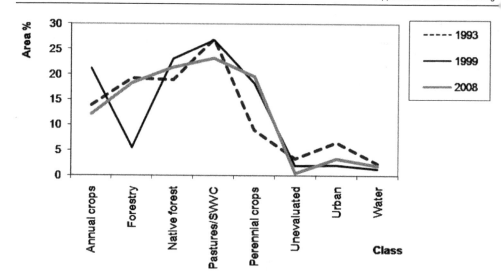

Fig. 8. Evolution of the land cover classes considered in this chapter in the municipality of Montenegro, Rio Grande do Sul state, Brazil.

## 5. Comparison and evaluation of errors

Two strategies were used to analize the errors of the classification methods: (a) Duplication of process with new test areas for the same classes in Caiuba region. (b) Confusion matrixs by truth of field for Montenegro municipality. Each strategy is related to a different error condition, that is, error of processing and error of the method respectively.

The maximum error for case (a) was for the class "Soils without vegetal cover", more or less 2% when the overall area was evaluated (Table 6). The interference of the clouds was of the same order (value insered with the class "Beach/dunes/clouds"). A confusion between dunes and water (sediments) occurred in the Caiuba region (central part of Fig. 9). Other sectors appear very similar for both images.

| Class | Image A | Image B |
|---|---|---|
| Water, % | 20.3 | 21.3 |
| Wetland, % | 18.9 | 18.5 |
| Pastures, % | 12.2 | 12.4 |
| Rice crops, % | 7.7 | 9.1 |
| Sandy fields, % | 13.4 | 13.6 |
| Beach/dunes/clouds, % | 8.4 | 7.8 |
| Forestry, % | 3.6 | 3.6 |
| SWVC,% | 15.6 | 13.8 |

Table 6. Errors derived from classification with new test areas in the Caiuba region for the image of Jan. 20th, 2002 in a total area of 204088 hectares. SWVC = Soil without vegetal cover.

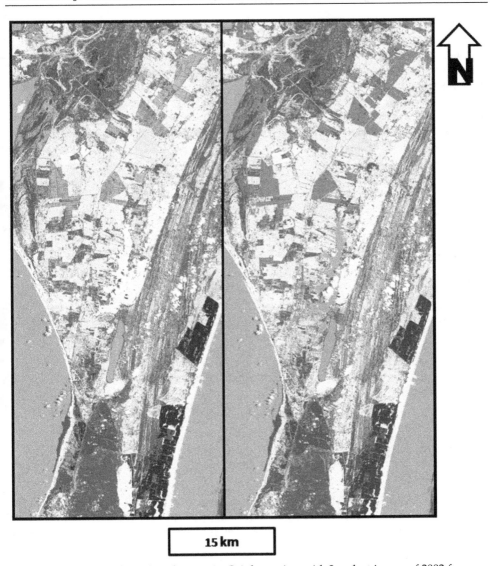

Fig. 9. Comparison of the classification in Caiuba region with Landsat image of 2002 for different test areas.

Data from Montenegro municipality considered the confusion matrix constructed with the truth of field for the Landsat 5 image of april 13, 2009. Thus, 48 control points were selected in the image, trying a "randomly - homogeneous" distribution on the territory of the municipality. Each point was verified at field in september-november 2009 and historical informations were collected with the local farmer when posible.

The accuracy of the method was moderate, that is, 42% for the full process (Table 7). Forestry, Pastures/SWVC, Perennial crops and Urban/outcrops showed the better results, with values greater or equal than 50%. The rest of the classes presented low accuracy with values in the interval 0 - 25%. The correlation coefficient of the quantity of control points and the accuracy was 0.43, suggesting few dependence between both variables. Anyway, a critical case occurred with the class "Native forest" with 11 control points and only two hits. An explanation for the low accuracy of the classification process and some specific classes is the shadow derived from the steep topography, causing confusion among classes and inconsistent results. A improvement of the results is obtained when principal components are considerer before classification, with a potential increment of accuracy of 10% (Schroder & Filippini-Alba, 2010b).

|  | Control points | SCC | Accuracy |
|---|---|---|---|
| Forestry | 9 | 5 | 56% |
| Pastures/SWVC | 11 | 6 | 55% |
| Water | 2 | 0 | 0% |
| Perennial crops | 7 | 4 | 57% |
| Native forest | 11 | 2 | 18% |
| Urban/outcrops | 4 | 2 | 50% |
| Anual crops | 4 | 1 | 25% |
| **Total** | **48** | **20** | **42%** |

Table 7. Results of the confusion matrix for the process of classification in the municipality of Montenegro, Rio Grande do Sul state, Brazil (Schroder & Filippini-Alba, 2010b). SCC = samples correctly classified.

## 6. Conclusion

Two categories of change detection techniques (Lu et al., 2004) were considered in this chapter, all of them including classification methods: Post-classification comparison and strategy with GIS.

The strategy with GIS isolated the poligon corresponding to the "Potential area for agriculture", then, the interference between some pair of classes was eliminated, by instance, wetlands and rice crops. The post-classification comparison allowed a rapid approach about the region with minor accuracy (preliminary results). Definition of the method used depends on the ratio between cost and efficiency according to the designed objectives.

Errors associated to classification methods are mainly due to the spectral answer, by undefinition of classes or occurrence of pixels of transition, because the errors derived from digitalization were insignificant. Atmospheric conditions and the regional topography also influence the process of classification.

Land cover changes in a dynamic way, sometimes with significant transformation rates of one class to another, as the discussed cases confirm. Truth of field appears as an optimal method to improve results, but the cost of process, in time, financial and human resources is incremented.

# 7. Acknowledgments

The authors are grateful to Dr. Angela Campos Diniz (Embrapa) and Dr. Cleber Palma (Federal University of Rio Grande), leaders of the projects that funded research activities. Partial support was obtained through the Foundation for Research Support in the State of Rio Grande do Sul (FAPERGS).

# 8. References

Bolfe, L.E., Siqueira, J.O.W. de, Pereira, R.S., Filippini -Alba, J.M. & Miura, A.K. (2009). Uso, ocupação das terras e banco de dados geográficos da metade sul do Rio Grande do Sul. *Ciência Rural*, Vol. 39, No. 6, (September 2009), pp. 1729 – 1737, ISSN 0103-8478.

Crist, E.P. & Kauth, R.J. (1986). The Tasseled Cap De-Mystified. *Photogrammetric Engineering and Remote Sensing*, Vol. 52, No. 1, (January 1986), pp. 81-86, ISSN 5201-0081.

Crosta, A. P. (1993). *Processamento digital de imagens de sensoriamento remoto*. State University of Campinas, ISBN 85-85369-02-7, Campinas, Brazil.

Davis, J. (2002). *Statistics and data analysis in geology*. Wiley, ISBN 04-71172-75-8, New York.

Embrapa (2009). Sistemas orbitais de monitoramento e gestão territorial. In: *Embrapa Monitoramento por Satélite*, 27/06/2011, Available from http://www.sat.cnpm.embrapa.br.

ER-Mapper (1995). *ER Mapper Reference*. Earth Resource Mapping, West Perth, Australia.

Filippini-Alba, J.M. & Siqueira, O.W. de. (1999). Caracterização da ocupação das terras em escala municipal, através do processamento de imagens Landsat. Área piloto, Pelotas, RS. *Agropecuária Clima Temperado*,Vol. 2, No. 2, (December 1999), pp. 223 – 231, ISSN 1415-6822.

Guild, L.S.; Cohen, W.B.; Kauffman, J.B. (2004). Detection of deforestation and land conversion in Rondônia, Brazil, using change detection techniques. *International Journal of Remote Sensing*, Vol. 25, No. 4, (February 2004), pp. 731-750, ISSN 0143-1161.

Inpe (n.d.). Catálogo de imagens. In: *National Institute of Spatial Research (Inpe)*, 10/jun/2010b. Available from <http://www.dgi.inpe.br/CDSR/>.

Inpe (n.d.). Marlin. In: *National Institute of Spatial Research (Inpe)*, 01/03/2010a. Available from <http://www.dgi.inpe.br/CDSR/marlin_PT.php>

Jensen, J.R. (1996). *Introductory digital image processing: a remote sensing perspective*. Prentice Hall , ISBN 0-13-205840-5, Upper Saddle River, USA.

Lillesand, T. M.; Kiefer, R.W. *Remote sensing and image interpretation*. Wiley, ISBN 0-471-57783-9, New York: 1994. 750p.

Lu, D.; Mausel, P.; Brondízio, E.; Moran, E. (2004). Change detection techniques. *International Journal of Remote Sensing*, Vol. 25, No. 12, pp. 2365-2407, ISSN 0143-1161.

Miranda, E. E. (2002). Brasil visto do espaço: Rio Grande do Sul. In: *Embrapa Monitoramento por Satélite*, 1 CD-rom.

Miranda, E. E. (2005). Brasil em Relevo. In: *Embrapa Monitoramento por Satélite*, 02/08/2010. Available from <http://www.relevobr.cnpm.embrapa.br>.

Moreira, M. A. (2005). *Fundamentos de sensoriamento remoto e metodologias de aplicação*. Universidade Federal de Viçosa, ISBN, Viçosa, Brazil.

Nasa (2004). ASTER: Advanced spaceborne thermal emission and reflection radiometer. In: *Nasa/JPL*, 27/06/2011. Available from < http://asterweb.jpl.nasa.com>.

Pontara, R.C.P. (1998). *Análise do comportamento espectral dos filitos carbonosos para interpretação de imagens*. Brasilia University, Academic dissertation, Brasília, Brazil.

Schroder, V.F.; Filippini-Alba, J.M. (2010). Classificacção de imagens orbitais com auxílio da análise por componentes principais no município de Montenegro – RS, 2009. In: *Infoteca/Embrapa*, 11/10/2011. Available from http://www.infoteca.cnptia.embrapa.br/handle/doc/884278

Schroder, V.F.; Filippini-Alba, J.M. (2010). Potencialidade do uso de imagens orbitais para detecção de mudanças temporais: estudo de caso no município de Montenegro – RS, 1993 – 2008. In: *Infoteca/Embrapa*, 11/10/2011. Available from http://www.infoteca.cnptia.embrapa.br/handle/doc/884267

Tanac (n.d.). Histórico. In: *Tanac Cia*. 08/07/2010, Available from <http://www.tanac.com.br/PT/institucional.php?codCategoriaMenu=148&nomA rea=Hist%F3rico&codDado=2&menu=138>.

# 5

# Remote Sensing Based Modeling of Water and Heat Regimes in a Vast Agricultural Region

A. Gelfan[1], E. Muzylev[1], A. Uspensky[2], Z. Startseva[1] and P. Romanov[3]
[1]Water Problem Institute of Russian Academy of Sciences, Moscow
[2]State Research Center of Space Hydrometeorology Planeta, Moscow
[3]City College of City University of New York, New York, NY
[1,2]Russia
[3]USA

## 1. Introduction

In many parts of the world, solution of the problem of water security is closely associated with possibility and predictability of access to soil water which accounts, globally, for approximately 60% of total rainfall. Soil water, so-called "green water", is fundamentally significant for maintaining sustainability of ecosystems and providing water for agriculture (in average, agriculture demands about 70% of global water use, and over 90% in some regions (Global Water Security, 2010). In addition, changes in the store of soil water due to evapotranspiration, infiltration of surface water and other water cycle processes strongly affect land-atmosphere interactions on intraseasonal to interannual timescales (Yang, 2004) and understanding physical mechanisms of these processes is central for effective modeling mesoscale atmospheric circulation.

Taking into account a vital role of "green water" both for ecosystems and agriculture, there is a need for physically based models allowing to describe land-surface processes in their interaction with atmosphere and to predict soil water availability under changing environment and climate. Processes of mass and energy exchange at the land surface, linking the atmosphere and soil water are of interest to several geophysical disciplines; particularly, the possibility of developing adequate models of these processes was studied intensively by hydrological and meteorological communities for a few decades. Recently, a huge contribution to our understanding these processes has been made by remote sensing community. Combination of these efforts results in intensive developing so called land-surface models (LSMs), considering as a key tool to predict successfully the likely future states of the terrestrial systems under anthropogenic pressure and climate change. There are a lot of reviews providing a detailed and comprehensive discussion of LSMs (e.g. Sellers et al., 1997; Pitman, 2003; Overgaard et al., 2006). Most of them are concentrated on the development of LSMs designed for use in weather and climate models. Below a review of LSMs will be done with emphasis on their availability to capture processes controlling soil water dynamics, and primarly, evapotranspiration. Additionally, focus will be on utilization of remotely sensed data in LSMs.

According to Sellers et al. (1997), the LSMs could be broadly divided onto three generations distinguishing largely on the level of complexity of the evapotranspiration process description. The first generation, developed in the late 1960s and 1970s, used simple bulk aerodynamic transfer formulations. Land surface considers as homogeneous one in the first-generation LSMs and spatially uniform surface parameters (water-holding capacity, albedo, roughness length etc.) are used, i.e. these models do not discriminate soil evaporation and transpiration. Examples are the "bucket" model by Manabe (1969) (probably, the earliest first-generation LSM) and more recent TOPUP model by Schultz et al. (1998), PROMET model by Mauser & Schädlich (1998). Use of the single value of the aerodynamic resistance regardless of the surface type is one of the common simplifications and major conceptual limitations of the first-generation LSMs (Pitman, 2003). Another one is that these models include a single soil layer for soil moisture and temperature simulations. According to the findings of the Project for the Intercomparison of Landsurface Parameterisation Schemes (PILPS; Henderson-Sellers et al., 1995), soil moisture dynamics simulated by the first-generation LSMs can not be often reproduced adequately by these models because of aforementioned conceptual limitations.

The second generation of the LSMs, developed in the early 1980s, explicitly represented, in contrast to the first-generation models, the influence of vegetation on the interaction between land surface and atmosphere. Taking into account difference between soil and vegetation provided an opportunity to begin integrating satellite data into LSMs (Pitman, 2003). Also, these models improve soil moisture representation replacing the simple conceptual scheme of Manabe (1969) by more sophisticated parametrizations (particularly, the vertically distributed Richards equation is often used). Such a new direction in land surface modeling took roots from a pioneering work of Deardorff (1978) developed a model for simulating soil moisture and heat transfer in two layers and vegetation as a single layer controlling heat balance on a land surface, as well as from the works of Dickinson (1986) and Sellers et al. (1986) developed the Biosphere Atmosphere Transfer Scheme (BATS) and Simple Biosphere Model (SiB) based on the ideas of Deardorff (1978). During the subsequent decades, a huge number of the second-generation LSMs have been developed; examples include SVAT (Kuchment & Startseva, 1991), VIC (Wood et al., 1992), BASE (Desborough & Pitman, 1998), CLM (Dai et al., 2002), SWAP (Gusev & Nasonova, 2002), LaD (Milly & Shmakin, 2002).

One of the principal advantages of the second-generation LSMs is their ability to consider snow processes affecting a major part of terrestrial water balance in cold regions. Snow sub-models of different vertical discretization have been implemented in LSMs and an intensive work on their evaluation and intercomparison (e.g. the recent SnowMIP2 Project (Rutter et al., 2009)) has been made. Additionally, some LSMs include parametrization of hydrothermal processes in a frozen soil (Wood et al., 1992; Gusev & Nasonova, 2002 among others) but an adequate description of these processes strongly affected soil moisture content before vegetation season still remains problematic (Pitman, 2003).

A step forward of the third-generation LSMs is in explicit description of a canopy physiology including biophysical mechanisms of stomatal conductance, photosynthesis, plant growth, etc. This ability opens up new opportunities not only to improve reproduction of evapotranspiration but, importantly, to address carbon exchange by plant. Description of the other land surface processes (soil moisture and temperature dynamics, snow processes, etc.) is similar to one utilized in the second-generation LSMs. One of the first LSMs of third generation were developed by Collatz et al. (1991) and Sellers et al. (1992), some recent examples representing improvements of the aforementioned second-generation LSMs include modifications of the SVAT-model (Kuchment et al., 2006), the CLM-model (Oleson et al., 2008), the SiBcrop model (Lokupitiya et al., 2009).

One of the major concerns restricting availability of LSMs, particularly on a regional scale, is the issue of spatial heterogeneity of land surface characteristics (soil, vegetation, topography) required for assigning the model parameters. Regional (mesoscale) heterogeneity is not captured by the existing ground-based observational network that leads to an excessive aggregation of the parameters and, as a rule, to decrease in accuracy of reproducing spatial distribution of the desired processes. Significant improvement of performance of LSMs can be reached by assimilation of information that is additional to the ground observations, first of all, information on land surface provided by satellite remote sensing. Remote sensing allows substituting the missed ground observations by measurements of the incoming and outgoing land surface radiation fluxes conversed into physical distributed parameters. No exaggeration to say that spatial data on land surface derived from remote sensing is the only source of the distributed parameters for LSMs at regional scale. Additionally, these data can be used for model evaluation purposes. Applicability of satellite remote sensing for improvement of the land surface modeling is reviewed, e.g. by Overgaard et al. (2006), including results of a number of field experiments (FIFE, HAPEX, KUREX and others). However, in spite of the fact that quantity and quality of satellite products have largely increased for the last decade and they have recognized as a potentially valuable source of distributed information, the majority of satellite products still needs considerable improvement and applicability of these products, even being improved, should be verified both for the specific region and for the used LSM.

This paper has three major objectives. First, the existing satellite-derived data of land surface and snow cover characteristics will be overviewed in brief and the specific developed technologies of the satellite data thematic processing will be presented for the study region located within the agricultural Black Earth area of the European Russia. Secondly, structure of the physically based distributed Remote Sensing Based Land Surface Model (RSBLSM) developed for simulation of vertical water-and-heat transfer in vegetation, unfrozen and frozen soil, snow cover will be described and the results of its testing against the available ground-based observations will be shown. Finally, the results of utilizing satellite-derived land surface and snow characteristics as the parameters and input variables of RSBLSM, as well as for the model evaluation, will be demonstrated and discussed.

## 2. Case study

The case study has been carried out for agricultural Black Earth region of the European Russia of 227,300 km$^2$ located in the steppe-forest physiographic zone (Fig. 1)

Relief of the region is low middle-hilly plain dissected by broad river valleys, ravines and gullies. Dominant slopes are ~1-4°. Absolute elevation marks of surface in the region are in the range of 150-260 m. Annual net radiation is 27-32 kcal/cm$^2$, and the sum of incoming solar radiation during the summer months is 41-44 kcal/cm$^2$. Annual precipitation is 519 mm, over 40% of which falls as snow. Snow water equivalent significantly varies over the area. At the northern part of the region snow cover deceases in the middle of April, southern parts are snow-free in March. Maximum rainfall (60-70 mm) falls as a rule in July. Ground water level lies at the depth of 15-30 m between rivers and of 3-5 m in river valleys.

Soils are mainly chernozem (podzolized, leached and typical), small part of the territory is occupied by floodplain meadow and gray soil. In texture, soils mostly relate to the loam. Sandy loam and sand are found rarer. In the north-western part of the region gray forest-steppe soils with patches of degraded chernozem are located. When moving to the south gray soils give place to leached chernozem often occurring in combination with powerful chernozem. The southeastern part of the region is occupied by ordinary chernozem. Most of the region territory is under cultivation, the natural vegetation is preserved mainly in river valleys and on slopes of gullies and ravines. Plough-lands compose the most part of the region (78 %), forests occupy about 5%; pastures take up about 16%; urbanized lands occupy less than 1%. The main crops are cereals (spring wheat and barley, winter wheat, less corn, buckwheat and rye) as well as sugar beets, potatoes and forage grasses. Ratio of crops in different households differs substantially. Region-averaged grain wedge is about 60% of all plough-lands. In the region there are 48 agricultural meteorological stations at which observations on meteorological characteristics and soil moisture under different crops, as well as snow cover characteristics are conducted. At several of them measurements of the evaporation pans are also made.

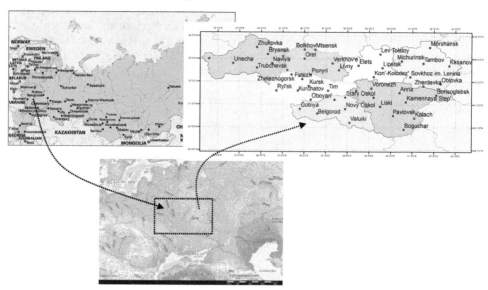

Fig. 1. Study region (points at the right upper map represent location of meteorological stations)

To assign the model parameters, most of which are the measured soil, vegetation and snow cover characteristics as well as to calibrate and validate the model the data have been attracted of above observations at agricultural meteorological stations for 21 years from 1971 to 2010.

## 3. Remote sensing of agricultural regions: products and algorithms of data processing

### 3.1 Remote sensing of land surface and vegetation characteristics

This section contains a brief overview of satellite instruments and methods for remote sensing of various land surface parameters, valuable in particular for hydrological applications. During the past decades, a series of sensors have been developed and launched, such as the Advanced Very High Resolution Radiometer (AVHRR) and the MODerate resolution Imaging Spectroradiometer (MODIS), which are respectively onboard the polar-orbiting satellites NOAA and NASA EOS Terra and Aqua. The AVHRR instrument has 6 VIS and IR channels with spatial resolution of 1 km (NOAA KLM Users Guide, 2005) that are informative with respect to different land surface and vegetation parameters. The MODIS instrument has 36 channels in VIS and IR (Justice et al., 1998) with 1 km spatial resolution, which provides information on even more geophysical parameters of land surface, vegetation, atmosphere, etc. In recent years new geostationary satellites have been developed, such as European Meteosat Second Generation (MSG), i.e. Meteosat-8 and - 9. Meteosat main payload is the optical imaging radiometer, the so called Spinning Enhanced Visible and Infrared Imager (SEVIRI). SEVIRI has 12 spectral channels from VIS to IR (Schmetz et al., 2002). It has channels similar to AVHRR, and the benefit is that SEVIRI provides measurements of the Earth-disc every 15 minutes comparing to the coverage of the polar-orbiting satellites (2 times a day for the same territory).

Among the above mentioned geophysical parameters we will focus on: land surface temperature ( $LST$ or $T_{ls}$ ), air foliage temperature ( $T_a$ ), emissivity ( $E$ ), as well as normalized difference vegetation index ( $NDVI$ ), vegetation cover fraction ( $B$ ) and leaf area index ( $LAI$ ). $LST$ is one of the important geophysical parameters. Together with the land surface spectral emissivity ( $LSE$ ) the $LST$ affect the heat and water transport between the surface and the atmosphere. There is a strong need in the remote sensing $LST$, since the conventional surface temperature observations are rather sparse (in space and time).

The possibilities of extracting $LST$ and $LSE$ information from thermal IR multichannel measurements in the "atmospheric window" spectral range (3.7-4.0, 10.5-12.5 μm) has been the subject of numerous investigations during last 20 years, see (Becker & Li, 1995).

Because of the land surface heterogeneity, the satellite measurements usually come from mixed pixels. At satellite pixel scale, $LSE$ refers to the area-weighted and channel-averaged emissivity ( $E$ ) and $LST$ refers to the radiometric surface temperature corresponding to the Field Of View (FOV) of a radiometer (Becker & Li, 1995).

### 3.1.1 AVHRR- and MODIS-based remote sensing products

The methods have been developed and tested for AVHRR/NOAA data processing (Muzylev et al, 2002, 2005) that provide the retrieval of two types of $LST$ (efficient radiation temperature $T_{s.eff}$ and land skin temperature $T_g$ ) and emissivity $E$ as well as the derivation of air-vegetation temperature $T_a$ and three vegetation characteristics, namely, vegetation index $NDVI$, fraction $B$ and leaf area index $LAI$. The algorithms for AVHRR-based estimation of $T_{s.eff}$, $T_g$, $T_a$ utilize cloud-free measurements in the split window channels 4 and 5 and linear statistical regression similar to well-known split window technique (Becker et al., 1995; PUM

LST, 2008; Wan et al., 1996). The values of emissivity $E_4$ and $E_5$ are specified *a priori* using one of alternate approaches. The required ancillary information is extracted from the classification-based emissivity model (Snyder et al., 1998) as well as from the empirical relationships between the emissivity and/or $NDVI / B$, see (Muzylev et al., 2002, 2005). To estimate $LAI$, the empirical relationships between $LAI$ and $NDVI$ (established for different land covers) have been applied (see Section 5.1, formulas (27) and (28)).

The threshold technique of cloud detection in the AVHRR FOV has been used that has allowed increasing the reliability of cloud-free fragment detection (Volkova & Uspenskii, 2007). The developed software package has been applied for AVHRR/NOAA cloud-free data thematic processing to generate named remote sensing products and cloud/precipitation parameters for various dates of the 1999-2010 vegetation periods. The error statistics of $T_a$, $T_{sg}$ and $T_{s.eff}$ derivation has been investigated for various samples using comparison with synchronous collocated in-situ measurements that has given root-mean square (RMS) errors in the range of 1.5-2.0, 3.5-4.5, and 2.5-3.5°C respectively (Muzylev et al., 2005, 2006, 2010). The archive of synchronous AVHRR/NOAA measurements, remote sounding data, and in-situ hydro-meteorological observations has been compiled for the study area and its separate parts for 1999-2010 vegetation seasons.

The dataset of MODIS-based remote sensing products has also been compiled on the base of special technology using LP DAAC web-site https://lpdaac.usgs.gov, that includes estimates of land surface temperature (LST) $T_{ls}$, $E$, $NDVI$, $LAI$ for the region of interest and 2003-2010 vegetation seasons. Two types of MODIS-based $T_{ls}$ and $E$ estimates have been extracted (for separate dates of the named time period): LST/E Daily L3 product (MOD11B1) with spatial resolution ~ 4.8 km and LST/E 5-Min L2 product (MOD11_L2) with spatial resolution ~ 1 km. The verification of $T_{ls}$ estimates has been performed by the comparison against analogous and collocated AVHRR-based ones (Muzylev et al, 2010).

### 3.1.2 SEVIRI-based land surface and land air surface temperature estimates

In the recent years there were a lot of studies on $LST$ derivations from SEVIRI/MSG data, see (PUM LST, 2008; Solovjev et al., 2009, 2010). In the State Research Center of Space Hydrometeorology "Planeta" (Moscow, Russia) the new methodology has been developed for the derivation of $LST$ and $LSE$ from cloud-free brightness temperatures measured in the SEVIRI channels 9 (10.8 μm) and 10 (12.0 μm) at three different times based on combination of two well-known techniques, i.e. split-window method (Wan et al., 1996) and two temperature method, see (Faysash et al., 2000) with additional hypothesis that the emissivity values $E_9$ and $E_{10}$ remain constant during the time interval between the first and the last image cycles used. The detailed description of the proposed technique can be found in (Uspensky et al., 2009; Solovjev et al., 2010). It is important to mention that unlike the technique from (PUM LST, 2008), this method does not require the accurate knowledge of emissivity ($E$) in the split-window channels.

The method described has been used for the period of 2009-2011 to produce $LST$ estimates over Europe and, in particular, for Black Earth zone of Russia. An example image of this $LST$ mapping is presented in Fig. 2.

The comparison with synchronous and collocated $LST$ products from Satellite Application Facility on Land Surface Analysis, SAF LSA (Lisbon, Portugal) (PUM LST, 2008) has been

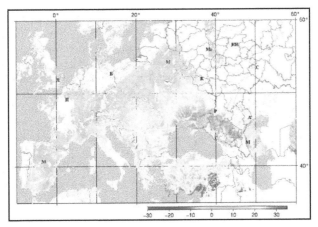

Fig. 2. *LST* map, 27.09.2011, 07:00 UTC

performed to validate the results over the central Europe. Good level of correlation has been reached, which could be treated as indirect proof of the method's efficiency. RMS deviation between the above mentioned LST estimates is in the range of 0.9-3.0°C. The upper limit could be shifted down to 2.4°C by subtracting systematic biases.

An additional validation has been performed through the inter-comparison with MODIS-based *LST* for the study region, see Fig. 3.

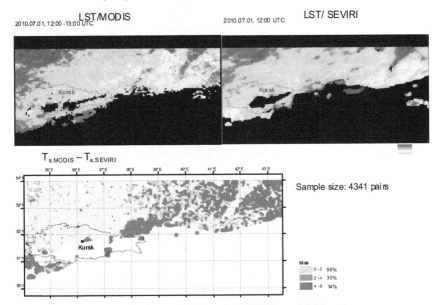

Fig. 3 Inter-comparison results for MODIS- and SEVIRI-based *LST* estimates

It is obvious that both *LST* maps are similar to each other. The discrepancy is rather small (0.0-2.0°C) for the most of the territory.

Along with this, a very important parameter such as land surface-air temperature $T_a$ can also be derived from SEVIRI data. A new method of $T_a$ derivation has been developed in (Uspensky et al., 2011). A multiple linear regression model has been constructed that estimates air temperature from satellite-observed $LST$, solar zenith angle (or related day of the year number), and land elevation. Land surface temperatures estimates from above were used in this scheme to calculate regression coefficients, as well as time-synchronous collocated in-situ measurements from the ground observation network. The development and validation experiments were carried out for the study area for the vegetation seasons of 2006-2009. Data from the above 48 agricultural meteorological stations were used for 8 standard synoptic times a day. All the data were subdivided on separate training and validation data sets. For vegetation season of 2009, an overall bias and standard deviation of calculations are approximately 0 and 1.9-2.1°C, respectively. The performance of the method is similar to the one presented in (Good, 2009).

## 3.2 Remote sensing of snow cover

Seasonal snow cover is among the most important factors for mid-latitude agriculture. Water accumulated in the snow pack in winter and released through the snowmelt is critical for crop development in early spring. For winter grain crops that are widely grown in Canada, Russia and Ukraine, snow pack presents an insulation material protecting the plants from freezing temperatures. Accurate information on the snow cover extent and variations during the winter season helps to identify areas of potential winterkill and to predict adverse conditions for crops development (e.g. Romanov, 2011).

Although most information on the snow cover distribution and properties for agriculture is traditionally obtained from in-situ observations at ground-based meteorological stations, the use of satellite-based snow products is becoming increasingly popular. First, this section presents a brief overview of current satellite-based snow mapping techniques and of snow products that can be used in agricultural applications. Then, a new technique used in this study and based on synergy of optical and microwave snow cover products available from MODIS and Advanced Microwave Sounding Radiometer (AMSR-E) instruments onboard EOS satellites Terra and Aqua is described.

### 3.2.1 Existing snow mapping techniques and products

To derive information on snow cover from satellite data a number of different techniques, both interactive and automated has been developed and is actively used. The most popular interactive snow product based on visual inspection of satellite optical imagery is Northern Hemipshere snow charts that have been generated by NOAA since 1972. Interactive maps of snow cover are currently produced within a computer-based Interactive Multisensor Snow and Ice Mapping System (IMS) that provides software tools and access to various datasets to facilitate the image analysis and map drawing by human analysts (Helfrich et al, 2007). Daily IMS snow maps are generated at 4 km resolution and are available from the National Snow and Ice Data Center (NSIDC) at http://nsidc.org/data/g02156.html. Despite some weaknesses associated primarily with subjectivity in the image analysis and interpretation NOAA IMS snow maps present a robust and consistent product. The overall accuracy of daily interactive snow maps is quite high with the yearly mean rate of agreement with surface observations data exceeding 90% (e.g., Brubaker et al., 2005). Most errors occur during fast and large-scale snow advance or retreat or when persistent cloudiness obscures the land surface.

In contrast to interactive techniques the interest to automated algorithms is attracted due to their low exploitation costs and ability to better utilize potentials of satellite data, particularly their high spatial resolution and multi-spectral capability. The two major techniques for mapping and monitoring snow cover from satellites are based, correspondingly on passive observations in the microwave and in the optical (visible to infrared) spectral range.

The advantage of using microwave sensors consists in their ability to "see" through most clouds and to provide information (although quite limited) on snow depth and snow water equivalent (SWE). The primary limitations of microwave measurements are associated with their coarse spatial resolution of 25-50 km, poor sensitivity to shallow and melting snow (Walker and Goodison, 1993) and difficulty to distinguish between snow and frozen rocks and soil (Grody and Basist, 1996). Snow depth and SWE estimates from observations in the microwave are limited only to dry snow packs while corresponding retrieval errors range typically between 50 and 100% (e.g. Kelly et al., 2003). Global monitoring of snow with microwave sensors data started in 1978 with the launch of Nimbus-7 with Scanning Multichannel Microwave Radiometer (SMMR) onboard and continued with a number of other sensors, including in particular, Special Sensor Microwave Imager (SSM/I) on Defense Meteorological Satellite Program (DMSP) satellites since 1987 (Armstrong and Brodzik 2005) and AMSR-E onboard EOS Aqua satellite since 2002 (Kelly et al, 2003).

As compared to satellite passive microwave measurements, observations in the optical spectral range allow for more accurate mapping of snow cover at higher spatial resolution. The mean accuracy of snow identification in optical bands usually exceeds 90%, but drops to 80-90% over dense coniferous forests (Simic, et al, 2004, Hall & Riggs, 2007). Daily global snow cover maps routinely generated with data from MODIS onboard NASA EOS Terra and Aqua satellites at 500 m spatial resolution (Hall et al, 2002) and from AVHRR onboard NOAA satellites at 4 km resolution (http://www.star.nesdis.noaa.gov/smcd/emb/snow/HTML/snow.htm). It is important that snow retrievals in the optical spectral bands are possible only under clear sky conditions. Partial improvement in the map area coverage can be achieved with geostationary satellites which provide multiple observations per day and hence increase the chance to see the land surface cloud clear (e.g, De Wildt et al., 2007). With geostationary satellites, however, the map coverage is only regional and is limited to the area within ~ 65° N and S.

Because of physical limitations of both principal snow remote sensing techniques, snow products generated with single sensor data lack either continuity or sufficient accuracy and spatial resolution and thus are hard to use in numerical model applications. In an attempt to improve satellite-based snow cover characterization several techniques have been proposed that combine snow cover observations in the optical and microwave spectral bands (e.g., Romanov et al. (2000), Brodzik et al. (2007), Foster et al. (2011)). The objective of these techniques is to maximize advantages offered by optical and microwave observations, to compensate for their weaknesses and to generate continuous snow maps at the highest possible spatial and temporal resolution. Most often in these algorithms optical snow retrievals are used in clear sky conditions, whereas microwave retrievals complement the optical data when cloudy.

### 3.2.2 Snow cover mapping through synergy of optical and microwave products from EOS satellites

A new technique used in this study is based on synergy of optical and microwave snow cover products available from MODIS and AMSR-E instruments onboard EOS satellites Terra and Aqua. The objective was to generate an advanced product providing continuous (gap free) characterization of the global snow cover distribution at 5 km spatial resolution at daily time step.

The algorithm utilizes two NASA daily snow products, MODIS snow cover map on a latitude-longitude grid at 5 km resolution (labeled by NASA as MOD10C1 and MYD10C1 correspondingly for MODIS Terra and Aqua) and Aqua AMSR-E–based snow water equivalent product AE_DySno. In the developed blending technique we took a cautious approach to the microwave data: microwave retrievals indicating no snow as well as retrievals over mountains were disregarded due to frequent omission of melting snow and shallow snow and frequent overestimates of snow cover in the mountains by microwave algorithms. The remaining microwave retrievals were used to complement snow cover distribution mapped with MODIS data in clear sky conditions. Within this approach some pixels in the daily map may remain undetermined. To eliminate these gaps in the coverage and to achieve continuity in the derived snow cover distribution pixels that remain undetermined in the current day snow map were filled in with the data from the previous day's blended snow map.

All available MODIS and AMSR-E snow products since 2002 have been reprocessed to derive almost 10-years-long time series of daily global snow cover maps. Snow maps in binary format are available at ftp://www.orbit.nesdis.noaa.gov/ pub/ smcd/ emb/ snow/ eos/. An example of a snow cover map generated through synergy of MODIS and AMSR-E data is presented in Fig. 4. To estimate the accuracy of the new snow product we compared it with available surface observations of snow cover. The comparison made for 9 consecutive winter seasons from 2002-2003 to 2010-2011 has shown that the yearly mean agreement of the blended MODIS and AMSR-E snow product to surface observations was 87%. This is only about 3% less than the accuracy of NOAA interactive snow maps estimated using the same method. The accuracy of the EOS blended product dropped to 80-85% in the middle of the winter season and increased to close to 100% in late spring, summer and early fall.

In order to assess consistency of satellite snow retrievals derived from the different satellites, we compared estimates of the derived snow covered area (SCA) from daily snow cover maps of MODIS Terra with ones of MODIS Aqua. The analysis has shown that the SCA derived from MODIS Terra and from MODIS Aqua changes synchronously.

Additionally, we compared dates of snow melt off as determined from satellite data with the dates of snow melt off as determined from the available ground-based observations. The comparison was performed for 48 meteorological stations. The results show that in most cases these dates differ by 1 to 10 days, however in some cases the difference exceeded two weeks (results of 2003 are shown in Fig. 5 as an example). Most probably the primary reason for the difference between satellite and surface estimates of the snow melt-off date consists in the difficulty of detecting shallow wet snow from satellites. However, the observed dates of snow cover decease averaged over the whole area are appeared to be very close to satellite-derived estimates of these dates.

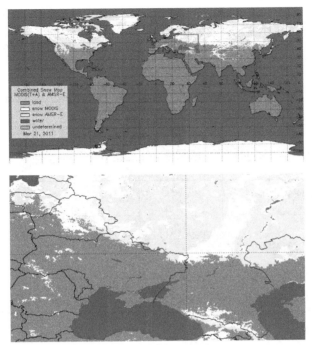

Fig. 4. Example of a blended global daily snow cover map derived from combined data of MODIS and AMSR-E instruments onboard Aqua and Terra satellites (upper) and zoomed in portion of the map covering Eastern Europe. Light blue color and white color represents snow cover identified correspondingly with microwave and optical satellite data (lower).

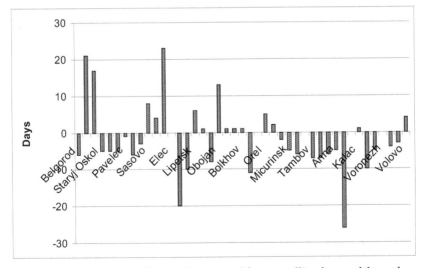

Fig. 5. Difference of snow melt off dates determined from satellite data and from the available ground-based observations

We consider the blended automated snow products as the most advanced ones providing a reliable and robust characterization of snow cover distribution at the satisfactory accuracy. Potential of using these products in conjunction with land surface model for reproducing snow cover distribution over the agricultural area will be demonstrated in section 5.2.

List of the satellite-derived data, both on land surface characteristics and snow, used in this study is shown in Table 1.

| Land Surface Characteristic | Name of Product | Sensor/ Satellite | Latitude-Longitude Resolution | Time Resolution | Time Period |
|---|---|---|---|---|---|
| Land Surface Temperature | NASA_MOD11_B1 | MODIS/ TERRA&AQUA | 0.05°x0.05° | Twice per day | 1 Apr to 31 Oct 2003-2010 |
| | NASA_MOD11_L2 | | 0.01°x0.01° | Twice per day | 1 Mar to 30 May 2002, 1 Mar to 31 Oct 2003-2010 |
| | Thematic processing from | AVHRR/NOAA | 1'x1.5' | Twice per day | 1 Apr to 31 Oct 1999-2010 |
| | | SEVIRI/ METEOSAT-9 | 0.06°x0.06° | Every 15min (at cloud-free condition) | 1 Apr to 31 Oct 2009-2011 |
| Land Surface Albedo | NASA_MOD043_C1 | MODIS/TERRA | 0.05°x0.05° | 16-day product | 1 Mar to 30 May 2003-2010 |
| Snow Water Equivalent | NASA_AE-DySno | MODIS/AQUA | 0.20°x0.20° | Once per day | 20 Jan to 30 May 2003-2010 |
| Snow Covered Area | NASA_MOD10_L2 | MODIS/TERRA | 0.01°x0.01° | Once per day | 20 Jan to 30 May 2002-2010 |
| Vegetation Cover Fraction | Thematic processing from | AVHRR/NOAA | 1'x1.5' | Twice per day | 1 Apr to 31 Oct 1999-2010 |
| Leaf Area Index | Thematic processing from | AVHRR/NOAA | 1'x1.5' | Twice per day | 1 Apr to 31 Oct 1999-2010 |
| | NASA_MOD15_A2 | MODIS/ TERRA&AQUA | 0.01°x0.01° | 8-day product | 1 Apr to 31 Oct 2003-2010 |
| Land Cover Classification | Land Cover Type | AVHRR/NOAA | 0.01°x0.01° | Static data generated from AVHRR data | |
| Tree Cover Fraction | Tree Cover Fraction | | | | |
| Fraction of Evergreen Tree Cover | Evergreen Tree Cover | | | | |

Table 1. Satellite-derived products used in this study

## 4. Remote sensing based land surface model: Structure, calibration and validation by the ground-based data

The developed comprehensive Remote Sensing Based Land Surface Model (RSBLSM) contains as major components the model of vertical water and heat transfer in the "Soil-Vegetation-Atmosphere" system (SVAT) for vegetation season as well as the model of vertical water and heat transfer in the "Soil-SNow-Atmosphere" system (SSNAT) for cold season. The first versions of these models were developed in the 1990s (Kuchment et al., 1989; Kuchment & Startseva, 1991; Kuchment & Gelfan, 1996). Later, conceptualization of the model was improved through accounting for additional processes (Gelfan et al., 2004; Kuchment et al., 2006), and, importantly, the methods of the model adaption to remote sensing data on land surface were developed (Muzylev et al., 2002; 2005; 2010; Kuchment et al., 2010). Below RSBLSM components used in this paper are described in brief.

### 4.1 SVAT component of the RSBLSM system

The SVAT model is intended for simulating evaporation from bare soil, transpiration by vegetation, vertical latent and sensible heat fluxes, vertically distributed soil water and heat content, soil surface and foliage temperatures, land surface radiation temperature as well as other variables characterized water and heat regimes of soil-vegetation system during a warm season. The model flowchart is shown in Fig. 6.

Land surface is considered as a two-component soil-vegetation system. Water/heat fluxes incoming to and outgoing from bare soil and vegetation cover are accounted for separately. So evapotranspiration ( $Ev$ ) is described as sum of two fluxes: bare soil evaporation ( $E_g$ ) and transpiration by vegetation ( $E_f$ ) as

$$E_g = \rho_a \cdot (r \cdot q^*(T_g) - q_a) / r_{ag} \tag{1a}$$

$$E_f = \rho_a \cdot (q^*(T_f) - q_a) \cdot LAI / (r_a + r_s) \tag{1b}$$

Sensible heat fluxes from surface of bare soil $H_g$ and from vegetation cover $H_f$ are calculated as:

$$H_g = \rho_a \cdot c_p \cdot (T_g - T) / r_{ag} , \tag{2a}$$

$$H_f = \rho_a \cdot c_p \cdot (T_f - T) / r_a \tag{2b}$$

where $T_g$ and $T_f$ are the soil surface and the foliage temperatures, respectively, $T$ and $q_a$ are the air temperature and specific air humidity at 2 m height, respectively; $q^*(T_g)$ and $q^*(T_f)$ are the specific air saturation humidity at the temperatures $T_g$ and $T_f$, correspondingly, $r$ is the relative air humidity near the soil surface, $r_{ag}$ and $r_a$ are the aerodynamic resistance between soil surfaces and foliage and between foliage and atmosphere, respectively,

$$r_{ag} = (C_g \cdot U)^{-1} \tag{3}$$

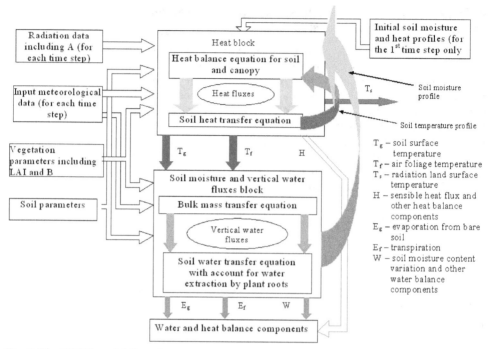

Fig. 6. The SVAT model flowchart

$$r_a = (C_e \cdot U)^{-1}, \qquad (4)$$

where $C_e$ and $C_g$ are the vapour transfer coefficients at the canopy level and at the ground one, respectively, depending on the surface roughness; $U$ is the wind velocity at 10 m height, $r_s$ is the stomatal resistance calculated by formula

$$r_s = r_0 \frac{\psi(\theta_{wp}) - \varsigma \psi_f}{\psi(\theta_{wp}) - \psi_f} \qquad (5)$$

$r_0$ is the minimum stomatal resistance, $\psi(\theta_{wp})$ is the soil matrix potential corresponding to wilting point $\theta_{wp}$, $\varsigma$ is the coefficient, $LAI$ is the leaf area index, $\rho_a$ is the air density, $c_p$ is the specific heat at constant pressure.

The relative air humidity $r$ near the soil surface in (1a) is defined by formula

$$r = \exp \frac{Mg\psi(\theta)}{R(T_g + 273)} \qquad (6)$$

where $\theta$ is the volumetric soil moisture content, $\psi(\theta)$ is the soil matrix potential, $M$ is the molecular mass of water, $g$ is the acceleration of gravity, $R$ is the universal gas constant.

To describe water transfer in the soil, the equation of soil moisture diffusion is applied taking into consideration water extraction by plant roots

$$\frac{\partial \theta}{\partial t} = \frac{\partial}{\partial z}\left[D(\theta)\frac{\partial \theta}{\partial z} - K(\theta)\right] - R(\theta,z) \tag{7}$$

where $K(\theta)$ is the hydraulic conductivity of soil, $D = K(\partial \psi / \partial \theta)$ is the soil moisture diffusion coefficient, $R(\theta,z)$ is the intensity of water extraction by plant roots

$$R(\theta,z) = -K(\theta)\cdot[\psi_r - \psi(\theta)]\cdot b_r \cdot \rho_r(z) \tag{8}$$

$\psi_r$ and $\rho_r(z)$ are the rootage water potential and the density, respectively, $b_r$ is the coefficient.

The soil matrix potential $\psi(\theta)$ and the soil hydraulic conductivity $K(\theta)$ can be assigned using different parametrizations. We compared some of them (Brooks & Corey, 1964; Clapp & Hornberger, 1978; van Genuhten, 1980) and found that they give similar simulation results. Hereafter, van Genuhten's formulas are used:

$$\psi(\theta) = -\frac{\left(S^{-1/m} - 1\right)^{1/n}}{\alpha} \tag{9}$$

$$K = K_0 S^{0.5}\left[1 - \left(1 - S^{1/m}\right)^m\right]^2 \tag{10}$$

where $S = (\theta - \theta_r)(\theta_s - \theta_r)^{-1}$ is the relative saturation; $\theta_s$ and $\theta_r$ are the residual and the saturated water contents respectively; $K_0$ is the saturated hydraulic conductivity; $\alpha > 0$ is the parameter, which is related to the inverse of the air entry pressure; $m = 1 - n^{-1}$, $n > 1$ is the parameter, which is a measure of the pore-size distribution.

The foliage water potential $\psi_f$ is assumed to be expressed in terms of $\psi_r$ using the relationship

$$\psi_f = \psi_r - r_r \rho_w \int\limits_0^{Z_{max}} R(\theta,z)dz \tag{11}$$

where $r_r$ is the rootage resistance, $Z_{max}$ is the maximum length of roots, and $\rho_w$ is the water density.

Heat transfer within a soil layer is described by

$$C_{eff}(\theta)\frac{\partial T}{\partial t} = \frac{\partial}{\partial z}(\lambda(\theta)\frac{\partial T}{\partial z}) \tag{12}$$

where $C_{eff}(\theta)$ and $\lambda(\theta)$ are the effective soil heat capacity and the soil heat conductivity.

The soil surface temperature $T_g$ is calculated from Eq. (12). The foliage temperature $T_f$ is obtained from the heat balance equation for the vegetation cover neglecting its heat content. Both $T_g$ and $T_f$ are used for assessing latent and sensible heat fluxes from bare soil and vegetation surfaces (Eqs. (1a) – (2b)).

The land surface radiation temperature $T_s$ is calculated from the long-wave radiation balance equation

$$\sigma \cdot T_s^4 = R_a - R_{lf} - R_{lg} \qquad (13)$$

where $R_a$ is the atmosphere counterradiation, $R_{lf}$ and $R_{lg}$ are long-wave components of radiation balance for vegetation and bare soil, respectively, calculated as functions of the measured meteorological variables, $\sigma$ is the Stefan-Boltzmann constant.

The input variables of the SVAT model are incoming radiation, air temperature, humidity and pressure, cloudiness, precipitation, wind speed assigned from the standard meteorological observations. Initial conditions for Eqs. (7) and (12), namely soil moisture and temperature profiles at the beginning of the vegetation season, are calculated by Eqs. (25), (26) (see the following section) through spin-up simulations of soil moisture and heat regimes during a winter previous to vegetation season.

Eqs. (7), (12) are numerically integrated by an implicit, four-point finite difference scheme with the time and spatial steps of 3 hours and 10 cm, respectively.

Spatial heterogeneity of land surface characteristics is taking into account in the model by the mosaic approach, i.e. by selection of plots on the studied territory with different soils, land-use and vegetation types which correspond to specific parameter values. Soils are characterized by bulk density, maximum hygroscopicity, porosity, field capacity, and saturated hydraulic conductivity. The heterogeneity of vegetation is represented by the minimum stomatal resistance, the leaf area index LAI, vegetation cover fraction $B$, and the aerodynamic resistance $r_a$.

The principal model parameters have been adjusted by calibration against ground-measured soil water content $W$, evapotranspiration $Ev$, vertical soil moisture profiles, and soil surface temperature. Also, the values of several parameters have been retrieved from the specific measurements at agricultural meteorological stations, some of them have been estimated using satellite data, and the values of certain parameters have been derived from literature sources. The spatial distributions of the most meteorological parameter values (being input model variables) have been built using interpolation procedures. The water and heat balance components for the entire considered territory have been calculated as weight-averages accounting for size of the area occupied by specific soil and vegetation.

In developing version of the model designed for utilizing satellite estimates of the land surface characteristics (built in a quasi-regular grid nodes) the uniform grid with 3x3 AVHRR/NOAA pixel cells ($\sim$ 7x5 km$^2$) has been superimposed on the entire investigated territory divided into plots with different soils and vegetation. Grid size for other sensors was assigned as close to one of AVHRR/NOAA. (Note, that pixel size for the IR channels of the AVHRR radiometer is 1' in latitude and 1.5' in longitude, for similar MODIS channels resolution is equal to 1 and 4.8 km, and for the same channels of SEVIRI it is 0.05° in latitude and 0.06° in longitude) . For nodes of the grids there have been built AVHRR-derived estimates of $T_a$, $T_{sg}$, $T_{s.eff}$, $NDVI$, $E$, $B$, and $LAI$, MODIS-derived estimates of $T_{ls}$, $E$, $NDVI$, $B$ and $LAI$, and SEVIRI-derived estimates of $T_{ls}$. Maps of some AVHRR-derived land surface characteristics for part of the study area are shown in Fig. 7. There have been also defined model parameters and input variables as well as there have been calculated the values of $Ev$, $W$ and other water and heat balance components together with $T_f$, $T_g$ and $T_s$.

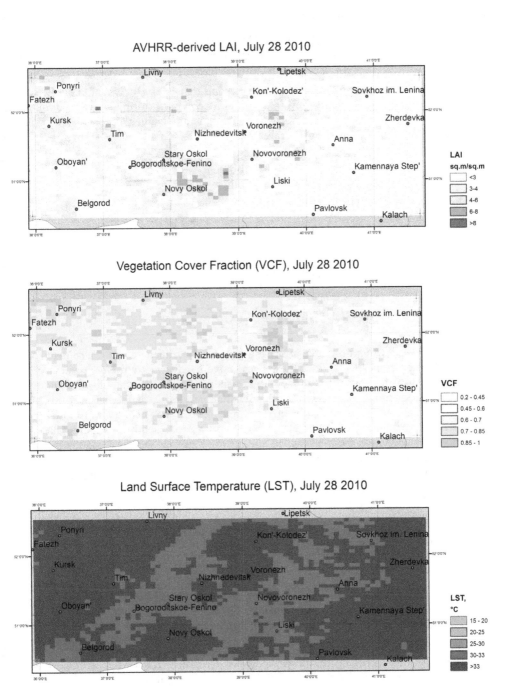

Fig. 7. Maps of some AVHRR-derived land surface characteristics

## 4.2 SSNAT component of the RSBLSM system

SSNAT model describes hydrothermal processes in soil and snow cover during a cold season, including processes of snow accumulation and melt, changes of soil moisture and temperature during soil freezing and thawing, as well as snowmelt water infiltration.

To simulate temporal changes of the snow depth and density, heat content of snow, water phase transformation and other processes within a snowpack during snow accumulation and melt, one-layer snow model proposed in (Kuchment, Gelfan, 1996) has been applied. The main equations of the model are written as:

$$\frac{dH}{dt} = \rho_w \left[ X_s \rho_0^{-1} - (S + E_s)(\rho_i i)^{-1} \right] - V \tag{14}$$

$$\frac{d}{dt}(iH) = \frac{\rho_w}{\rho_i}(X_s - S - E_s) + S_i \tag{15}$$

$$\frac{d}{dt}(wH) = X_l + S - E_l - R - \frac{\rho_i}{\rho_w} S_i \tag{16}$$

$$c_s \frac{d}{dt}(T_s H) = Q_a - Q_g - \rho_w L S + \rho_i L S_i \tag{17}$$

where $H$ is the snow depth; $i$ and $w$ are the vertically averaged volumetric contents of ice and liquid water, respectively; $T_s$ is the vertically averaged temperature of snowpack; $S$ is the melt rate; $S_i$ is the rate of freezing of liquid water in snow, $E_l$ is the rate of liquid water evaporation from snow; $E_s$ is the rate of snow sublimation, $Q_a$ is the net heat flux at the snow surface; $Q_g$ is the ground heat flux; $X_s$ and $X_l$ are the snowfall and rainfall rates, respectively (partitioning of the total precipitation, $X$, into solid and liquid phase is a function of the air temperature); $V$ is the snowpack compression rate; $R$ is the snowmelt outflow from snowpack; $c_s$ is the specific heat capacity of snow; $\rho_w$, $\rho_i$, and $\rho_0$ are the densities of water, ice, and fresh-fallen snow, respectively; $L$ is the latent heat of ice fusion.

The melt rate $S$ is found from the energy balance of the snowpack at zero snow temperature as:

$$S = \begin{cases} (Q_a - Q_g)(\rho_w L)^{-1} = (Q_{sw} + Q_{lw} - Q_{ls} + Q_T + Q_E + Q_P - Q_g)(\rho_w L)^{-1}, Q_a - Q_g > 0 \\ 0, Q_a - Q_g < 0 \end{cases} \tag{18}$$

where $Q_{sw}$ is the net short wave radiation; $Q_{lw}$ is the downward long wave radiation; $Q_{ls}$ is the upward long wave radiation from snow; $Q_T$ is the sensible heat exchange; $Q_E$ is the latent heat exchange; $Q_P$ is the heat content of liquid precipitation.

The heat flux components of $Q_a$ for an open agricultural site are calculated by the empirical relationships using the observed meteorological variables (air temperature, air humidity, wind speed, precipitation, and cloudiness) as the inputs.

The rate $S_i$ of freezing of liquid water in snowpack is calculated as:

$$S_i = \begin{cases} H\dfrac{dw}{dt}, & T_s = 0°C \wedge Q_a - Q_g < 0 \wedge \dfrac{|Q_a - Q_g|}{\rho_i L} \geq H\dfrac{dw}{dt} \\[2mm] \dfrac{|Q_a - Q_g|}{\rho_i L}, & T_s = 0°C \wedge Q_a - Q_g < 0 \wedge \dfrac{|Q_a - Q_g|}{\rho_i L} < H\dfrac{dw}{dt} \\[2mm] 0, & T_s = 0°C \wedge Q_a - Q_g \geq 0 \\[2mm] X_I, & T_s < 0°C \end{cases} \tag{19}$$

The snowpack compression rate $V$ (in cm s$^{-1}$) is found from:

$$V = \frac{v_1 \rho_s}{\exp(v_2 T_s + v_3 \rho_s)} \frac{H^2}{2} \tag{20}$$

where $\rho_s$ is the density of snowpack (in g cm$^{-3}$) equal to $\rho_s = \rho_i i + \rho_w w$; $v_1$, $v_2$, and $v_3$ are the coefficients equal to 2.8×10$^{-6}$ cm$^2$ s$^{-1}$ g$^{-1}$; -0.08 $^{\circ}$C$^{-1}$; 21 cm$^3$ g$^{-1}$, respectively.

The outflow of liquid water from snow is calculated as:

$$R = \begin{cases} X_I + S - E_I - w_{max}\dfrac{dH}{dt}, & w = w_{max} \\[2mm] 0, & w < w_{max} \end{cases} \tag{21}$$

where $w_{max}$ is the holding capacity of snowpack related to its density $\rho_s$ as

$$w_{max} = 0.11 - 0.11\frac{\rho_s}{\rho_w} \tag{22}$$

Numerical integration of the Eqs. (14) – (17) is carried out by an explicit finite-difference scheme with the time-step of 3 hours.

To calibrate the snow pack model for the study area, the meteorological data from 48 stations for the period from 1 November 2001 to 31 May 2002 were used. Then the model was validated against snow depth observations during the winter-spring seasons of 2002-2003 and 2003-2004. The comparison of snow modeling results with the observed snow depth at the ground-based stations has demonstrated a good correspondence between the two datasets. The standard error of the simulated snow depth was 5.8 cm during the calibration stage and 7.7 cm at the validation stage.

Water and heat transfer in a soil during the processes of soil freezing, thawing and infiltration of water are described by the following equations (Gelfan, 2006):

$$\frac{\partial W}{\partial t} = \frac{\partial}{\partial z}\left(D_\theta \frac{\partial \theta}{\partial z} + D_I \frac{\partial I}{\partial z} - K\right) \tag{23}$$

$$c_T \frac{\partial T}{\partial t} - \rho_w L \frac{\partial W}{\partial t} = \frac{\partial}{\partial z}\left(\lambda \frac{\partial T}{\partial z}\right) + \rho_w c_w \left(D_\theta \frac{\partial \theta}{\partial z} + D_I \frac{\partial I}{\partial z} - K\right)\frac{\partial T}{\partial z} \tag{24}$$

where $W$ and $I$ are the total water content and ice content of soil, respectively ($W = \theta + \rho_i I / \rho_w$); $c_T = c_{eff} + \rho_w L(\partial \theta / \partial T)$; $c_{eff}$ is the effective heat capacity of soil equals $c_{eff} = \rho_g c_g (1 - P) + \rho_w c_w \theta + \rho_i c_i I$; $\rho$ and $c$ are the soil density and the specific heat capacity, respectively (indexes $w$, $i$ and $g$ refer to water, ice and soil matrix, respectively); $P$ is the soil porosity; $D_\theta = K(\partial \psi / \partial \theta)_I$; $D_I = K(\partial \psi / \partial I)_\theta$; $\psi = \psi(\theta, I)$ is the matrix potential of soil.

If soil is frozen ($I(z,t) \neq 0$), than hydraulic and thermal characteristics of soil are functions of ice content, i.e. $K = K(\theta, I)$, $\lambda = \lambda(\theta, I)$. One can see that for an unfrozen condition ($I(z,t) = 0$), Eq. (23) reduces to Eq. (7) (neglecting water extraction by plant roots) and Eq. (24) reduces to Eq. (12).

The matrix potential, $\psi = \psi(\theta, I)$, and the hydraulic conductivity, $K = K(\theta, I)$ are determined from (Gelfan, 2006):

$$\psi(\theta, I) = -\frac{\left(S_f^{-1/m} - 1\right)^{1/n}}{\alpha} \times \left[\frac{\theta_0 - \theta_r}{\theta_0 - I - \theta_r} + \frac{\theta_r}{\theta}\left(1 - \frac{\theta_0 - \theta_r}{\theta_0 - I - \theta_r}\right)\right](1 + 8I)^2 \qquad (25)$$

$$K(\theta, I) = K_0 S_f^{0.5}\left[\frac{1 - \left(1 - S_f^{1/m}\right)^m}{(1 + 8I)}\right]^2 \qquad (26)$$

where $S_f = (\theta - \theta_r)(\theta_0 - I - \theta_r)^{-1}$ is the relative saturation of frozen soil.

Note, that under $I(z,t) = 0$ formulas (25), (26) reduce to the van Genuchten's formulas (9), (10) for an unfrozen condition.

The values of $\theta_0$ and $\theta_r$ are assumed to be equal to the measured soil porosity and maximum hydroscopicity, respectively. The values of $a$, $n$, $c_T$, and $\lambda$ are calculated from the measured soil characteristics, such as bulk density, field capacity, and wilting point by the formulas presented in Gelfan (2006).

Equations (23) and (24) are numerically integrated by an implicit, four-point finite difference scheme with the time and spatial steps of the finite difference scheme of 1 hour and 10 cm, respectively.

## 5. Results

### 5.1 Satellite based modeling water and heat regimes of the study region during a vegetation season

The model has been verified by comparing the calculated and measured vertical soil moisture and temperature profiles, values of land surface temperature, radiation balance, soil water content of one-meter soil layer, and evapotranspiration. Results of such a collation for several characteristics above are shown in the Figs. 8-10 by the example of some meteorological stations located in the study area.

Moreover, there has been executed comparison of modeled values of $T_f$, $T_g$ and $T_s$ with their satellite-derived analogues $T_a$, $T_{sg}$ and $T_{s.eff}$ for AVHRR and $T_{ls}$ for MODIS and

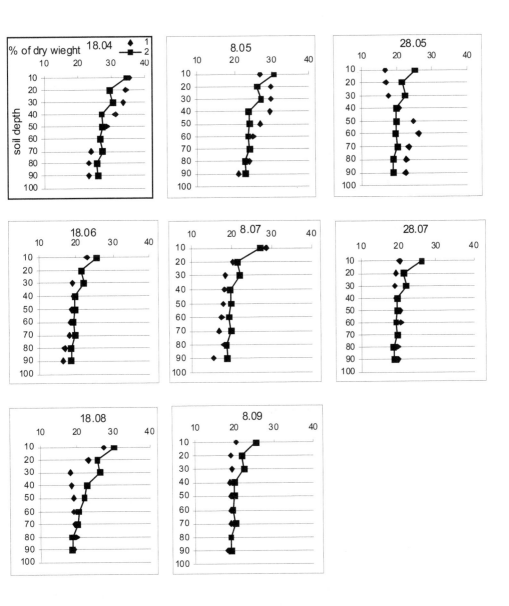

Fig. 8. Modeled (1) and measured (2) vertical soil moisture profiles for perennial grasses at water balance station Nizhnedevitsk for 2003 vegetation season

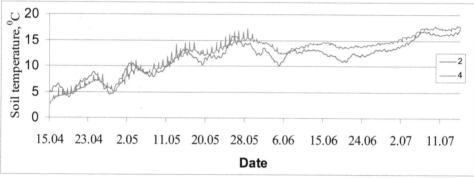

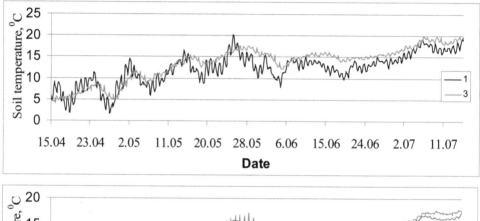

Fig. 9. Soil temperature under vegetation cover modeled for the depth 20 (1) and 40 (2) cm and measured at the same depths for perennial grasses at Nizhnedevitsk water balance station in 2003

SEVIRI. For most terms of each considered vegetation season the differences $T_{s.eff} - T_s$, $T_a - T_f$, and $T_{sg} - T_g$ have not exceed the standard errors of AVHRR-derived estimates of $T_{s.eff}$, $T_a$ and $T_{sg}$ as one can see from Fig. 11 illustrating this result for the part of the study area. Separate local spots with a difference of 20°C on these figures correspond to the clouds above given plots.

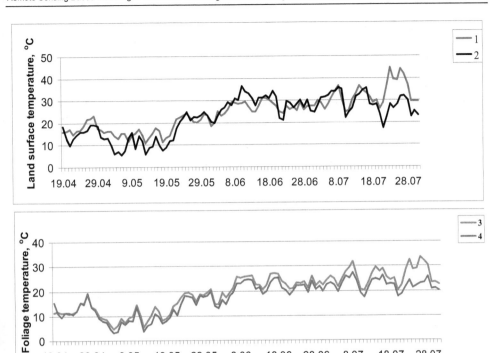

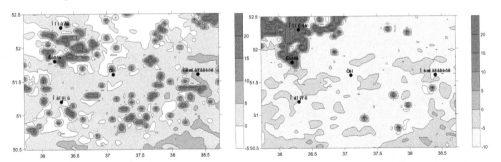

Fig. 10. Modeled (1) and measured (2) soil surface temperatures, modeled foliage temperature (3) and measured air temperature (4) for winter wheat at agricultural meteorological station Petrinka for vegetation season 1999

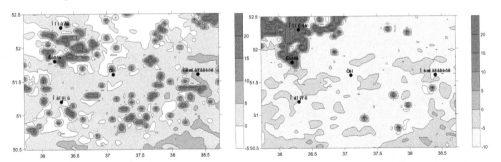

Fig. 11. LST difference map ($\Delta T = T_{s.eff}$ (AVHRR) – $T_s$(MODEL)) for part of the study region of 23,500 km$^2$ with spacing of ~ 5 km, 11.00 GMT 27 May (left) and 12 August (right) 2009

Similar distributions have been also built for the difference of $T_{ls}(MODIS) - T_s(MODEL)$. Results of comparing SEVIRI-derived temperature $T_{ls}$ defined by continuous measurements during 6-day interval of the vegetation season 2009 with three-hour ground-based observation data on air temperature $T$ at agricultural meteorological station Livny and with modeled temperature $T_f$ calculated using the same ground-based data are shown in Fig. 12. As seen from this figure, the temporal behavior of these variables is quite close.

To discover relevance of the model parameters (stomatal resistance $r_0$, leaf area index $LAI$, soil and vegetation albedos ($A_g$, $A_f$) and emissivities ($E_s$, $E_p$) for describing the water and heat exchange processes, sensitivity of the model (in particular, of the quantities $Ev$ and $T_s$) to these parameter variations has been investigated. In (Kuchment & Startseva, 1991) it has been shown that $Ev$ is strongly influenced by $r_0$. The present study has confirmed this effect. Particularly, changing the values of $r_0$ for winter wheat and perennial grasses by 20 and 60 % results in changing the values of $Ev$ by 4-5 and 8-10 mm per decade, correspondingly. Strong sensitivity of quantities $Ev$ and $T_s$ to variations of $LAI$ was also found. It was particularly notable in periods of rapid plant growth, their yellowing, mowing, as well as in the beginning of vegetation season.

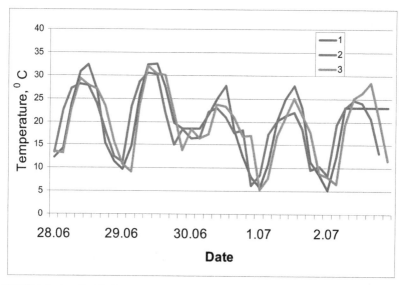

Fig. 12. SEVIRI-derived LST (1), air temperature (2), and air vegetation temperature modeled using three-hour ground observation data (3) on 28.06.-03.07.2009 at agricultural meteorological station Livny

Changing the relative values of $LAI$ by 0.1 and 0.2 resulted in changing the values of $Ev$ by 2-2.5 and 3-3.5 mm per decade and $T_s$ by 1.0-1.2 and 1.4-1.6°C, respectively, and changing the maximum values of $LAI$ from 3 to 5 leads to increase of $Ev$ 7-8 mm per decade and decrease of $T_s$ ~ 2°C. Numerical experiments with a shift of time mowing led to changing values of $Ev$ for perennial grasses to 15-17 mm per decade and values of $T_s$ to more than 3°C (fig. 13).

Such sensitivity makes it possible to select appropriately the values of $LAI$ at the specific site, that, in turn, leads to close to actual assessment of vegetation cover fraction $B$ that is shown below. The influence of the integral soil and vegetation albedos $A_g$ and $A_f$ on the values of $Ev$ and $T_s$ was different. Evapotranspiration was varied more when changing $A_f$ whereas temperature $T_s$ was changed significantly by variation of $A_g$. Particularly, change of $A_g$ by 0.2 led to changes of $Ev$ for several crops by 1.5-2.5 mm per decade and corresponding

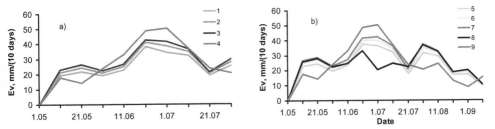

Fig. 13. Modeled values of evapotranspiration Ev for 1997 vegetation season at agricultural meteorological station Petrinka for winter wheat (a); perennial grasses (b) and different values of $LAI_{max}$: 2.5 (1); 3.5 (2); 4.5 (3); 3.0 (5); 4.0 (6); 5.0 (7); (8) 4.5 with shift of mowing day for 3 weeks. (4) and (9) corresponds to measured values of Ev for perennial grasses.

variations of $T_s$ reached 3°C and more. Similar variations of $A_f$ led to changes of $Ev$ by 4-5 mm and more per decade and also of $T_s$ by 1.5-2°C. Direct effect of soil and vegetation emissivities $E_s$ and $E_p$ on $Ev$ and $T_s$ was negligible.

The main conclusion from all experiments described above is that the key parameters affecting evapotranspiration and soil water content under wet soil conditions are $r_0$ and shading parameters $LAI$ and $B$. At the same time the influence of the latter two is often more substantial than of the first one. Under dry soil conditions (usually occurring when increasing land surface temperature), this effect becomes less noticeable and the value of evapotranspiration is mainly determined by soil water content of the upper one meter soil layer.

High sensitivity of $Ev$ and $T_s$ to $LAI$ variations as well as possibility to control current values of $LAI$ by comparing modeled $T_f$, $T_g$ and $T_s$ with satellite-derived $T_a$, $T_{sg}$ and $T_{s.eff}$ allowed specifying time behavior of $LAI$ for several crops in the absence of phytometry data. Satellite-based values of $LAI$ were estimated using empirical relationships between $LAI$ and $NDVI$ for grasslands (Biospheric Aspects, 1993) (27) and for agricultural crops (Biftu & Gan, 2001) (28):

$$LAI = NDVI \cdot 1.71 + 0.48 \qquad (27)$$

$$LAI = -2.5 \cdot \ln(1.2 - 2 \cdot NDVI) \qquad (28)$$

Specified in that way time behavior of $LAI$ is presented at Fig. 14. Here sudden changes correspond to time intervals of plant mowing.

In initial versions of the model the fraction B was calculated as follow

$$B = 1 - \exp(-p \cdot LAI) \qquad (29)$$

Here $p$ is empirical coefficient that was adjusted by comparing modeled magnitudes of $T_f$, $T_g$, and $T_s$ with satellite-derived values of $T_a$, $T_{sg}$, and $T_{s.eff}$ using $LAI$ estimates determined from both satellite and ground data. Numerical experiments were carried out under three scenarios of estimating $LAI$ and $B$: 1) values of $B$ were calculated by (29) using $LAI$ determined by ground observation data on phenological stage changes and plant heights for different land-use; 2) values of $B$ were also calculated by (29) using satellite-

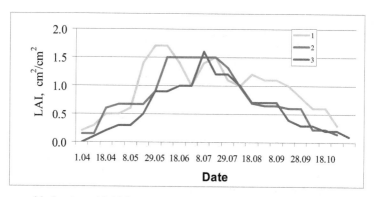

Fig. 14. Temporal behavior of LAI for vegetation season 2004 built by data of AVHRR/3 (1), MODIS (3) and by ground observation data at water balance station Nizhnedevitsk for perennial grasses (1, 2) and without allocation of cultures (3)

derived $LAI$ ; 3) estimates of $B$ and $LAI$ were generated from satellite data. Values of $T_a$ and $T_f$ under these scenarios are close to each other: for the most of observation times their differences do not exceed 2.5 and 3.5°C respectively (i.e. they are close to RMSE for satellite-derived $T_a$). These results are confirmed by rather high (0.65-0.75 at different seasons) $T_f$ and $T_a$ correlation coefficient. Similar results were also obtained when comparing $T_s$ with $T_{s.eff}$ and $T_{sg}$ with $T_g$ under all the scenarios. Examples of comparing AVHRR-derived and ground-based estimates of $B$ for perennial grasses are presented in Fig. 15.

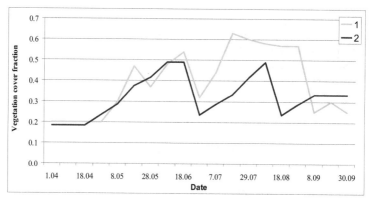

Fig. 15. Comparing AVHRR-derived (1) and ground-based estimates (2) of vegetation cover fraction $B$ for perennial grasses at the station Kursk for vegetation season 1999

Described results gave impetus to research possibility of direct use of satellite-based estimates of vegetation cover and LSTs in the model. High desirability of such use while model simulating water and heat balance components for vast territories is due to the necessity of distributed estimates of the land surface characteristics, especially under the lack of ground-based observation data. To assimilate satellite-based estimates of vegetation and meteorological characteristics the updating of the SVAT model has been performed including:

1. Replacing the ground point-wise estimates of the model parameters *LAI* and *B* by their AVHRR- or MODIS-based analogues. The efficiency of such approach has been proved through comparisons: between satellite-derived and ground-based data on *LAI* and *B* behavior during vegetation season; between satellite-derived, modeled, and in-situ measured temperatures; between modeled and actual values of evapotranspiration *Ev* (Fig. 16) and of soil water content *W* for one-meter soil layer (Fig. 17). The

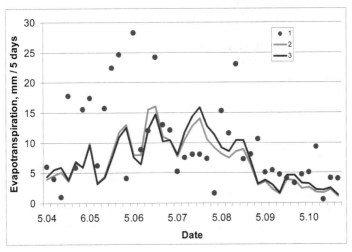

Fig. 16. Evapotranspiration Ev for vegetation season 2008 measured on grassland at water balance station Nizhnedevitsk (1), modeled using AVHRR-derived LAI for perennial grasses (3) and MODIS-derived LAI (2)

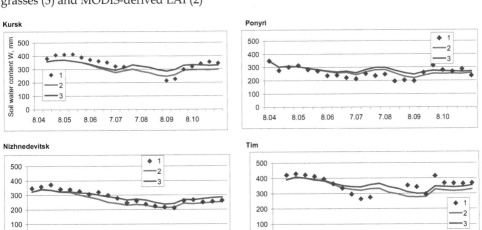

Fig. 17. Soil water content: ground measured for corn at four agricultural meteorological stations (1), modeled using AVHRR-derived LAI and B for corn (2) and MODIS-derived LAI (3) for vegetation season 2008.

discrepancies between $T_a$, $T_s$, $T_{s.eff}$ and $T_f$, $T_g$, $T_s$ as well as between $T_{ls}$ and $T_s$ and ground-measured ones do not exceed standard errors of satellite-derived estimates $T_a$, $T_{sg}$ and $T_{s.eff}$ respectively while the modeled and measured values of $Ev$ and $W$ are found close to each other within a standard error of their estimation.

2.  Entering the AVHRR- or MODIS-based LST estimates as the input SVAT model variables instead of their standard ground-based estimates if the time-matching of satellite and ground-based observations takes place. The SEVIRI-derived $T_{ls}$ estimates can also play the same role. Permissibility of such replacement has been verified while comparing remote sensed, modeled and ground-based temperatures as well as calculated and measured values of $W$ and $Ev$ (Fig. 18) The SEVIRI-based $T_{ls}$ estimates are found to be very informative and useful due to their high (up to 15 min) temporal resolution.

3.  Inputting AVHRR- and MODIS-derived $LAI$ and $B$, AVHRR-, MODIS-, and SEVIRI-based LSTs in each grid cell of the model in order to account for the space variability of vegetation cover parameters and meteorological characteristics. Ground-based data on precipitation, air temperature and humidity prepared by Inverse Distance Squared (IDS) interpolation method are also inserted into the model in each grid node. The calculations of vertical water and heat fluxes, soil water and heat contents and other water and heat balance components for above region of interest have been carried out using the described updated SVAT model and the fields of AVHRR/3- or MODIS-derived $LAI$ and $B$ estimates together with AVHRR/3- or MODIS- or SEVIRI-derived $T_{ls}$ retrievals for 1999-2010 vegetation seasons. Fig.19 shows the results of calculation of latent heat flux and soil water content at the same date of 2010 as shown in Fig. 7. The acceptable accuracy levels for above values assessment have been achieved under all scenarios of parameter and input model variable specification using satellite- and ground-based data.

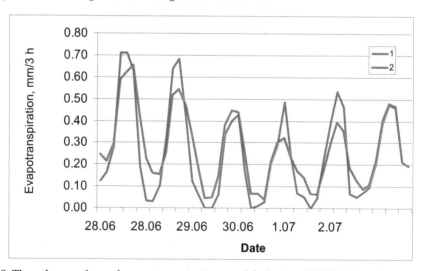

Fig. 18. Three-hour values of evapotranspiration modeled using SEVIRI-derived data (1) and three-hour observation data at the agricultural meteorological station Livny (2) on 28.06.-03.07.2009

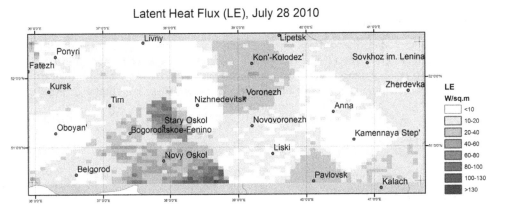

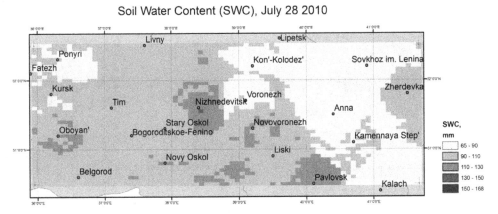

Fig. 19. Calculated latent heat flux and soil water content distributed over the study region

As follows from the above, the presented approach can be used for vast territories under the lack or absence of ground observations. The most promising in this case is the utilization of SEVIRI data due to their frequent occurrence.

### 5.2 Satellite based modeling of snow cover

The calibrated snow model (14) – (22) was applied to calculate snow cover characteristics in each 0.01° pixel of the spatial grid within the study areas. The meteorological data interpolated from the available meteorological stations to each pixel by the IDS method were used as the input to the snow cover models. In pixels where MODIS data were available the simulated land surface temperature and albedo were replaced by corresponding satellite-derived variables. Fields of snow cover characteristics based on satellite observations and the snow pack model were generated for the study area for the time period from January 1st to May 31th of the years 2002 – 2005. The initial snow cover distribution in the model was assigned following AMSR-derived SWE data on 1 January for all pixels. For the pixels where the initial SWE values were unavailable because of the lack of coverage or for some other reasons, these values were interpolated from neighboring pixels by the IDS interpolation method.

Maps of the simulated distributions of these snow characteristics were compared with the corresponding satellite maps. The simulated changes of SCA were appeared to be in satisfactory correspondence with the satellite-derived SCA. As an example of this correspondence, simulated and satellite-derived dynamics SCA are shown in Fig. 20 for several sub-regions adjoining the meteorological stations of the study area.

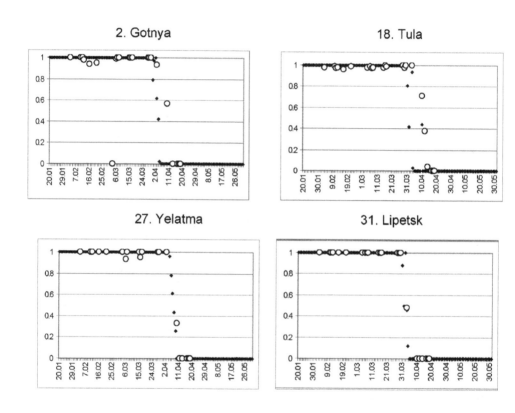

Fig. 20. Simulated (solid dots) and MODIS-derived (circles) dynamics of SCA for some polygons within the study area (winter-spring season of 2004)

However, the simulated maps of SWE substantially differ from the corresponding AMSR-derived SWE maps as it is illustrated by Fig. 21. We assume that this difference is resulted from by the fact that the accuracy of the SWE estimated from the radiometric satellite measurements noticeably decreases during melt period when snowpack is saturated by melted water so the AMSR-derived SWE maps may not adequately represent the real SWE distribution for this period of the year.

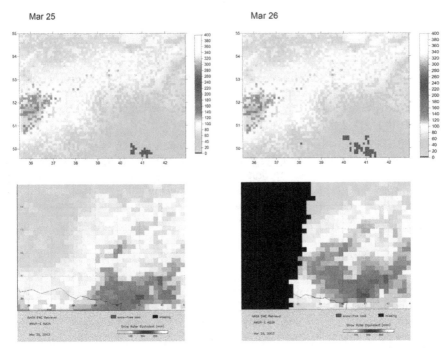

Fig. 21. Simulated (upper raw) and AMSR-derived (lower raw) distribution of SWE (mm) on 25-26 March, 2003.

## 6. Conclusion

The paper highlights the importance of satellite remote sensing data used in conjunction with a land surface model in describing water and heat regimes of vast agricultural regions. Accuracy and robustness of the corresponding quantified assessments by LSM is strongly restricted by insufficiency, both in space and time, of ground observations of highly heterogenic soil, vegetation and snow characteristics, soil temperature, etc., which can be used as the model parameters or input variables, as well as for calibration and validation of the model. Satellite remote sensing is not only an additional source of land surface data allowing substitution of the missed ground observations into the model, but, for majority of regions, it is a unique data source. From the other side, in spite of the diversity of satellite data on land surface characteristics and technologies of the data thematic processing, opportunity of using these data in LSMs is often problematic because of many reasons, e.g. irregularity of data, uncertain data accuracy, etc.

Moreover, there is no general approach allowing reliable choice of appropriate satellite data and/or processing technique; this choice depends on both specific features of the region of interest and the used model. In this study, we tried, firstly, to demonstrate opportunities of utilizing land-surface/snow remote sensing products obtained from the different satellites (NOAA, EOS Terra and Aqua, METEOSAT-9) and sensors (SEVIRI, AMSR-E, AVHRR, MODIS) in the developed LSM. Secondly, we analyzed sensitivity of the simulation results to different satellite remote sensing data. It is important that most of the used products were derived by the originally developed processing techniques, especially new technique for processing SEVIRI measurement and the technology based on synergy of optical and microwave snow products. Other main emphasis of the study is to develop a new physically based distributed Remote Sensing Based Land Surface Model (RSBLSM). The model is aimed to simulation of vertical water-and-heat transfer and, importantly, it takes into account hydrothermal processes in the "frozen soil-snow-atmosphere" system. These processes are critical for cold region agriculture, as they define crop development in early spring before the vegetation season beginning.

The obtained results allow us to conclude that utilization of the differently derived satellite-based estimates in the developed physically based model, intensively calibrated and validated against the available ground observations, provides an opportunity for reproducing spatial fields of evapotranspiration, soil moisture and temperature at different soil depths, temperature of soil/vegetation surfaces, snow and other water and heat characteristics for the vast agricultural region.

## 7. Acknowledgment

We thank E.Volkova, A.Kukharsky, S.Uspensky, V.Solovjev from SRC "Planeta" for their participation in the development of satellite data processing methods, as well as M.Alexadrovich from Water Problem Institute of RAS for technical assistance in modeling and data processing. The presented study was carried out under support of the Russian Foundation of Basic Researches (grant № 10-05-00807).

## 8. References

Armstrong, R. L., Brodzik, M. J. (2005) Northern Hemisphere EASE-Grid Weekly Snow Cover and Sea Ice Extent Version 3. *Digital media, http://nsidc.org/data/nsidc-0046.html*, National Snow and Ice Data Center, Boulder, Colorado, USA.

Becker F, Li, Z.-L. (1995) Surface temperature and emissivity at various scales: definition, measurement and related problems. *Rem. Sens. Rev.*,Vol. 12, pp. 225-253.

Biospheric Aspects of the Hydrological Cycle (BAHC). (1993) Report № 27, 103 p. Ed. by BAHC Core Project Office, Institut für Meteorologie, Freie Universitat Berlin, Germany.

Biftu, G.F., Gan, T.Y. (2001) Semi-distributed, physically based, hydrologic modeling of the Paddle River basin, Alberta, using remotely sensed data. *J. Hydrol.,* Vol. 244, pp. 137-156.

Brodzik, M. J., Armstrong, R., Savoie, M. (2007) Global EASE-Grid 8-day Blended SSM/I and MODIS Snow Cover. *Digital media.* *http://nsidc.org/data/docs/daac/nsidc0321_8day_ssmi_modis_blend/index.html*, National Snow and Ice Data Center, Boulder, Colorado, USA.

Brooks, R.H., Corey, A.T. (1964) *Hydraulic properties of porous media.* Hydrol. Pap. Colorado State Univ. № 3. 27 p.

Brubaker, K. L., Pinker, R. T., Deviatova, E. (2005) Evaluation and Comparison of MODIS and IMS Snow-Cover Estimates for the Continental United States Using Station Data. *J. Hydrometeor.,* Vol. 6, pp. 1002–1017. doi: 10.1175/JHM447.1

Clupp, R.B., Hornberger, G.M. (1978) Empirical equations for some hydraulic properties. *Water Resources Research,* Vol. 14, № 4, pp. 601-604.

Collatz, G.J., Ball. J.T., Grivet, C., Berry, J.A. (1991) Physiological and environmental regulation of stomatal conductance, photosynthesis and transpiration: a model that includes a laminar boundary layer. *Agricultural and Forest Meteorology,* Vol. 54, pp. 107–136.

Dai, Y., et al. (2003) The Common Land Model (CLM) version 1.0. *Bull. Am. Meteorol. Soc.,* Vol. 84, pp. 1013– 1023.

de Wildt, M. D., Gabriela, S., Gruen, A. (2007) Operational snow mapping using multitemporal Meteosat SEVIRI imagery. *Remote Sensing of Environment,* Vol. 109, pp. 29-41.

Deardorff ,J.W. (1978) Efficient prediction of ground surface temperature and moisture with inclusion of a layer of vegetation. *Journal of Geophysical Research,* Vol. 83, pp. 1889–1903.

Desborough, C.E., Pitman, A.J. (1998) The BASE land surface model. *Global and Planetary Change,* Vol. 19, pp. 3-18.

Dickinson, R.E., Henderson-Sellers, A., Kennedy, P.J., Wilson, M.F. (1986) *Biosphere Atmosphere Transfer Scheme (BATS) for the NCAR Community Climate Model,* NCAR Tech. Note, TN-275 + STR, 69 pp.

Faysash, A., Smith, E.A. (2000) Simultaneous Retrieval of Diurnal to Seasonal Surface Temperatures and Emissivities over SGP ARM-CART Site Using GOES Split Window. *J. Appl. Meteor.,* Vol. 39, pp. 971-982.

Foster, J.L., Hall, D. K., Eylander, J.B., Riggs, G.A., Nghiem, S.V., Tedesco, M., Kim, E., Montesano, P.M., Kelly, R. E.J., Casey, K.A., Choudhury, B. (2011) A blended global snow product using visible, passive microwave and scatterometer satellite data. *International Journal of Remote Sensing,* Vol. 32, pp. 1371-1395.

Gelfan, A. N., Pomeroy, J. W., Kuchment, L.S. (2004) Modelling forest cover influences on snow accumulation, sublimation, and melt. *J. Hydrometeorology,* Vol. 5, No. 5, pp. 785–803.

Gelfan, A. N. (2006) Physically based model of heat and water transfer in frozen soil and its parametrization by basic soil data. In: *Predictions in Ungauged Basins: Promises and Progress. Proceedings of symposium S7 held during the Seventh IAHS Scientific Assembly at Foz do Iguazu, Brazil, April 2005,* Eds: M. Sivapalan,. IAHS Publ., Vol. 303, pp. 293-304.

Global Water Security – an engineering perspective. *Report of the The Royal Academy of Engineering,* Publ. by The Royal Academy of Engineering, London, UK, April 2010.

Good, E. (2009) Blending in situ and satellite data for monitoring land air temperatures. *Proc. of the 2009 Eumetsat Meteorol. Sat. Conf, Bath, UK, 21 - 25 September 2009,* 5 p.

Grody, N.C., Basist, A.N. (1996) Global identification of snow cover using SSM/I Measurements. *IEEE Trans. Geosci. Remote Sens.,* Vol. 34, N 1, pp. 237-249.

Gusev, E. M., Nasonova, O.N. (2002) The simulation of heat and water exchange at the land–atmosphere interface for the boreal grassland by the land-surface model SWAP. *Hydrol. Processes,* Vol. 16, pp. 1893-1919.

Hall, D.K., Riggs, G., Salomonson, V., DiGirolamo, N.E., Bayr, K.J. (2002) MODIS snow cover products. *Remote Sensing of Environment,* V. 83, pp. 181-194.

Hall, D.K, Riggs, G.A. (2007) Accuracy assessment of the MODIS snow-cover products. *Hydrological Processes*, Vol. 21, pp. 1534–1547.

Helfrich, S.R., McNamara, D., Ramsay, B.H., Baldwin, T., Kasheta, T. (2007) Enhancements to, and forthcoming developments in the Interactive Multisensor Snow and Ice Mapping System (IMS). *Hydrological Processes*, Vol. 21, pp. 1576–1586.

Henderson-Sellers, A, Pitman, A.J., Love, P.K., Irannejad, P., Chen, T. (1995) The project for intercomparison of land surface parameterization schemes (PILPS) phases 2 and 3. *Bulletin of the American Meteorological Society*, Vol. 76, pp. 489–503.

Justice, C. O., Vermote, E., Townshend, J. R. G., et al. (1998) The Moderate Resolution Imaging Spectroradiometer (MODIS): land remote sensing for global change research. *IEEE Trans. Geosci. Remote Sens.*, Vol. 36, pp. 228-1249.

Kelly, R. E., Chang, A. T., Tsang, L., Foster, J.L. (2003) A prototype AMSR-E global snow area and snow depth algorithm. *IEEE Transactions on Geoscience and Remote Sensing*, Vol. 1, pp. 230-242.

Kuchment, L.S., Motovilov, Yu., G., Startseva, Z.P. (1990) Modeling of moisture transport in the soil-vegetation-surface boundary layer system for hydrological problems. *Water Resources*, Vol. 16, N 2, pp. 121-128. (Translated from Russian, *Vodnye Resursy* (1989), N 2, pp.32-39, Plenum Publishing Corporation, Consultants Bureau, New York, USA).

Kuchment, L.S., Startseva, Z.P. (1991) Sensitivity of evapotranspiration and soil moisture in wheat fields to changes in climate and direct effects of carbon dioxide. *Hydrol. Sci. J.*, Vol. 36, N 6, pp. 631-643.

Kuchment, L. S., Gelfan, A.N. (1996) The determination of the snowmelt rate and the meltwater outflow from a snowpack for modeling river runoff generation. *J. Hydrology*, Vol.179, pp. 23-36.

Kuchment, L.S., Demidov, V.N., Startseva, Z.P. (2006) Coupled modeling of the hydrological and carbon cycles in the soil–vegetation–atmosphere system. *J. Hydrology*, Vol.323, pp.4-21

Kuchment, L.S., Romanov, P., Gelfan, A.N., Demidov, V.N. (2010) Use of satellite-derived data for characterization of snow cover and simulation of snowmelt runoff through a distributed physicall basedmodel of runoff generation *Hydrol. Earth Syst. Sci.*, Vol. 14, pp. 339–350.

Lokupitiya, E., et al. (2009) Incorporation of crop phenology in Simple Biosphere Model (SiBcrop) to improve land-atmosphere carbon exchanges from croplands. *Biogeosciences*, Vol.6, pp.969–986.

Manabe, S. (1969) Climate and the ocean circulation. The atmospheric circulation and hydrology of the Earths surface. *Mon.Wea. Rev.*, Vol. 97, pp. 739–774.

Mauser, W., Schädlich, S. (1998) Modelling the spatial distribution of evapotranspiration on different scales using remote sensing. *J. Hydrol.*, Vol. 212–213, pp. 250–267.

Milly, P.C.D., Shmakin, A.B. (2002) Global modeling of land water and energy balances. Part I: the Land Dynamics (LaD) model. *Journal of Hydrometeorology*, Vol. 3, pp. 283–299.

Muzylev, E.L., Uspensky, A.B., Volkova, E.V., Startseva, Z.P. (2002) Simulation of hydrological cycle of river basins using synchronous high resolution satellite data. *Russian Meteorology and Hydrology*, N 5, pp. 52-63.

Muzylev, E.L., Uspensky, A.B., Volkova, E.V., Startseva, Z.P. (2005) Using satellite information for modeling heat and moisture transfer in river watersheds. *Earth Research from Space*, N 4, pp. 35-44. (In Russ.)

Muzylev, E., Uspensky, A., Startseva, Z., Volkova, E. (2006) Modeling vertical heat and water fluxes from river basin area with AVHRR/NOAA-based information on land surface characteristics. *Proc. of the 7th International Conference on Hydroinformatics, Nice, France, 4-8 September 2006.* Ed. P.Gourbesville, J.Cunge, V.Guinot, S.-Y.Liong, Vol. 2, pp. 1163-1170, Research Publishing, Chennai, India.

Muzylev, E.L., Uspenskii, A.B., Startseva, Z.P., Volkova, E.V., Kukharskii, A.V. (2010) Modeling water and heat balance components for the river basin using remote sensing data on underlying surface characteristics. *Russian Meteorology and Hydrology*, Vol. 35, N 3, pp. 225-235, DOI: 10.3103/S1068373910030106

NOAA KLM User's Guide. (2005) *http://www.ncdc.noaa.gov/oa/pod-guide/ncdc/docs/klm/html/c1/sec1-2.htm.*

Oleson, K.W., Niu, G.-Y., Yang, Z.-L., Lawrence, D. M., Thornton, P. E., Lawrence, P. J., Stöckli, R., Dickinson, R. E., Bonan, G. B., Levis, S., Dai, A., Qian, T. (2008) Improvements to the Community Land Model and their impact on the hydrological cycle. *J. Geophys. Res.*, Vol. 113, G 01021, doi:10.1029/2007JG000563.

Overgaard, J., Rosbjerg, D., Butts, M. B. (2006) Land-surface modelling in hydrological perspective – a review. *Biogeosciences*, Vol. 3, pp. 229–241.

Pitman, A.J. (2003) The evolution of, and revolution in, land surface schemes designed for climate models. *Int. J. Climatol.*, Vol. 3, pp. 479–510.

PUM LST (Product User Manual. Land Surface Temperature) (2008) *SAF/LAND/IM/PUM_LST/2.1*, 49p.

Romanov, P., Gutman. G., Csiszar, I. (2000) Automated monitoring of snow over North America with multispectral satellite data. *Journal of Applied Meteorology*, Vol. 39, pp. 1866–1880.

Romanov, P. (2011) Satellite-Derived Information on Snow Cover for Agriculture Applications in Ukraine, in Use of Satellite and In-Situ Data to Improve Sustainability. *NATO Science for Peace and Security Series C: Environmental Security.* Part 2, pp.81-91, DOI: 10.1007/978-90-481-9618-0_9

Rutter, N. et al. (2009), Evaluation of forest snow processes models (SnowMIP2), J. Geophys. Res., 114, D06111, doi:10.1029/2008JD011063

Schmetz, J., Pili, P., Tjemkes, S., Just, D. et al., (2002) An introduction to Meteosat Second Generation (MSG). *Bull. Amer. Meteor. Soc.*, Vol. 83, pp. 977-992.

Schultz, K., Franks, S., Beven, K. (1998) *TOPUP – A TOPMODEL based SVAT model to calculate evaporative fluxes between the land surface and the atmosphere, Version 1.1, Program documentation,* Department of Environmental Science, Lancaster University, UK.

Sellers, P.J, Mintz, Y., Sud, Y.C., Dalcher, A. (1986) A Simple Biosphere model (SiB) for use within general circulation models. *Journal of the Atmospheric Sciences*, Vol. 43, pp. 505–531.

Sellers, P.J., Berry, J.A., Collatz, G.J., Field, C.B., Hall, F.G. (1992) Canopy reflectance, photosynthesis and transpiration. III. A reanalysis using improved leaf models and a new canopy integration scheme. *Remote Sensing of the Environment*, Vol. 42, pp. 187–216.

Sellers, P. J., Dickinson, R. E., Randall, D. A., Betts, A. K., Hall, F. G., Berry, J. A., Collatz, G. J., Denning, A. S., Mooney, H. A., Nobre, C. A., Sato, N., Field, C. B., Henderson-Sellers, A. (1997) Modeling the exchanges of energy, water, and carbon between continents and the atmosphere. *Science*, Vol. 275, pp. 502-509.

Simic, A., Fernandes, R., Brown, R., Romanov, P., Park, W. (2004). Validation of VEGETATION, MODIS, and GOES C SSM/I snow-cover products over Canada based on surface snow depth observations. *Hydrological Processes*, Vol. 18, No. 6, pp. 1089-1104.

Snyder, W.C., Wan, Z., Zhang, Y., Feng, Y.-Z. (1998) Classification-based emissivity for land surface temperature measurement from space . *Int. J. Rem. Sens.*, Vol. 19, No. 14, pp. 2753-2774.

Solovjev, V.I., Uspensky, S.A. (2009) Monitoring of land surface temperatures based on second generation geostationary meteorological satellites, *Earth Studies from Space*, No. 3, pp.17-25, (in Russian).

Solovjev, V.I., Uspensky, A.B., Uspensky, S.A. (2010) Derivation of land surface temperature using measurements of IR radiances from geostationary meteorological satellites, *Russian Meteorology and Hydrology*, Vol. 35, No. 3, pp. 159-167, DOI 10.3103/ S1068373910030015.

Uspensky, S., Solovjiev, V., Uspensky, A.,(2009) Monitoring of land surface temperatures based on SEVIRI/METEOSAT-9 measurements. *Proceedings of 2009 EUMETSAT Meteorological Satellite Conference, 2009, EUMETSAT*, pp 55-60.

Uspensky, S., Uspensky, A., Rublev, A. (2011) Land surface air temperature retrieval capabilities with geostationary meteorological satellite data. *Proc. "Intern. Symp. on Atmospheric Radiation and Dynamics", St.Petersburg, June 21-24, 2011,*pp. 37-38.

van Genuchten, M.Th. (1980) A closed form equation for predicting the hydraulic conductivity of unsaturated soils. *Soil Sci. Soc. Am. J.*, No. 44, pp.892-898.

Volkova, E.V., Uspenskii, A.B. (2007). Detection of clouds and identification of their parameters from the satellite data in the daytime. *Russian Meteorology and Hydrology.* Vol. 32, No. 12, pp. 723-732, DOI: 10.3103/S1068373907120011.

Walker, A.E., Goodison, B.E. (1993) Discrimination of a wet snow cover using passive microwave satellite data. *Annals of Glaciology*, Vol. 17, pp. 301-311

Wan, Z., Dozier, J. (1996) A generalised split-window algorithm for retrieving land surface temperature from space. *IEEE Trans. Geosci. Rem. Sens.*, Vol. 34, No. 4, pp. 892-905.

Wood, E.F, Lettenmaier, D.P, Zartarian, V.G. (1992) A land-surface hydrology parameterization with subgrid variability for general circulation models. *Journal of Geophysical Research* Vol. 97, pp. 2717-2728.

Yang, Z.-L. (2004) Modeling land surface processes in short term weather and climate studies, In: *Observation, Theory and Modeling of Atmospheric Variability*, Ed. X. Zhu, X. Li, M. Cai, S. Zhou, Y. Zhu, F.-F. Jin, X. Zou, and M. Zhang, pp. 288-313, World Scientific Series on Meteorology of East Asia, World Scientific, New Jersey.

# 6

# Mapping Soil Salinization of Agricultural Coastal Areas in Southeast Spain

Ignacio Melendez-Pastor, Encarni I. Hernández,
Jose Navarro-Pedreño and Ignacio Gómez
*Department of Agrochemistry and Environment,*
*University Miguel Hernández of Elche*
*Spain*

## 1. Introduction

Soil salt content is a key factor that determines soil chemical quality together with soil reaction, charge properties and nutrient reserves (Lal et al., 1999). An adequate salt supply is essential for an optimum development of photosynthetic mechanism and other biochemical processes in plants (Sitte et al., 1994). Soil salt content constitutes an environmental problem when salt accumulation generates drastic changes in soil physical and chemical properties, adversely affecting soil productivity and plant growth (Richards, 1954; Qadir et al., 2000).

Salinization affects about 30% of the irrigated land of the world, decreasing this area approximately 1-2% per year due to salt-affected land surfaces (FAO, 2002). In Europe, about 1-3 million hectares of the land are affected by salinization (European Commission, 2003), and most of these areas are situated in the Mediterranean basin. In Spain, about 18% of the 3.5 million hectares of irrigated land are severely affected or at serious risk of soil salinization (European Commission, 2002). Soil salinization is a frequent problem in arid and semiarid regions like Southeast Spain (Hernández Bastida et al., 2004). In these areas, agriculture with a great water requirement combined with high water tables and an adverse climate (increased occurrence of extreme drought events) have forced irrigation with poor quality water, causing processes of soil degradation and salinization, limiting crop growth and the production capacity (Pérez-Sirvent et al., 2003; Acosta et al., 2011).

Evaluating the spatial variability of basic soil properties in saline soils, and mapping spatial distribution patterns of these soil properties helps to make effective site-specific management decisions (Ardahanlioglu et al., 2003). Accordingly, remote sensing techniques and geographic information systems (GIS) have introduced a new era for soil resources assessment and monitoring in terms of information quality (Mermut and Eswaran, 2001). A priori knowledge of spectral characteristics of remotely sensed materials is fundamental to any valuable quantitative analysis (Ben-Dor et al., 1997). The variety of absorption processes occurring in the soil and their wavelength dependence allow us to derive information about the chemistry of the minerals composing it from the reflected or emitted light (Clark, 1999). Reflectance spectra of soils are attributed to numerous soil properties. There are no narrow absorption bands linked to soil salinity status, since it is determined by soil properties such as pH, electrical conductivity, salt content and exchangeable sodium percentage (Csillag et al., 1993; Farifteh et al., 2008). In this sense, soil reflectance is derived from the particular spectral behaviour of the heterogeneous combination of minerals, organic matter and soil water (Ben-Dor and Banin, 1994). Salt-affected soils cations ($Na^+$, $Mg^{2+}$, $K^+$, and $Ca^{2+}$) and anions ($Cl^-$, $SO_4^{2-}$, $CO_3^{2-}$ and $HCO_3^-$) can be detected by optical spectrometers since salt minerals have diagnostic spectral features occurring in the visible and near infrared (VNIR) and short-wave infrared (SWIR) spectral regions (Farifteh et al., 2008). Saline soils usually have evaporate minerals, which spectral features that can be explained by vibrational absorption due to water molecules chemically bound as part of the crystal structure (Howari et al., 2000). In this sense, the spectral differences of evaporates of single salt compounds are determinant of the type and mineralogy of the soils (Howari et al., 2000).

Remote sensing has been extensively employed in soil salinity studies. Data from aerial photography, videography, and optical, thermal, microwave or geophysical sensors has been used in soil salinity mapping (Metternich and Zinck, 2003). Perhaps, the most widely used remote sensing data in recent decades have been provided by multispectral (Landsat, SPOT, IRS, ASTER) or hyperspectral (DAIS, HyMap, AVIRIS, Hyperion) sensors in the spectral range approximately between 400 and 2500 nm. Researchers have frequently employed remote sensing data to map soil salinity with multispectral (Metternich and Zinck, 1997; Dwivedi et al., 2001; Melendez-Pastor et al., 2010a) and hyperspectral images (Dehaan and Taylor, 2002, 2003; Schmid et al., 2009, Ghrefat and Goodell, 2011). Pioneering studies in the 1970s employed air-borne and satellite-borne multispectral scanners to detect soil salinity, indicating the better capability of infrared bands over visible bands to locate saline soils and the low contribution of thermal bands to improve the delineation of saline areas (Richardson et al., 1976; Dalsted et al., 1979). Nowadays, imaging spectroscopy techniques are employed for the automatic detection of soil salinization with airborne or satellite sensor (Dehaan and Taylor, 2002, 2003; Dutkiewicz et al., 2009; Schmid et al., 2009; Weng et al., 2009; Melendez-Pastor et al., 2010a; Ghrefat and Goodell, 2011). Imaging spectroscopy deals with the mapping of ground materials by detecting and analysing reflectance/absorbance features in hyperspectral (or multispectral) images (Clark, 1999). Imaging spectroscopy adds a new dimension of remote sensing by expanding point spectrometry into a spatial domain and under field conditions, which is a very good approach for the study of soil properties (Ben-Dor et al., 2009).

The aim of this chapter is the application of remote sensing for the study of soil salinity of an agricultural area in southeast coast of the Iberian Peninsula. Different digital image processing techniques were applied to satellite multispectral images (Landsat TM). 'Conventional' hard classification techniques were combined with spectral mixture analysis and soil properties to achieve a better understanding of the soil salinization process in the study area.

Multispectral satellite images such as those obtained by the Landsat program provide low or free cost worldwide coverage for four decades. Moreover, salinization problems are concentrated in arid and semi-arid regions, often in developing countries with few economic resources. Although there are more advanced sensors that can provide a more precise quantification of the extent of soil salinity (e.g. hyperspectral), their high cost difficult its extensive use. Therefore, it is necessary to continue investigating the application of multispectral image repositories as a tool to assist in the monitoring and management of saline soils.

## 2. Material and methods

This study will evaluate the applicability of various remote sensing techniques for studying salinization processes in an agricultural coastal area. One of the greatest difficulties in the application of remote sensing techniques to the study area is the fragmentation of the territory by the existence of small plots and buildings that create a dispersed mixture of spectral signals to the scale of a moderate spatial resolution multispectral remote sensing image as those acquired by the Landsat Thematic Mapper sensor. This difficulty motivates the need to evaluate various techniques and methodological approaches to carry out this study as necessary to help monitoring the processes of salinization.

Representative soils of the area were sampled and their properties were characterized at the laboratory by standard methods. Predominant land cover classes at the soil sampling plots and at additional land cover validation points were identified. Land cover is a fundamental variable that impacts on and links many parts of the human and physical environments (Foody, 2002) with a great influence on soil properties (Caravaca et al., 2002; Majaliwa et al., 2010; Biro et al., 2011). Both kinds of information in a GIS database were included. In this sense, the effect of land cover on soil properties was statistically evaluated. Then, multispectral images were employed for a hard land cover mapping with a supervised approach using the k-nearest-neighbour classifier. Accuracy assessment methods highlighted the need to employ a mixed pixel focus to deal with the particularities of the study area. Spectral unmixing techniques allowed the identification of representative spectral endmembers and the obtainment of their corresponding fraction images. Finally, fraction endmembers were employed to characterize land cover classes and to predict soil properties with various statistical methods.

### 2.1 Description of the study area

The study area is located in a coastal zone of Southeast Spain, in the province of Alicante. It is located around 38.14°N and 0.73°W, at the south of the cities of Elche and Alicante. The study area (Figure 1) comprises alluvial plains resulting from the accumulation of sediments from the Segura and Vinalopó rivers. During most of the Holocene (~10,000 years ago to present) the study area was a large lagoon (Blázquez, 2003). In the last centuries, the ancient lagoon was transformed into an irrigated agricultural land draining the wetland. Nowadays, this area is a mixture of small-size cities, coastal urban areas, scattered residential houses, irrigated crops and isolated and scattered wetlands. The perimeter of the study area was delimited according to natural or man-made features in order to enclose a large coastal plain area primarily occupied by irrigated agricultural activities. The study area lies in the north with the natural parks of *El Hondo* and the *Salinas de Santa Pola*. Both natural areas are wetlands included in the RAMSAR list of wetlands of international importance. The east and south boundaries are the *Sierra del Molar* and the *Segura River* respectively. Urban areas and sclerophyllous vegetation mainly occupy the Sierra del Molar, while the Segura River is the most important watercourse in southeast of Iberian Peninsula providing water for irrigation agriculture and to fill the reservoirs that currently comprise

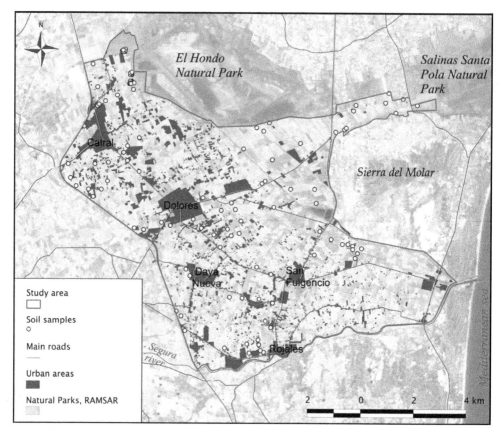

Fig. 1. Study area with the Landsat scene (false-colour composite RGB:742) and superimposed cartographic information (soil samples, urban areas, natural parks and roads).

the wetland of *El Hondo*. The western boundary of the study area is a motorway that cuts north to south the floodplain.

This coastal region has a semiarid Mediterranean climate, with a mean annual rainfall of less than 300 mm and a mean annual temperature of 17 °C and defined by the Köppen climate classification system as *Bsk* class (dry climate with a dry season in summer and a mean annual temperature about 18 °C). The climate is arid or semiarid according to the aridity index of Martonne (De Martonne, 1926) and the aridity index of UNEP (1997) respectively. Figure 2 shows the daily climatic diagram of mean temperature, precipitation and evapotranspiration (by the Penman-Monteith method) for the hydrological year 2010-2011 (from October to September) at Catral meteorological station. Mean daily temperature (blue line) varies from approximately 9°C in winter to more than 25°C in summer. Rain events (red bars) mainly occurred from December 2010 to May 2011 with total accumulated precipitation of 182 mm. This very scarce precipitation joint with an accumulated evapotranspiration of 1115 mm implied that the hydrological year was very dry.

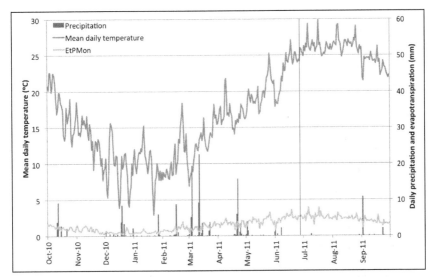

Fig. 2. Daily values of precipitation, mean temperature and evapotranspiration for the hydrological year 2010-11 at Catral station. Source data from the Spanish Ministry of Environment and Rural and Marine Affairs (MARM).

Predominant soil classes are Entisols according to the Soil Taxonomy (Soil Survey Staff, 2006) but affected by agriculture practices along years. They are characterized by a massive presence of carbonates and soluble salt content. In the studied area, irrigation is essential to support agriculture. The water deficit during several months requires irrigation while low quality water is used in the poorly drained soils of these coastal plains, being soil salinization an environmental problem. Thus, the study area soils are subjected to severe risk of physical, chemical and biological degradation (De Paz et al., 2006) that endanger agriculture sustainability.

## 2.2 Field survey

Field survey was done in the late spring and summer months of the hydrological year 2010-2011 to collect soil samples and identify land cover classes. An extensive soil sampling was done, and 116 samples were collected and geographically referenced. Samples were obtained from the upper 5 cm as solar radiation in VNIR spectral range has limited penetration capabilities. Soil samples were dried at room temperature and a 2 mm sieve was used to separate the fine fraction to be analysed. Analysed soil characteristics included in the study were electrical conductivity (EC) (1:5 w/v water extraction), pH and organic carbon (OC) by wet chemical oxidation (Walkley and Black, 1934) with potassium dichromate oxidation (Nelson and Sommers, 1982).

A land cover validation campaign was also conducted along with the soil survey in order to allow accuracy assessment of generated land cover maps. Land cover validation points were randomly generated in a GIS and a database with the land cover category generated. A total of 205 land cover validation points were identified, combining field observation and recent aerial orto-photography (0.5 m of spatial resolution). Land cover classes identified in the study area were: water bodies, seasonal or permanent crops, saltmarshes and misused agricultural field that tends to be saltmarshes, palm groves, marshes with almost permanent inundation, and anthropic areas (Table 1).

| Land cover | ID | Features |
|---|---|---|
| Water | 1 | Wetlands water tables and irrigation ponds |
| Arable land | 2 | Herbaceous (e.g. alfalfa, barley) and horticultural (e.g. melon, broccoli) crops |
| Permanent crops | 3 | Fruit trees (e.g. orange, lemon, pomegranate) |
| Fallow/abandoned | 4 | Fallow or recently abandoned agricultural land. |
| Saltmarsh | 5 | Halophyte vegetation (e.g. *Salicornia sp.*, *Suaeda sp.*, *Limonium sp.*, *Halocnemum sp.*) |
| Palm groves | 6 | Palm trees plantations and nurseries, mainly from *Phoenix dactilifera* |
| Marsh | 7 | *Phragmites australis* dominated wetland vegetation |
| Man-made/urban | 8 | Urban areas, roads, farms or industrial areas |

Table 1. Descriptions of land cover classes identified in the study area.

Land cover categories at soil sampling points were also identified and included along with soil properties in a GIS database for the land cover classification training stage and for further spatial analyses. Note that land cover (i.e. biophysical materials found on the land) and land use (i.e. how the land is being used by human beings) (Jensen, 2007) are different terms but often used together or interchangeably. In this chapter, we adopt the term land cover because we are interested in knowing about the biophysical characteristics of the study area, but the knowledge of both land use and land cover are important for land planning and land management activities (Lillesand et al., 2003).

## 2.3 Satellite imagery preprocessing

Remote sensing data were acquired by the Thematic Mapper (TM) sensor on-board the Landsat 5 satellite. Meteorological conditions and the satellite pass over the study area conditioned the date of image acquisition. A scene acquired on 28th June 2011 (path 199 row 33) was employed for analyses. A vertical black line on Figure 2 indicates the time of acquisition of the scene. No rain events happened 16 days prior to the scene acquisition date. Typically summer meteorological conditions without cloud coverage and high temperature were registered on the date of image acquisition, and thus the image quality was optimal.

Satellite image preprocessing included geometric and atmospheric corrections with the aim to ensure the spatial comparability with other data sources and to obtain at-ground reflectance pixel spectra, respectively. Various georeferenced data types were used for the geometric correction: aerial orthophotos (0.5 m of pixel resolution) and digital cartography (scale = 1:10000). The Landsat 5 TM scene was geometrically corrected using Ground Control Points (GCP) identified on the orthophotos and cartographic maps. A quadratic mapping function of polynomial fit and the nearest neighbour resampling method were used for the correction. The nearest neighbour resampling method was selected because it ensures that the original (raw) pixel values are retained in the resulting output image, which is an important requirement in any change detection analysis (Mather 2004). The maximum allowable root mean square error (RMSE) of the geometric correction was less than half a pixel, a reference value frequently cited (Townsend and Walsh 2001; Jensen, 2005).

Atmospheric correction involves the estimation of the atmospheric optical characteristics at the time of image acquisition before applying the correction to the data (Kaufman, 1989). This type of correction is a pre-requisite in many remote sensing applications such as in classification and change detection procedures (Song et al., 2001). Radiometric calibration was applied prior to the atmospheric correction. The conversion of raw digital numbers ($DN_{raw}$) of Landsat level 1 (L1) image products to at-satellite radiance values ($L_{sat}$) required the application of current re-scaling values (Chander et al., 2010) by applying the following expression (Chander and Markham, 2003; Chander et al., 2010):

$$L_{sat} = \left( \frac{L_{MAX\lambda} - L_{MIN\lambda}}{255} \right)(DN) + L_{MIN\lambda} \qquad (1)$$

Where $L_{sat}$ is at-satellite radiance [W/(m² sr µm)]; $L_{MIN\lambda}$ is the spectral radiance that is scaled to $Q_{calmin}$ [W/(m² sr µm)] ($Q_{calmin}$ is the minimum quantized calibrated pixel value, i.e. DN=0, corresponding to $L_{MIN\lambda}$; $L_{MAX\lambda}$ is the spectral radiance that is scaled to $Q_{calmax}$ [W/(m² sr µm)] ($Q_{calmax}$ is the maximum quantized calibrated pixel value, i.e. DN=255, corresponding to $L_{MAX\lambda}$); and DN are digital numbers of the L1 image product. Surface reflectance values ($\rho$) were computed by using the image based COST method (Chavez Jr, 1996). Path radiance ($L_p$) values were computed by using the equation reported in Song et al. (2001) that assumes 1% surface reflectance for dark objects (Chavez Jr, 1989, 1996; Moran et al., 1992). The optical thickness for Rayleigh scattering ($\tau_r$) was estimated according to the equation given in Kaufman (1989).

## 2.4 Land cover classification

Image classification procedures aim to automatically categorize all pixels in an image into land cover classes or themes (Lillesand et al., 2003). Thematic mapping from remotely sensed data can be defined as grouping together cases (pixels) by their relative spectral similarity (unsupervised component) with the aim of allocating cases based on their similarity to a set of predefined classes that have been characterized spectrally (supervised component) (Foody 2002). Multispectral images (like Landsat TM scenes) are frequently used to perform the classification based on spectral pattern recognition methods that exploits the pixel-by-pixel spectral information as the basis for automated classification (Lillesand et al., 2003). In this study, a supervised land cover classification of the Landsat TM image was performed with a $k$-nearest-neighbour clustering algorithm to obtain a discrete or 'hard' categorical land cover map for the study area. $K$-nearest-neighbour (KNN) classifier searches away from the pixel to be classified in all directions of the spectral space until it encounters $k$ user-specified training pixels and then assigns the pixel to the class with the majority of pixels encountered (Jensen, 2005). KNN algorithm has been successfully applied for land cover classification with remote sensing data (Franco-Lopez et al., 2001; Haapanen et al., 2004; Blanzieri and Melgani, 2008). Land cover classes assigned to the soil plots were employed in the training stage of the algorithm. Major urban areas were digitized with a GIS and masked-out of the supervised classification procedure since urban areas induce a great spectral confusion. Water areas training points were also included in the training dataset.

The land cover validation database was employed to evaluate the performance of the classification. Land cover map accuracy assessment was quantified with statistical methods such as the error matrix and the kappa statistic. The error matrix is a square array of numbers organized in rows and columns that express the number of sample units (i.e. pixels) assigned to a particular category relative to the actual category as indicated by the reference data (Congalton, 2004). Reference data are in the columns while the rows indicate the map categories to be assessed. This form of expressing accuracy as an error matrix is an effective way to evaluate both errors of inclusion (commission errors) and errors of exclusion (omission errors) present in the classification as well as the overall accuracy (Congalton et al., 1983). In addition to the error matrix, the Kappa coefficient developed by Cohen (1960) was employed to quantify the accuracy of the land cover map. Cohen's Kappa (or KHAT) is a measure of agreement for nominal scales based on the difference between the actual agreement of the classification (i.e., agreement between computer classification and reference data as indicated by the diagonal elements) and the chance agreement, which is indicated by the product of the row and column marginal (Congalton et al., 1993).

## 2.5 Spectral unmixing

A mixed pixel results when a sensor's Instantaneous Field of View (IFOV) includes more than one land cover type on the ground (Lillesand et al., 2003). The spectrum of a single pixel is a complex measurement that integrates the radiant flux from all the spatially unresolved materials in the IFOV, regardless of whether or not we know their identities (Adams and Gillespie, 2006). Spectral mixture analysis (SMA) has been developed as a method to transform the reflectance in the bands of multispectral images to fractions of reference endmembers, which are reflectance spectra of well-characterized materials that mix to produce spectra equivalent to those of pixels of interest in the image (Adams et al., 1995). As part of SMA techniques, linear spectral unmixing (LSU) models tread the radiation recorded by a sensor as the result of a linear mixture of spectrally pure endmember radiances (Small and Lu 2006). This method is based on the assumptions that: 1) the recorded radiation by the sensor for each pixel is limited to the sensor's IFOV, and assumes no influences by reflected radiation from neighbouring pixels (Settle and Drake 1993), 2) the overall global radiance is proportional to the surface occupied by each land cover type, and 3) the spectrally pure endmembers are valid for the whole study area (Quarmby et al. 1992). LSU models describe radiation reflected by an individual pixel $(i,j)$ of a band $k$ as the result of the product of reflectance for each land cover type by their respective mixture fraction plus an additional associated error for each pixel. The general expression of the model is presented in the following equation:

$$\rho_{i,j,k} = \sum\nolimits_{m=1,p} F_{i,j,m}\, \rho_{m,k} + e_{i,j} \qquad (2)$$

Where $\rho_{i,j,k}$ is the observed reflectance of a pixel for row $i$, column $j$, and band $k$ ; $F_{i,j,m}$ is the proportion of component $m$ of a pixel for row $i$, column $j$, for each one of the pure components; $\rho_{m,k}$ is the characteristic reflectance for component $m$ in band $k$ ; and $e_{i,j}$ is the error associated to the estimation of proportions for each pixel $i, j$. The Least Square Mixing Model proposed by Shimabukuro and Smith (1991) is commonly used to resolve linear spectral mixture models. The method proposed by Shimabukuro and Smith (1991) assumes two initial restrictions for the computation of the proportions of spectrally pure endmembers. The first one implies that pure endmember proportions must range between 0 and 1. This means that the proportions of the components are normalized to a common range of potential values. The following expression summarizes this first restriction:

$$0 \leq F_{i,j,m} \leq 1 \tag{3}$$

The second restriction is that the sum of the fractions for every component is equal to the total pixel surface. In this way, it is quite simple to express the individual contribution or fraction of an endmember in relation to the total reflectance of the pixel.

$$\sum_{m=1,p} F_{i,j,m} = 1 \tag{4}$$

The choice of a LSU model must consider both the landscape of the test site and the ability of the model to depict the structure, shape and distribution of the basic landscape components (Ferreira et al. 2007). Well-chosen endmembers not only represent materials found in the scene, but provide an intuitive basis for understanding and describing the information in the image (Adams and Gillespie 2006). Endmembers were obtained after applying a spatial and spectral remote sensing data dimensionality reduction with the minimum noise fraction (MNF) and pixel purity index (PPI) techniques, respectively. The MNF is used to detect the inherent dimensionality of image data, segregating noise from the signal in the data and reducing computational requirements for subsequent processing tasks (Boardman and Kruse, 1994). The MNF as modified from Green et al. (1988) consists in two steps: 1) applying a transformation, based on an estimated noise covariance matrix to decorrelate and rescale the noise in the data (noise has unit variance and no band-to-band correlations); and 2) performing a standard principal component transformation of the noise-whitened data. A final dataset of coherent and almost noise-free bands are selected from the MNF output and can be used for subsequent processing steps. Pixel Purity Index (PPI) is a procedure for finding the most spectrally pure (extreme) pixels that typically correspond to mixing endmembers in multispectral and hyperspectral images (ITT VIS, 2008). PPI is computed by repeatedly projecting $n$-dimensional scatterplots onto a random unit vector; the extreme pixels in each projection (those pixels that fall onto the ends of the unit vector) are recorded and the total number of times each pixel is marked as extreme is noted. The selection of extreme pixels corresponding to analogous surface features is complex due to the great number of pixels typically found in remote sensing image data. The $n$-dimensional visualizer implemented in ENVI software (ITT Visual Information Solutions) is a tool to locate, identify, and cluster the purest pixels and most extreme spectral responses in a data set. The distribution of these points in $n$-space can be used to estimate the number of spectral endmembers and their pure spectral signatures (Boardman, 1993). Three endmembers were used in the LSU model of the study area, namely green vegetation (GV), non-photosynthetic vegetation (NPV) and shade (S). The GV endmember represents the signature of green dense vegetation, the NPV endmember is the signature of bare soil or sparse non-photosynthetic vegetation, and the shade endmember represents the signature of dark pixels and water bodies.

## 2.6 Statistical methods

The possible existence of differences in soil properties based on the land cover classes was determined by the use of the analysis of variance (ANOVA). ANOVA is used to evaluate significant differences between means of independent variables. The observed variance of independent variables is partitioned into components by several explanatory variables (factors). Land cover class was the factor employed in the analysis. Post-hoc analysis was performed using Tukey method.

Relationships between fraction endmembers and soil properties were studied by the principal component analysis (PCA). PCA is a technique of data dimensionality reduction that performs an orthogonal transformation to convert potentially correlated input variables into uncorrelated variables or principal components. The first components accumulate most of the variance and therefore, the most useful information about the variables.

Regression analyses between soil properties and fraction endmembers were performed for quantitative estimation of soil properties. A linear regression analysis applying a stepwise method for variable entry and removal was the selected statistical technique. Model selection was based on the lower typical error of the estimation and minimum collinearity.

## 3. Results and discussion

The relationship between various soil properties and land cover classes was analysed. Two approaches to the study of land cover are presented. A first categorization based on discrete land cover classes and another based on mixture fractions. Finally, statistical models for predicting soil properties of interest in the study of soil salinity through the use of mixture fractions are presented.

### 3.1 Soil properties

The study was focused on soil electrical conductivity (EC), pH and organic carbon (OC) (Table 2). These properties are important in chemical and biological quality of soils (Lal et al., 1999). Previous studies in semiarid areas combining remote sensing and soil analyses have indicated significant differences in these properties in different land cover classes (Biro et al.,

| Land covers | EC (mS/cm) | pH | OC (%) |
|---|---|---|---|
| Arable land | 1.38 ± 1.01 *ab* | 8.23 ± 0.26 *ab* | 1.92 ± 0.85 *a* |
| Permanent crops | 0.70 ± 1.00 *b* | 8.40 ± 0.26 *b* | 1.29 ± 0.24 *a* |
| Fallow/abandoned | 3.52 ± 2.43 *ac* | 8.14 ± 0.25 *ab* | 1.46 ± 0.36 *a* |
| Saltmarsh | 3.98 ± 2.86 *c* | 8.05 ± 0.07 *ab* | 3.61 ± 0.81 *b* |
| Palm groves | 3.82 ± 2.42 *ac* | 8.20 ± 0.20 *ab* | 2.11 ± 1.00 *a* |
| Marsh | 4.71 ± 2.04 *c* | 7.82 ± 0.09 *a* | 5.24 ± 1.39 *c* |
| P-value | <0.001*** | <0.001*** | <0.001*** |

Table 2. Descriptive statistics (mean ± standard deviation) of soil properties based on land cover classes. The *p*-value and homogeneous subgroups (lower case letters; Tukey test, $P < 0.05$) resulting from the ANOVA test are included.

2011). Cultivated areas (i.e. arable land and permanent crops, homogenous subgroup *b*) have lower electrical conductivity values than natural or semi-natural vegetation. The construction of drainage systems at agricultural areas to encourage the leaching for salinity control has been a traditional amelioration strategy (Qadir et al., 2000). This fact explains that marshes and saltmarshes soils have higher EC values than the other land cover classes, since they are areas with poor drainage and temporally flooded. Salinity increases when farming finishes (i.e. fallow/abandoned) because irrigation water is not available to promote salts leaching.

The pH values were slightly alkaline but significantly different for marshes. Wetland soils are characterized by the permanent or seasonal inundation of the land, promoting anaerobic conditions and thus reduced redox conditions (high concentration of $H^+$ which implies low pH)(Reddy et al., 2000). The organic carbon content was also different depending on the type of land cover. Arable land and permanent crops soils have organic carbon content ranging from 1.46 to 2.11% that is not very high (Pérez-Sirvent et al., 2003). Opposite, wetland soils (i.e. stable saltmarshes and saltmarshes) exhibited the highest organic carbon contents. Compared to upland areas, most wetland soils show an accumulation of organic matter by the higher rates of photosynthesis in wetlands than other ecosystems and the lower rates of decomposition due to anaerobic conditions (Reddy et al., 2000).

All soil properties are significantly correlated (P<0.01) according to the Pearson bivariate correlation test applied to the full dataset. Figure 3 shows two scatterplots of the land cover classes average values (error bars represent the standard deviation) of pH and organics carbon versus electrical conductivity (EC). EC is negatively correlated with pH (R=-0.61) and positively correlated with organic carbon (R=0.34), while pH and organic carbon are negatively correlated (R=-0.32). Two sets of distinct land uses mainly dependent on the EC values are distinguished: 1) active cultures: with low EC and OC values, and 2) natural vegetation and crops of low requirements (palm groves): with high EC and OC values, increasing as the land cover is more similar to the wetland. Palm groves are the most halotolerant crop and require little tillage.

### 3.2 Land cover classification

A land cover map was obtained with the *k*-nearest-neighbours algorithm (Figure 4). Optimum results were obtained with *k*=4. The area occupied by the land cover classes (hectares and percentage of the total area) was quantified (Table 3).

The study area is mainly agricultural but largely occupied by urban areas. Urban/man-made areas represent 14.4% of the study area. There is a clear distinction between the northeast portion (area between the two natural parks and the *Sierra del Molar*) with large fields and less presence of buildings, and the rest of the study area, with numerous buildings scattered, villages, small-size towns and smaller parcels. Dominant land cover classes are also different at these two sectors. Close to the natural parks, there are many saltmarshes (7.09% of the study area), marshes (1.95% of the study area), palm groves (2.21% of the study area) and arable land (45.76% of the area and mainly forage, barley and melons). The other sector has a massive presence of permanent crops (16.3% of the study area and mainly citrus trees such as orange and lemon trees).

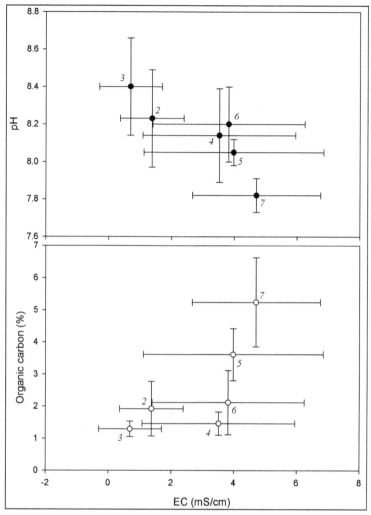

Fig. 3. Scatterplots with the average values pH and organic carbon versus electrical conductivity for land cover classes (numbers in italics are the ID number of the class). X and Y bars represent one standard deviation.

The distribution of land cover classes can be explained by the characteristics of the soils. Generally, the closest soils to the wetland areas of the natural parks are more saline. These soils have a poor drainage due to its lower altitude and very high-water tables, largely due to the horizontal flow of water and salts from the nearby water bodies. Permanent crops class dominates in areas that are close to the towns, being better drained and less saline. Fallow/abandoned areas (12.11% of the study area) are spread throughout the study area as a result of the abandonment of farming on individual fields. However, abandoned land is more present in the proximal portion of the natural parks since the conditions of salinization of soils led to their abandonment.

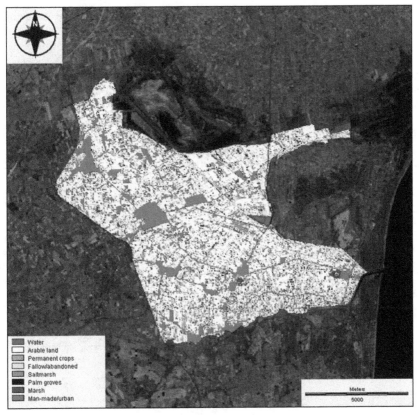

Fig. 4. Land cover map of the study area.

| Land cover classes | Area (ha) | Area (%) |
|---|---|---|
| Water | 17.37 | 0.20 |
| Arable land | 4061.52 | 45.76 |
| Permanent crops | 1446.93 | 16.30 |
| Fallow/abandoned | 1074.51 | 12.11 |
| Saltmarsh | 629.01 | 7.09 |
| Palm groves | 196.02 | 2.21 |
| Marsh | 172.71 | 1.95 |
| Man-made/urban | 1277.73 | 14.40 |
| TOTAL | 8875.80 | 100 |

Table 3. Area occupied by land cover classes according to the map obtained by k-nearest neighbour.

The land cover map accuracy was evaluated with the data set of validation points. Overall accuracy was a 68%, and KHAT value was 0.56. According to Landis and Koch (1977),

KHAT values ranging from 0.4 to 0.8 exhibit a moderate agreement. Inter-class confusion was detected analysing the error matrix. A portion of arable land (78% of producer's accuracy, 65% of user's accuracy) was wrongly classified as permanent crops, fallow/abandoned land or saltmarshes. A great portion of palm groves (21% of producer's accuracy and 75% of user's accuracy) was classified as arable land. The performance of the automatic classification for marshes (90% of producer's accuracy and 75% of user's accuracy) and water areas (100% of producer's accuracy and 80% of user's accuracy) was highly satisfactory. The performance of the KNN algorithm for our land cover classification approach was enough good and comparable with the accuracy obtained by Franco-Lopez et al. (2001) classifying a forest stand (52% of overall accuracy with $k$=10), and the results of the experiment carried on by Samaniego and Schulz (2009) classifying crop types (47% of overall accuracy with $k$=5).

### 3.3 Spectral unmixing and land covers

Spectral mixture analysis was applied to obtain fraction images of green vegetation (GV), non-photosynthetic vegetation (NPV) and shade endmembers. Spectral signatures of selected endmembers are highly distinctive (Figure 5). These endmembers had optimal spectral separability as measured with the transformed divergence method. GV endmember is associated with vigorous vegetation, NPV endmember is associated with bare soil and dry halophytic vegetation, and shade endmember is associated with water bodies and low illuminated areas.

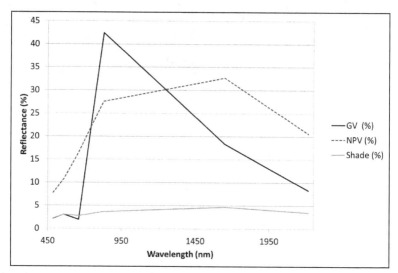

Fig. 5. Plot showing the spectral signatures of selected green vegetation (GV), non-photosynthetic vegetation (NPV) and shade endmembers.

Fraction images of the three endmembers and the residual fraction of the spectral mixture analysis were obtained (Figure 6). Values range from 0 for low high membership to the image fraction (black colour) to 1 for high membership (white colour). Fractions images are continuous variables that are graphically represented with a greyscale colour ramp. High values of the shade fraction image are present in the wetland areas of the natural parks and in a triangular area in the middle-right boundary of the study area that corresponds with a small wetland. A white area in the right of the image corresponds with the Mediterranean Sea. High green vegetation fraction values area scattered through the study area. They correspond with active crops at the time of the image acquisition. Indeed, the white areas in the NPV image fraction correspond with bare soil and saltmarshes which vegetation is quite dry in summer and have a great spectral confusion with background soil. Urban areas were also associated with this fraction image. Finally, high values of the residual fraction are located in industrial areas, whose spectral signature was notably different respect to the three endmembers of the unmixing model.

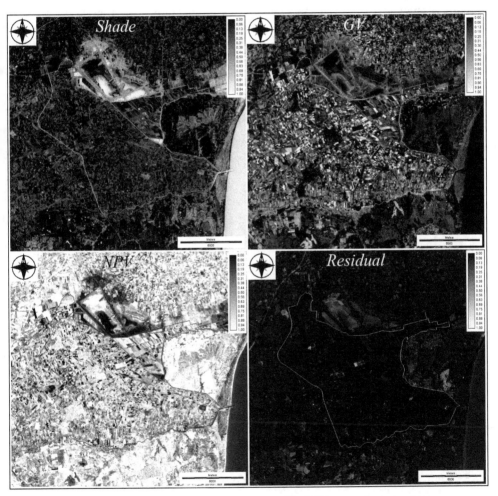

Fig. 6. Fraction images of shade/water, green vegetation (GV), non-photosynthetic vegetation (NPV) and the residual component of the linear spectral mixture analysis. White/black polygon represents the boundary of the study area.

Average values of the three fraction images for the land covers were computed and represented in a ternary diagram (Figure 7). Water land cover has high shade fraction values (>90%) and very low values for the GV and NPV fractions. Marshes have an important fraction of shade (>55%) and around 30% of the GV fraction. This mixture composition is highly indicative of the marshes structure with green *Phragmites australis* stands, growing on flooded or water-saturated soils. Shade fraction has a low contribution in the other land covers (<30%). Saltmarshes, permanent crops, arable land, fallow/abandoned and palm groves land cover classes have GV fraction values between 20-40% and NPV fraction values between 50-70%. This relative homogeneity in the mixture fractions values for different land cover classes could be attributed to the lower water availability in summer, that promotes a drying and browning of the vegetation and promotes spectral confusion. Melendez-Pastor et al. (2010b) previously observed this phenomenon in the study area. They also employed ternary diagrams, combining mixture fraction and land cover classes for a drought year and an average year. Soil or NPV fractions increase their contribution in a dry weather scenario (i.e. drought or summer) and the water and GV fractions have a lower contribution.

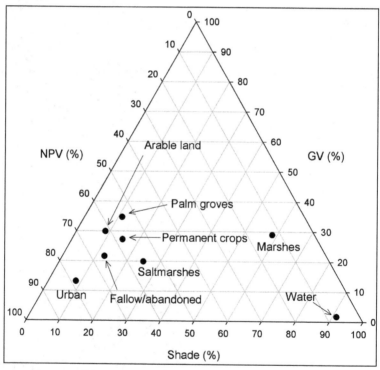

Fig. 7. Ternary diagram of the average mixture fraction values for the land cover classes.

## 3.4 Fraction endmembers to predict soil properties

Mixture fraction values were statistically related to soil salinity. Principal component analysis provided valuable information about the relationship among soil properties and spectral mixture analysis fractions. The first three principal components accumulated 75.8%

of total variance. PC1 was positively correlated with NPV fraction (factor loading = 0.988) and negatively correlated with GV fraction (factor loading = -0.902). PC1 might be used to separate vegetated from non-vegetated pixels. PC2 was positively correlated with electrical conductivity (factor loading = 0.778) and the shade fraction (factor loading = 0.604) and negatively correlated with the pH (factor loading = -0.785). PC2 might be used to differentiate soil salinity status.

Salinization status seems to be related to the abundance of the shade fraction. This result could be explained by the presence of water at the soil profile, which is an evidence of poor drainage that could lead to salt accumulation. Thus, monitoring shade fraction values along a year could be an indirect method to detect the evolution of soil electrical conductivity with remote sensing. PC3 was positively correlated with the residual fraction of the spectral unmixing (factor loading = 0.853) and organic carbon content (factor loading = 0.711). Evident negative correlations with the PC3 were not found.

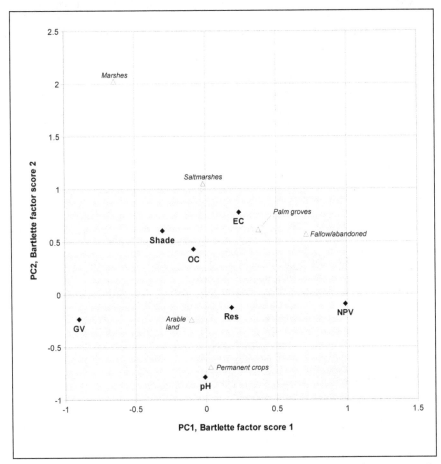

Fig. 8. Factor loadings plot for the measured soil properties and mixture fractions, and average values factor scores plot for the land cover classes.

Previous studies assessed the relationships between PC factor loadings of soil properties and PC factor scores of land cover classes (Biro et al., 2011). We also included the mixture fraction values for soil plots in the principal component analysis. Soil properties and mixture fractions factor loadings and average factor scores for land cover classes of the first two components were plotted to explore their relationship (Figure 8). Factor loading values range from -1 to 1. Factor scores of land cover classes were also included in the plot. Land cover classes were differentiated from each other along PC1, mainly because of the high positive factor loading of the NPV fraction and high negative factor loading of the GV fraction. Also, land cover classes were differentiated from each other along PC2, mainly because of the high positive loadings of the shade fraction, electrical conductivity and organic carbon content, and high negative loading value of pH.

Finally, we tested the usefulness of mixture fractions to predict EC. Stepwise linear regression was employed to model relationship among EC and mixture fractions. EC variable was normalised with the natural logarithm, while the other variables were normal. Table 4 summarizes the main parameters of two linear regression models. Moderate adjustment was obtained from the regression with R=0.338 for the first model and R=0.408 for the second model. The ANOVA test (data not included in the table) indicated the usefulness of the models with a $p$-value <0.001 for both cases. Model 1 included GV fraction as the unique mixed fraction predictor variable, while model 2 included GV and also the shade fraction. In both models, the coefficient B has a negative value for the variable GV, suggesting an inverse relationship between the green covers and EC. By contrast, the regression coefficient B was positive for the shade fraction, indicating a direct relationship between the presence of shadows/water and EC. This latter observation corroborates the interpretation given in the principal component analysis on the usefulness of the shade fraction to predict soil salinity. Collinearity statistics revealed the absence of collinearity problems as we obtained tolerance values much greater than 0 and VIF values much lower than 15 (SPSS, 2009).

| Model | R | R² | Adjusted R² | Predictors | Unstandardized Coefficients | | Sig. | Collinearity Statistics | |
|-------|---|----|----|----|----|----|----|----|----|
| | | | | | B | Std. Error | | Tolerance | VIF |
| 1 | 0.338 | 0.115 | 0.107 | (Constant) | 7.483 | 0.106 | 0.000 | | |
| | | | | GV | -1.081 | 0.279 | 0.000 | 1.000 | 1.000 |
| 2 | 0.408 | 0.167 | 0.152 | (Constant) | 7.302 | 0.123 | 0.000 | | |
| | | | | GV | -1.005 | 0.273 | 0.000 | 0.989 | 1.011 |
| | | | | Shade | 1.320 | 0.492 | 0.008 | 0.989 | 1.011 |

VIF: variable inflation factor

Table 4. Summary of results of linear regression models.

Regression models with mixture fractions did not show collinearity that could lead us to false predictions. A major constraint using proximal and remote sensing data for mapping salinity is related to the fact that there is a strong vertical, spatial and temporal variability of salinity in the soil profile (Mulder et al., 2011). Direct and precise estimation of the salt quantities is difficult by using satellite data with a low spectral resolution because these fail to detect specific absorption bands of some salt types, and the spectra interfere with other soil chromophores (Mougenot et al., 1993). More research will help to improve the prediction of soil properties with remote sensing data for a fast assessment of soil status over large areas and at low cost.

## 4. Conclusions

This chapter provides an interesting case of study on the application of remote sensing to soil salinity. The land use and management greatly affect soil salinity and land cover mapping helps to delineate areas with different severity of salinization. The use of spectral mixture analysis in combination with land cover maps and soil properties data is a more advanced technique. Mixture fractions help to know the spectral behaviour of land cover and their constituents using a simple three endmembers model. In addition, mixture fractions can be used as predictors in regression models to predict the electrical conductivity of soils. The results of regression models were encouraging but require further research to improve them. Since mixture fractions are sensitive to spectral changes due to changes in ground surface, they may be particularly useful for mapping the severity of soil salinization processes over time with low coast satellite images. The combined use of soil properties analytics, land cover maps and spectral mixture analysis is feasible for monitoring saline soils and land management over large areas with a reduced cost.

## 5. Acknowledgment

The authors acknowledge the Spanish Ministry of Science and Innovation for the financial support of the project '*Efectos de la aplicación de un compost de biosólido sobre la calidad de suelos con distinto grado de salinidad*' with the reference CGL2009-11194 that allowed this research.

## 6. References

Acosta, J.A.; Faz, A.; Jansen, B.; Kalbitz, K. & Martínez-Martínez, S. (2011). Assessment of salinity status in intensively cultivated soils under semiarid climate, Murcia, SE Spain. *Journal of Arid Environments*, Vol. 75, No. 11 (November 2011), pp. 1056-1066, ISSN 0140-1963

Adams, J.B. & Gillespie, A.R. (2006). *Remote Sensing of Landscapes with Spectral Images. A Physical Modeling Approach*. Cambridge University Press, ISBN 0521662214, Cambridge, UK

Adams, J.B.; Sabol, D.E.; Kapos, V.; Filho, R.A.; Roberts, D.A.; Smith, M.O. & Gillespie, A.R. (1995). Classification of Multispectral Images Based on Fractions of Endmembers: Application to Land-Cover Change in the Brazilian Amazon. *Remote Sensing of Environment*, Vol. 52, No. 2 (May 1995), pp. 137-154, ISSN 0034-4257

Ardahanlioglu, O.; Oztas, T.; Evren, S.; Yilmaz, H. & Yildirim, Z.N. (2003). Spatial variability of exchangeable sodium, electrical conductivity, soil pH and boron content in salt- and sodium-affected areas of the Igdir plain (Turkey). *Journal of Arid Environments*, Vol. 54, No. 3 (July 2003), pp. 495-503, ISSN 0140-1963

Ben-Dor, E. & Banin, A. (1994). Visible and near-infrared (0.4-1.1 μm) analysis of arid and semiarid soils. *Remote Sensing of Environment*, Vol. 48, No. 3 (June 1994), pp. 261-274, ISSN 0034-4257

Ben-Dor, E.; Inbar, Y. & Chen, Y. (1997). The reflectance spectra of organic matter in the visible near-infrared and short wave infrared region (400–2500 nm) during a controlled decomposition process. *Remote Sensing of Environment*, Vol. 61, No. 1 (July 1997), pp. 1-15, ISSN 0034-4257

Ben-Dor, E.; Chabrillat, S.; Demattê, J.A.M.; Taylor, G.R.; Hill, J.; Whiting, M.L. & Sommer, S. (2009). Using Imaging Spectroscopy to study soil properties. *Remote Sensing of Environment*, Vol. 113, No. SUPPL. 1 (September 2009), pp. S38-S55, ISSN 0034-4257

Biro, K.; Pradhan, B.; Buchroithner, M. & Makeschin, F. (2011). Land use/land cover change analysis and its impact on soil properties in the Northern Part of Gadarif Region, Sudan. *Land Degradation & Development*, Vol. in press DOI: 10.1002/ldr.1116, ISSN 1085-3278

Blanzieri, E. & Melgani, F. (2008). Nearest Neighbor Classification of Remote Sensing Images With the Maximal Margin Principle. *IEEE Transactions on Geoscience and Remote Sensing*, Vol. 46, No. 6 (June 2008), pp. 1804-1811, ISSN 0196-2892

Blázquez, A.M. (2003). *L'Albufera d'Elx: evolución cuaternaria y reconstrucción paleoambiental a partir del estudio de los foraminíferos fósiles*. PhD dissertation. Universitat de Valencia, Valencia, Spain.

Boardman, J.W. (1993). Automating Spectral Unmixing of AVIRIS Data Using Convex Geometry Concepts, *Summaries of the Fourth Annual JPL Airborne Geosciences Workshop, Volume 1: AVIRIS Workshop, JPL Publication 93-26*, pp. 11-14. Washington D.C., USA.

Boardman, J.W. & Kruse, F.A. (1994). Automated spectral analysis: A geological example using AVIRIS data, northern Grapevine Mountains, Nevada. *Proceedings, Environmental Research Institute of Michigan (ERIM) Tenth Thematic Conference on Geologic Remote Sensing*, pp. I-407 - I-418. Ann Arbor (MI), USA.

Caravaca, F.; Masciandaro, G. & Ceccanti, B. (2002). Land use in relation to soil chemical and biochemical properties in a semiarid Mediterranean environment. *Soil and Tillage Research*, Vol. 68, No. 1 (October 2002), pp. 23-30, ISSN 01671987

Chander, G. & Markham, B. (2003). Revised Landsat-5 TM Radiometric Calibration Procedures and Postcalibration Dynamic Ranges. *IEEE Transactions on Geoscience and Remote Sensing*, Vol. 41, No. 11 PART II, (November 2003), pp. 2674-2677, ISSN 0196-2892

Chander, G.; Haque, M.O.; Micijevic, E. & Barsi, J.A. (2010). A Procedure for Radiometric Recalibration of Landsat 5 TM Reflective-Band Data. *IEEE Transactions on Geoscience and Remote Sensing*, Vol. 48, No. 1 (January 2010), pp. 556-574, ISSN 0196-2892

Chavez Jr, P.S. (1989). Radiometric calibration of Landsat Thematic Mapper multispectral images. *Photogrammetric Engineering & Remote Sensing*, Vol. 55, No. 9 (September 1989), pp. 1285 -1294, ISSN 0099-1112

Chavez Jr, P.S. (1996). Image-based atmospheric corrections - Revisited and improved. *Photogrammetric Engineering & Remote Sensing*, Vol. 62, No. 9 (September 1996), pp. 1025-1036, ISSN 0099-1112

Clark, R.N. (1999). Chapter 1: Spectroscopy of Rocks and Minerals, and Principles of Spectroscopy. In: *Manual of Remote Sensing, Volume 3, Remote Sensing for the Earth Sciences*, A.N. Rencz, (Ed.), pp. 3-58. John Wiley and Sons, ISBN 0471294055, New York, USA

Cohen, J. (1960). A Coefficient of Agreement for Nominal Scales. *Educational and Psychological Measurement*, Vol. 20, No. 1 (April 1960), pp. 37-46. ISSN 0013-1644

Congalton, R.G. (2004). Putting the map back in map accuracy assessment. In: *Remote Sensing and GIS accuracy assessment*, R.S. Lunetta & J.G. Lyon, (Eds.), pp. 1-11. CRC Press, ISBN 156670443X, Boca Raton (FL), USA

Congalton, R.G.; Oderwald, R.G. & Mean, R.A. (1983). Assessing Landsat Classification Accuracy Using Discrete Multivariate Analysis Statistical Techniques. *Photogrammetric Engineering & Remote Sensing*, Vol. 49, No. 12 (December 1983), pp. 1671-1678, ISSN 0099-1112

Csillag, F.; Pasztor, L. & Biehl, L.L. (1993). Spectral band selection for the characterization of salinity status of soils. *Remote Sensing of Environment*, Vol. 43, No. 3 (March 1993), pp. 231-242, ISSN 0034-4257

Dalsted, K.J.; Worcester, B.K. & Brun, L.J. (1979). Detection of Saline Seeps by Remote Sensing Techniques. *Photogrammetric Engineering & Remote Sensing*, Vol. 45, No. 3 (March 1979), pp. 285-291, ISSN 0099-1112

Dehaan, R.L. & Taylor, G. R. (2002). Field-derived spectra of salinized soils and vegetation as indicators of irrigation-induced soil salinization. *Remote Sensing of Environment*, Vol. 80, No. 3 (June 2002), pp. 406-417, ISSN 0034-4257

Dehaan, R.L. & Taylor, G.R. (2003). Image-derived spectral endmembers as indicators of salinisation. *International Journal of Remote Sensing*, Vol. 24, No. 4 (February 2003), pp. 775-794, ISSN 0143-1161

De Martonne, E. (1926). L'indice d'aridité. *Bulletin de l'Association des géographes français*, Vol. 9, pp. 3-5, ISSN 0004-5322

De Paz, J.M.; Sánchez, J. & Visconti, F. (2006). Combined use of GIS and environmental indicators for assessment of chemical, physical and biological soil degradation in a Spanish Mediterranean region. *Journal of Environmental Management*, Vol. 79, No. 2 (April 2006), pp. 150-62, ISSN 03014797

Dutkiewicz, A.; Lewis, M. & Ostendorf, B. (2009). Evaluation and comparison of hyperspectral imagery for mapping surface symptoms of dryland salinity. *International Journal of Remote Sensing*, Vol. 30, No. 3 (February 2009), pp. 693-719, ISSN 0143-1161

Dwivedi, R.S.; Ramana, K.V.; Thammappa, S.S. & Singh, A. N. (2001). The Utility of IRS-1C LISS-Ill and PAN-Merged Data for Mapping Salt-Affected Soils. *Photogrammetric Engineering & Remote Sensing*, Vol. 67, No. 10 (October 2001), pp. 1167-1175, ISSN 0099-1112

European Commission. (2002). *Towards a Thematic Strategy for Soil Protection. COM(2002) 179 final*. European Commission, Brussels, Belgium.

European Commission. (2003). *Extent, causes, pressures, strategies and actions that should be adopted to prevent and to combat salinization and sodification in Europe*. Directorate General Environment, Directorate B, Erosion Working Group (Task 5; Topic: Salinization and Sodification), Brussels, Belgium.

FAO. (2002). *Crops and Drops, making the best use of water for agriculture*, Food and Agriculture Organization of the United Nations, Rome, Italy.

Farifteh, J.; Van Der Meer, F.; Van Der Meijde, M. & Atzberger, C. (2008). Spectral characteristics of salt-affected soils: A laboratory experiment. *Geoderma*, Vol. 145, No. 3-4 (June 2008), pp. 196-206, ISSN 0016-7061

Ferreira, M.E.; Ferreira, L.G.; Sano, E.E. & Shimabukuro, Y.E. (2007). Spectral linear mixture modelling approaches for land cover mapping of tropical savanna areas in Brazil. *International Journal of Remote Sensing*, Vol. 28, No. 2 (January 2007), pp. 413-429, ISSN 0143-1161

Foody, G.M. (2002). Status of land cover classification accuracy assessment. *Remote Sensing of Environment*, Vol. 80, No. 1 (April 2002), pp. 185-201, ISSN 0034-4257

Franco-Lopez, H.; Ek, A.R., & Bauer, M.E. (2001). Estimation and mapping of forest stand density, volume, and cover type using the k-nearest neighbors method. *Remote Sensing of Environment*, Vol. 77, No. 3 (September 2001), pp. 251-274, ISSN 0034-4257

Ghrefat, H.A. & Goodell, P.C. (2011). Land cover mapping at Alkali Flat and Lake Lucero, White Sands, New Mexico, USA using multi-temporal and multi-spectral remote sensing data. *International Journal of Applied Earth Observations and Geoinformation*, Vol. 13, No. 4 (August 2011), pp. 616-625, ISSN 0303-2434

Green, A.A.; Berman, M.; Switzer, P. & Craig, M.D. (1988). A transformation for ordering multispectral data in terms of image quality with implications for noise removal. *IEEE Transactions on Geoscience and Remote Sensing*, Vol. 26, No. 1 (January 1988), pp. 65-74, ISSN 0196-2892

Haapanen, R.; Ek, A.R.; Bauer, M.E. & Finley, A.O. (2004). Delineation of forest/nonforest land use classes using nearest neighbor methods. *Remote Sensing of Environment*, Vol. 89, No. 3 (February 2004), pp. 265-271, ISSN 0034-4257

Hernández Bastida, J.A.; Vela de Oro, N. & Ortiz Silla, R. (2004). Electrolytic Conductivity of Semiarid Soils (Southeastern Spain) in Relation to Ion Composition. *Arid Land Research and Management*, Vol. 18, No. 3 (July 2004), pp. 265-281, ISSN 15324982

Howari, F.M., Goodell, P.C. & Miyamoto, S. (2000). Spectral properties of salt crusts formed on saline soils. *Journal of Environmental Quality*, Vol. 31, No. 5 (September 2002), pp. 1453-61, ISSN 00472425

ITT VIS. (2008). *ENVI 4.5 User's Guide*. ITT Visual Information Solutions Boulder (CO), USA.

Jensen, J.R. (2005). *Introductory Digital Image Processing* (3rd Edition). Prentice Hall, ISBN 0131453610, Upper Saddle River (NJ), USA

Jensen, J. R. (2007). *Remote Sensing of the Environment: An Earth Resource Perspective* (2nd edition), Pentice Hall, ISBN 0131889508, Upper Saddle River (NJ), USA

Kaufman, Y.J. (1989). The atmospheric effect on remote sensing and its corrections. In: *Theory and Applications of Optical Remote Sensing*, G. Asrar, (Ed.), pp. 336-428, Wiley-Interscience, ISBN 0471628956, New York, USA

Lal, R., Mokma, D. & Lowery, B. (1999). Relation between soil quality and erosion. In: *Soil quality and soil erosion*, R. Lal, (Ed.), pp. 237-258, CRC Press, ISBN 1574441000, Boca Raton (FL), USA

Landis, J.R. & Koch, G. G. (1977). The measurement of observer agreement for categorical data. *Biometrics*, Vol. 33, No. 1 (March 1977), pp. 159-174, ISSN 0006341X

Lillesand, T.M.; Kiefer, R.W. & Chipman, J.W. (2003). *Remote Sensing and Image Interpretation* (5th edition). John Wiley and Sons, ISBN 9812530797, Hoboken (NJ), USA

Majaliwa, J.G.M.; Twongyirwe, R.; Nyenje, R.; Oluka, M.; Ongom, B.; Sirike, J.; Mfitumukiza, D.; Azanga, E.; Natumanya, R.; Mwerera, R. & Barasa, B. (2010). The Effect of Land Cover Change on Soil Properties around Kibale National Park in South Western Uganda. *Applied and Environmental Soil Science*, Article ID 185689, ISSN 1687-7667

Mather, P.M. (2004). *Computer Processing of Remotely-Sensed Images: An Introduction* (3rd edition). Wiley, ISBN 0470849185, Chichester, UK

Melendez-Pastor, I.; Navarro-Pedreño, J.; Koch, M., & Gómez, I. (2010). Applying imaging spectroscopy techniques to map saline soils with ASTER images. *Geoderma*, Vol. 158, No. 1-2 (August 2010), pp. 55-65, ISSN 0016-7061

Melendez-Pastor, I.; Navarro-Pedreño, J.; Koch, M., & Gómez, I. (2010). Multi-resolution and temporal characterization of land use classes in a Mediterranean wetland with land cover fractions. *International Journal of Remote Sensing*, Vol. 31, No. 20 (October 2010), pp. 5365-5389, ISSN 0143-1161

Mermut, A.R., & Eswaran, H. (2001). Some major developments in soil science since the mid-1960s. *Geoderma*, Vol. 100, No. 3-4 (May 2001), pp. 403 -426, ISSN 0016-7061

Metternicht, G.I. & Zinck, J.A. (2003). Remote sensing of soil salinity: Potentials and constraints. *Remote Sensing of Environment*, Vol. 85, No. 1 (April 2003), pp. 1-20, ISSN 0034-4257

Metternicht, G. & Zinck, J.A. (1997). Spatial discrimination of salt- and sodium-affected soil surfaces. *International Journal of Remote Sensing*, Vol. 18, No. 12 (June 1997), pp. 2571-2586, ISSN 0143-1161

Moran, M.S.; Jackson, R.D.; Slater, P.N. & Teillet, P.M. (1992). Evaluation of simplified procedures for retrieval of land surface reflectance factors from satellite sensor output. *Remote Sensing of Environment*, Vol. 41, No. 1-2 (August-September 1992), pp. 169-184. ISSN 0034-4257

Mougenot, B.; Pouget, M. & Epema, G.F. (1993). Remote sensing of salt affected soils. *Remote Sensing Reviews*, Vol. 7, No. 3-4, pp. 241-259, ISSN 02757257

Mulder, V.L.; De Bruin, S.; Schaepman, M.E. & Mayr, T. R. (2011). The use of remote sensing in soil and terrain mapping – A review. *Geoderma*, Vol. 162, No. 1-2 (April 2011), pp. 1-19, ISSN 0016-7061

Nelson, D.W. & Sommers, L. E. (1982). Total carbon, organic carbon, and organic matter. In: *Methods of Soil Analysis*. A.L. Page; R.H. Miller & D.R. Keeney (Eds.), pp. 539-579, American Society of Agronomy (ASA),Madison (WI), USA

Pérez-Sirvent, C.E.; Martínez-Sánchez, M.J.; Vidal, J. & Sánchez, A. (2003). The role of low-quality irrigation water in the desertification of semi-arid zones in Murcia, SE Spain. *Geoderma*, Vol. 113, No. 1-2 (April 2003), pp. 109-125, ISSN 0016-7061

Qadir, M.; Ghafoor, A., & Murtaza, G. (2000). Amelioration strategies for saline soils: A review. *Land Degradation & Development*, Vol. 11, No. 6 (November-December 2000), pp. 501-521, ISSN 1085-3278

Quarmby, N.A.; Townshend, J.R.G.; Settle, J.J.; White, K.H.; Milnes, M.; Hindle, T.L. & Silleos, N. (1992). Linear mixture modelling applied to AVHRR data for crop area estimation. *International Journal of Remote Sensing*, Vol. 13, No. 3 (February 1992), pp. 415-425, ISSN 0143-1161

Reddy, K.R.; D'Angelo, E.M. & Harris, W.G. (2000). Biogeochemistry of Wetlands. In: *Handbook of Soil Science*, M.E. Sumner, (Ed.), p. G-89-G-119, CRC Press, ISBN 0849331366, Boca Raton (FL), USA

Richards, L.A. (1954). *Diagnosis and improvement of saline and alkali soils. Agriculture Handbook No. 60*. United States Department of Agriculture (USDA), Washington D.C., USA

Richardson, A.J.; Gerbermann, A.H.; Gausman, H.W. & Cuellar, J.A. (1976). Detection of Saline Soils with Skylab Multispectral Scanner Data. *Photogrammetric Engineering & Remote Sensing*, Vol. 42, No. 5 (May 1976), pp. 679-684, ISSN 0099-1112

SPSS (2009). *PSAW 18.0 for Windows Help System*. IBM Corpotation, Armonk (NY), USA

Samaniego, L. & Schulz, K. (2009). Supervised Classification of Agricultural Land Cover Using a Modified k-NN Technique (MNN) and Landsat Remote Sensing Imagery. *Remote Sensing*, Vol. 1, No. 4 (December 2009), pp. 875-895, ISSN 2072-4292

Schmid, T.; Koch, M. & Gumuzzio, J. (2009). Applications of hyperspectral imagery to soil salinity mapping. In: *Remote Sensing of Soil Salinization: Impact and Land Management*, G. Metternicht & J.A. Zinck, (Eds.), pp. 113-140. CRC Press, ISBN 1420065025, Boca Raton (FL), USA

Settle, J.J. & Drake, N.A. (1993). Linear mixing and the estimation of ground cover proportions. *International Journal of Remote Sensing*, Vol. 14, No. 6 (March 1993), pp. 1159-1177, ISSN 0143-1161

Shimabukuro, Y.E. & Smith, J.A. (1991). The least-squares mixing models to generate fraction images derived from remote sensing multispectral data. *IEEE Transactions on Geoscience and Remote Sensing*, Vol. 29, No. 1 (January 1991), pp. 16-20, ISSN 0196-2892

Sitte, P.; Ziegler, H.; Ehrendorfer, F. & Bresinsky, A. (1994). *Strasburger. Tratado de Botánica* (8th Spanish edition), Omega, ISBN 8428209790, Barcelona, Spain

Small, C. & Lu, J.W.T. (2006). Estimation and vicarious validation of urban vegetation abundance by spectral mixture analysis. *Remote Sensing of Environment*, Vol. 100, No. 4 (February 2006), pp. 441-456, ISSN 0034-4257

Soil Survey Staff. (2006). *Keys to Soil Taxonomy* (10th Edition).United States Department of Agriculture (USDA), Washington D.C., USA

Song, C.; Woodcock, C.E.; Seto, K.C.; Lenney, M.P. & Macomber, S.A. (2001). Classification and Change Detection Using Landsat TM Data: When and How to Correct Atmospheric Effects? *Remote Sensing of Environment*, Vol. 75, No. 2 (February 2001), pp. 230-244, ISSN 0034-4257

Townsend, P.A. & Walsh, S.J. (2001). Remote sensing of forested wetlands: application of multitemporal and multispectral satellite imagery to determine plant community composition and structure in southeastern USA. *Plant Ecology*, Vol. 157, No. 2 (December 2001), pp. 129-151, ISSN 1385-0237

UNEP. (1997). *World Atlas of Desertification* (2nd edition). United Nations Environment Programme (UNEP), ISBN 0340691662, Nairobi, Kenya

Walkley, A. & Black, I.A. (1934). An Examination of the Degtjareff Method for Determining Soil Organic Matter, and A Proposed Modification of the Chromic Acid Titration Method. *Soil Science*, Vol. 37, No. 1 (January 1934), pp. 29-38, ISSN 0038-075X

Weng, Y.; Gong, P. & Zhu, Z. (2008). Reflectance spectroscopy for the assessment of soil salt content in soils of the Yellow River Delta of China. *International Journal of Remote Sensing*, Vol. 29, No. 19 (October 2008), pp. 5511-5531, ISSN 0143-1161

# 7

# High Resolution Remote Sensing Images Based Catastrophe Assessment Method

Qi Wen et al.*
*National Disaster Reduction Center of China
China*

## 1. Introduction

In recent years, there seems to be more and more occurrences of natural disasters happening around the world due to abnormal climate change. To deal with natural disasters, disaster assessment technology will provide technical support and help facilitate decision making for disaster relief, disaster prevention and reduction, post-disaster recovery and reconstruction (The six editing room of Press of China Standards, 2010). Airborne remote sensing and satellite remote sensing, which feature no time limitation, no geographical restriction, wide coverage and high accuracy, are widely used in disaster assessment scenario, because they can provide prompt and accurate information (Xie & Zhang, 2000). After several earthquakes happened around China (Xingtai, Haicheng, Tangshan, Longling, Datong) during 1970s and 1980s, China had widely implemented airborne remote sensing photography and seismic damage interpretation (Zhang, 1993). Remote sensing had also played an important role in disaster assessment of recent occurring disasters, including Wenchuan Earthquake, Yushu Earthquake, Zhouqu Debris flow and Yingjiang Earthquake (Chen et al., 2008; Shi et al., 2010). This chapter will mainly discuss the catastrophe assessment method and technical flow used by National Disaster Reduction Center of China (NDRCC) during Wenchuan Earthquake(2008), Yushu Earthquake(2010), Zhouqu Debris Flow(2010). Further discussion and advises are also given.

## 2. The flow of catastrophe assessment method

Traditionally, catastrophe assessment process can be divided into three major steps: disaster scope assessment, physical quantity assessment and direct economic losses assessment. In addition, two assessment processes: rapid disaster assessment and ground investigation process are needed to supplement the major steps. Rapid disaster assessment is usually carried out to preliminarily judge the disaster condition after the disaster occurred. The reported data from Rapid Disaster Assessment, combined with that from following Disaster Scope Assessment, are integrated to have an overview of disaster scope and extent. Another supplementary process, Ground Investigation Process, is usually carried out to cross-

* Yida Fan, Siquan Yang, Shirong Chen, Haixia He, Sanchao Liu, Wei Wu, Lei Wang, Juan Nie, Wei Wang, Baojun Zhang, Feng Xu, Tong Tang, Zhiqiang Lin, Ping Wang and Wei Zhang
*National Disaster Reduction Center of China, China*

validate disaster condition after Rapid Disaster Assessment process. The rapid disaster assessment result, ground investigation data, combined with high resolution remote sensing data will provide a detailed comprehensive assessment report of disaster condition. Figure 1 illustrates the flow of catastrophe assessment method.

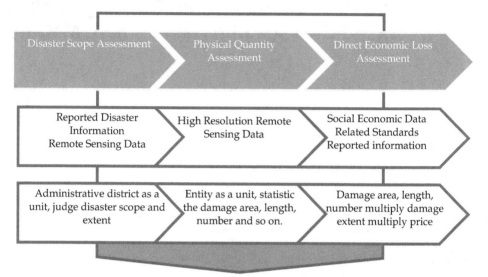

Fig. 1. The flow of catastrophe assessment method

## 2.1 Preliminary judgment of disaster condition

When the disaster happens, the remote sensing system will response immediately. By comparing pre-disaster and post-disaster high resolution remote sensing images, data on damage of transportation lines, houses, variation of terrain will be generated, and preliminarily overview of damage and disaster trend will be concluded based on these data.

## 2.2 Ground investigation

When the disaster situation is stable, according to preliminary judgment result, an expert investigation group will be sent to investigate in-field the disaster situation of sample region. Prior to ground investigation, the disaster region should be partitioned as multi-level grids in high-resolution remote sensing images. The first level partitioning is to partition the remote sensing images to functional zone according to administrative district, based on which the second level partitioning is to partition each functional zone to smaller unit, named "Entity", according to street in order to facilitate the ground investigation. The third level partitioning is to partition the "Entity" to even smaller grid unit according to house and parcel of rice field. All the grids are indexed, so that the investigation team members can fill out the corresponding forms. Investigation team uses Ground Disaster Information Collection System (Figure 2) to retrieve disaster information of 3rd level grid unit, including number of houses and their floors, estimated construction cost, and general information of 1st level unit, including affected population, damage extent and field photos. This information is collected and sent to the assessment team in NDRCC.

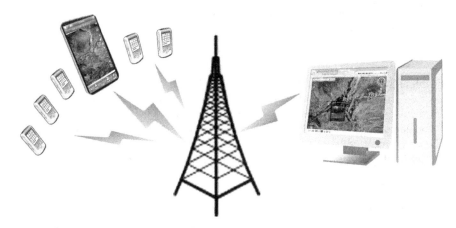

Fig. 2. Ground Disaster Information Collection System

Ground Disaster Information Collection System is a GIS software with the client-server architecture. The client side implemented in mobile phone is used to collect text and picture disaster information, label the position and send information to the sever side. The server side receives information from client side, interprets it and stores it to database.

## 2.3 Disaster scope assessment

When the disaster situation is stable, data reported by local government, assessment results generated by investigation expert group, interpreted data from remote sensing images are integrated and analyzed together to implement disaster scope assessment for sample regions. Regression relations between the sample regions and the entire disaster area are then established. Also, damage assessment data in adjacent regions are interpolated so as to generalize assessment of the whole area based on that of sample region.

In the case of an earthquake, disaster loss assessment is graded using Comprehensive Disaster Index. Disaster scope and extent of loss are assessed as a unit of administrative district(such as "county" in China). The Comprehensive Disaster Index is usually calculated as a weighted sum of the average seismic intensity, the toll of death and missing, the rate of number of death and missing per ten thousands persons, the number of collapsed houses, the rate of number of collapsed houses per ten thousands people, the geological disaster risk, and the rate of number of relocated person per ten thousand people.

$$DI = \sum (f_k \times DI_k) \qquad (1)$$

$DI_k$ is the k[th] normalized single indicator:

$$DI_k = [DI_k - \min(DI_k)] / [\max(DI_k) - \min(DI_k)] \qquad (2)$$

$f_k$ is the k[th] weight, and should be adjusted according to the specific condition of the disaster.

Average seismic intensity can be calculated by weighted sum of rates of different seismic intensity level for different area:

$$I = \sum (I_i \times S_i / S) \tag{3}$$

$I_i$ is the intensity level. $S_i / S$ is the rate of different seismic intensity level for different area; the toll of death and missing is the number of dead and missing persons in each county; number of collapsed houses is counted separately for different counties; geological disaster risk is calculated as a weighted sum of number of endangered residential areas, broken roads, blocked rivers, collapsed bridges and reservoirs. The rate of number of relocated per ten thousands of persons is calculated based on the base household population of each county.

## 2.4 Physical quantity assessment

The physical quantity assessment is carried out based on the previous data gathered in the first step of Disaster Scope Assessment. Based on extensive research and practical work experiences, NDRCC has established NNDDCA (the National Natural Disaster Damage Comprehensive Assessment Framework). Disaster loss is classified as five dimensions, and further divided to two hundred and twenty nine indicators. The five dimensions include: population loss, economic losses, residents' housing and property damage, national economy and industry loss, government offices and social undertakings loss. Several indicators, which are suitable to be assessed in remote sensing images, are first selected from the two hundred and twenty nine indicators and the damage of which is then assessed using artificial visual interpretation method and automatically machine recognition technology. The remote sensing index framework is then established.

One thing to notice is that, the affected population is not included in the remote sensing assessment index framework, and the economic losses can be statistically calculated from residents' housing and property damage, national economy and industry loss, government offices and social undertakings loss. The losses of residents' property are also not included in the remote sensing assessment index framework.

According to the characteristics of remote sensing technology, housing of three categories of statistical objects is classified as housing type; select specific object from agriculture, industry and services of state-owned enterprises; select water conservancy, environment, public facilities and other departments from government offices and social undertakings. Specific objects in the three categories can be divided into three new categories of remote sensing index of physical quantity damage assessment: housing and buildings, important infrastructure and natural resources. The first category, housing and buildings, includes building, bungalow, chimney and water tower. In the second category, important infrastructures include transportation, municipal, power, water, telecommunication and radio communication, pipeline facilities, etc., natural resources include land resources and water resources. Table 1 below shows a selection of assessment indicators.

In our case, data extracted from high-resolution remote sensing images in the step of Physical Quantity Assessment is usually brought to the decision-making chain as major information resource. By using pre-disaster and post-disaster UAV and airborne high-resolution remote sensing images collected from members of NDRCC, we have

| | | | Building |
|---|---|---|---|
| Remote sensing indexes of physical quantity damage assessment | Housing and buildings | | Bungalow |
| | | | Chimney and water tower |
| | Important infrastructures | Transportation infrastructure | National road |
| | | | Provincial road |
| | | | County road |
| | | | Bridge |
| | | | Railway |
| | | | airport pavement |
| | | Municipal infrastructure | Municipal road |
| | | | Public traffic facility |
| | | | Urban green land |
| | | | Street lamp |
| | | Power facility | Power generation, substation facility |
| | | | Overhead power line |
| | | Water conservancy facility | reservoir |
| | | | channel |
| | | | dam |
| | | | embankment |
| | | telecommunication and radio communication facility | emission station |
| | | | transmission line |
| | | Pipeline resources | Overhead and ground pipeline |
| | Natural resources | Land resources | Farmland |
| | | | Woodland |
| | | | Building land |
| | | Water area resources | River |
| | | | Lake |

Table 1. Remote sensing indexes of physical quantity damage assessment

implemented disaster targets recognition and change detection, through which disaster data is extracted and integrated. The data includes number and area index of houses, number and length of roads, number and area index of power facilities, water conservation facilities, and communication stations, municipal utilities, and land resources.

### 2.5 Direct economic losses assessment

Based on physical quantity assessment results and related standards, Direct Economic Loss is assessed for houses, infrastructure of government, industry, and personal property within disaster areas. For example, the economic loss of a building can be calculated by multiplying build-up area, damage rate, and unit price. The damage rate is usually calculated using damage extent grading method. Other indicators follow the same method in calculating that of building.

## 3. Case studies of remote sensing catastrophe assessment

Catastrophe afflicted areas usually have complex terrain and frequently changing weather in a short time after the disaster, which makes earth observation technology one of the best methods for disaster emergency monitoring and disaster assessment due to the acquisition capability of high resolution images. In this section, three cases in China will be discussed in detail to illustrate the Disaster Assessment Technology NDRCC had used: Wenchuan Earthqueake, Yushu Earthquake, and Zhouqu Debris Flow Disaster. During Wenchuan Earthquake in May 2008, we used high resolution airborne remote sensing images, reported data collected by local government and ground investigation data in combination for Disaster Scope Assessment (Fan et al., 2008a); during Yushu Earthquake in April 2010, we implemented housing damage assessment (Yang et al., 2011); during Zhouqu Debris Flow Disaster in August 2010, we implemented Physical Quantity Assessment and Direct Economic Loss Assessment (Qi et al., 2011). The Physical Quantity Assessment is implemented based on disaster scope assessment result with indicators like houses and building, important infrastructure, land resources, and the Direct Economic Loss Assessment is implemented based on physical quantity assessment result.

### 3.1 Wenchuan earthquake

On May 12th, 2008, an earthquake of magnitude 8.0 Richter occurred in the countryside of Wenchuan, Sichuan province, China. The maximum seismic intensity was 11 degree which is higher than that of Tangshan Earthquake (7.8 Richter). Aftershocks frequently occurred and the largest one was of magnitude 6.4 Richter. Four hundred and seventeen counties out of ten provinces (Sichuan, Gansu, Shanxi, Chongqing, Guizhou, Yunnan, Henan, Hubei, Shanxi, Hunan) suffered from this disaster, leading to 69197 deaths, 18289 missing and 374000 injuries by July 13th 2010. The highly affected areas are most in mountainous valley where traffic was not convenient. The earthquake and following aftershock wrecked the fragile transportation and communication facilities, making disaster relief a very difficult job since rescue personnel, supplies, vehicles and large-scale disaster rescue equipments couldn't access the disaster scene in time.

### 3.1.1 Preliminary judgment of disaster scope

A seismic intensity distribution map was immediately made after the earthquake using empirical data from USGS (Figure 3). Results showed that the maximum intensity was beyond 9 degrees. Then, data of population and areas of affected towns and counties in the affected regions is calculated according to the map.

### 3.1.2 Remote sensing assessment of housing damage

Short after the earthquake in Wenchuan, we rapidly collected 1277 remote sensing images captured by 24 satellites from 12 countries using the International Charter "SPACE AND MAJOR DISASTERS" and domestic satellites data sharing mechanism. Several airplanes of

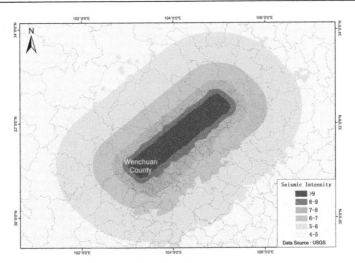

Fig. 3. Seismic intensity distribution map

National Administration of Surveying and Chinese Academy of Sciences flew to affected area to take airborne remote sensing image, in a resolution of 0.5m and 2m.

Take the indicator of houses for example. With the help of remote sensing images, damage condition of houses in satellite-covered regions was estimated. For uncovered regions, damage condition of houses was estimated by proportional spatial extrapolation and interpolation based on data from satellite-covered regions in a unit of village or town. Based on destruction probability of general fortified buildings, combined with the seismic intensity data, the relationship between collapse rate and damage rate was calculated. Damage condition of housing in affected region was then estimated. Figure 4 illustrates the flow of damage assessment process for the indicator of houses during Wenchuan Earthquake (Fan

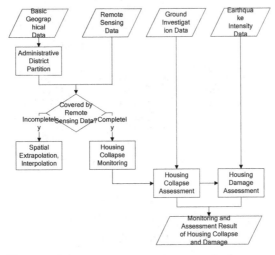

Fig. 4. The flow of collapse and damage assessment process for houses

et al., 2008b). Figure 5, 6, and 7 show the damage assessment map of sample regions: Maoxian County, Qingchuan County and the city of Dujiangyan. Figure 8 demonstrates the comprehensive assessment result of damage of the whole disaster area for single indicator of houses.

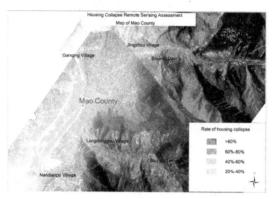

Fig. 5. The damage assessment result of Maoxian County for houses.

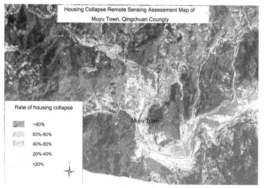

Fig. 6. The damage assessment result of Qingchuan County for houses.

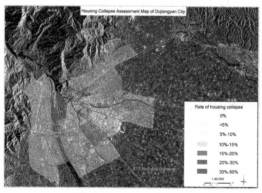

Fig. 7. The damage assessment result of Dujiangyan City for houses.

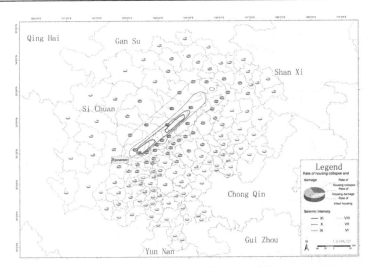

Fig. 8. Comprehensive assessment map of damage for houses

### 3.1.3 Disaster monitoring

In the following days after the quake, we continually and extensively monitored traffic, main national roads and settlements in disaster area to fast and accurately identify and locate traffic jam and landslide with the help of high-resolution remote sensing images.

EROS-B satellite image on May 15th illustrates several traffic jams occurred along 213 and 317 national highway in suburban area around Wenchuan due to secondary disaster landslide (Figure 9).

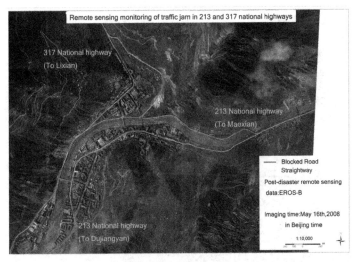

Fig. 9. Remote sensing image of the area along 213 and 317 national highways in the countryside of Wenchuan. Be noticed traffic jams are spotted with red lines

Quickbird satellite image on May 16th illustrates 5 landslides happened in a 2-kilometre diameter area around Yongan County, which seriously interrupted traffic (Figure 10).

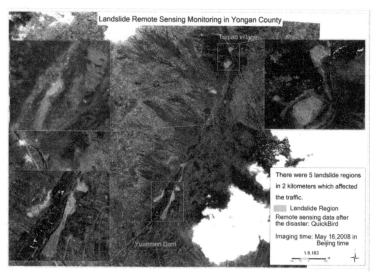

Fig. 10. Remote sensing image from Quickbird satellite about landslide around Yongan County

Comparing EROS-B satellite image of 0.7m on May 27th and airborne remote sensing image of 0.5m resolution on May 18th, about 10 large refugee settlements distributed along both sides of urban main road in Qingchuan County (Figure 11), which increased significantly in ten days.

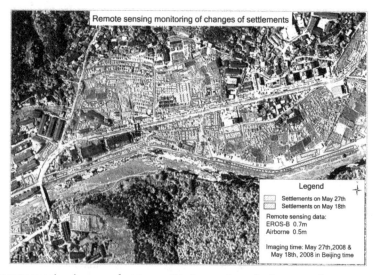

Fig. 11. Remote sensing images about quantity increasing of refugee settlements

### 3.1.4 Disaster scope assessment results

Combined with remote sensing interpretation results, ground investigation information and reported data from local government, we built Comprehensive Disaster Index (CDI) for Disaster Scope Assessment. The index will be demonstrated in detail as follows:

The weight of average seismic intensity was 0.3; the weight of rate of death and missing per ten thousands persons was 0.15, and the total weight was 0.3; the weight of houses collapsed rate per ten thousands was 0.1, and the total weight was 0.2; the weight for geological disaster risk was 0.1; the weight of relocated rate per ten thousands was 0.1.

Disaster area was partitioned into three sub region (Figure 12): serious region, severe region, and general region. The area and CDI of serious region was around 26000 km², and 0.4 above; the area and CDI of severe region was around 90000 km², and between 0.15 and 0.4; the area and CDI of general region was around 384000 km², and between 0.01 and 0.15.

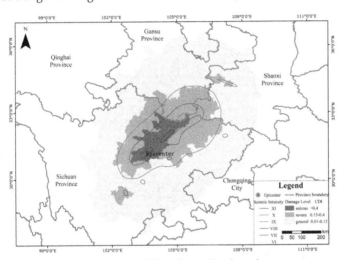

Fig. 12. Disaster scope assessment map of Wenchuan Earthquake

### 3.2 Yushu earthquake

On April 14th, 2010, a magnitude-7.1 earthquake hit Yushu County, Yushu Tibetan Autonomous Prefecture, in Qinghai Province of China. The epicentre (pointer on earth surface directly above) was located in 14-kilometre depth at 33.2 degrees north latitude, 99.6 degrees east longitude, with a seismic intensity 9-degree at maximum. Aftershocks frequently occurred with the largest one of magnitude-6.3. The earthquake wrecked Jiegu town and its surrounding area, resulted in 2220 deaths, 70 missing and 12135 injuries by p.m. 5, April 14th 2010. Many factors, such as high latitude, inconvenient traffic and cold weather, make disaster relief very difficult.

### 3.2.1 Preliminary assessment

Firstly, a seismic intensity distribution map was made by earthquake intensity empirical model according to earthquake magnitude, depth of epicentre and regional fracture zone distribution. Jiegu Town, Yushu County was in the hard-hit area with IX degree. Figure 13 illustrates contour of seismic intensity and terrain distribution.

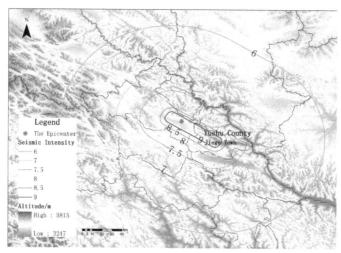

Fig. 13. Seismic intensity and terrain distribution map

By comparing pre-disaster airborne remote sensing images and post-disaster EROS-B high resolution satellite images, we analyzed the damage condition of housings, and estimated disaster scope and damage extent rapidly. Figure 14 illustrates the assessment result.

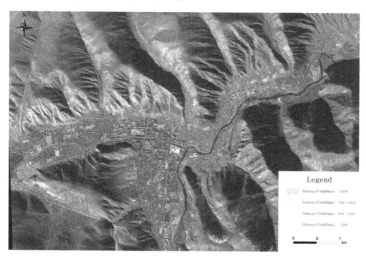

Fig. 14. Rapid assessment of disaster scope and extent of houses damage

### 3.2.2 Ground investigation and comprehensive assessment

Immediately after the quake, an airplane of Chinese Academy of Sciences (CAS) was sent to Yushu to take airborne remote sensing images. Images with a total storage of 409G and 0.4

meter spatial resolution were acquired, covering 4500 km² hard-hit area. Then combined with the result from preliminary assessment process, we partitioned residential region of Jiegu Town into 3-level grids according to street, function, and structure. About 685 girds of functional level were indexed.

First level (Figure 15) includes 4 categories: government office, community, industry area, and restricted area; second level (Figure 16) includes 6 categories: government office, school, general office, community, industry area, and restricted area; third level (Figure 17) includes 5 categories: building, bungalow, greenhouse, square, and stadium.

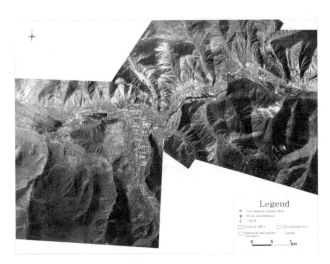

Fig. 15. Grids of first level

Fig. 16. Grids of second level

Fig. 17. Grids of third level

Hazard assessment expert group was then sent for field investigation. Damage extent, building size, architecture style, and actual pictures of each third level grid were collected and sent back to experts in NDRCC for interpretation and verification.

### 3.2.3 Disaster monitoring

At the same time, some key areas were monitored 24 hours a day using satellite and UAV images, including landslide region in Road 214, traffic jams in Road 308, refugee settlements area, terrain.

According to UAV monitoring images of 0.2m in Figure 18, 9861 refugee tents were placed in Jiegu urban region by April 19th, 5 days after the quake.

Fig. 18. Tent settlement monitoring

26 landslide regions were spotted, among which 5 were along Road 214 in the south of Jiegu town, which resulted in traffic jam. The other 21 were mainly located in mountain region and had little affect to traffic. Figure 19 illustrates traffic jam caused by landslide along Road 214.

Fig. 19. Traffic jam monitoring and assessment in landslide regions

### 3.2.4 Disaster scope assessment results

Figure 20 illustrates CDI about the disaster scope and extent of loss. CDI for this case is a little different compared to that of previous Wenchuan Earthquake. For serious regions covering around 992 km², CDI was 1; for severe regions of around 7030 km², CDI was [0.2,1]; for general regions of 27840 km², CDI was [0.01,0.2].

### 3.2.5 Housing damage assessment results

The housing damage scope and extent of loss were defined again by integrating field investigation data, data reported by local government (Figure 21). This is the first time satellite-airplane-ground integration assessment mode used in national disaster assessment. The damage scope was partitioned into three parts: collapsed building area, about 1,314,000m²; serious damaged building area, about 2,332,000 m²; and minor damaged building area, about 680,000 m². Collapsed building is defined as houses collapsed into ruins; serious damaged building as structures of houses been destroyed, thus it must be reconstructed; minor damaged building as structure been slightly affected, and it can still be used after reinforcement. Figure 8 and table 1 show housing damage assessment results. Figure 22 illustrates the remote sensing hazard monitoring and assessment flow in Yushu Earthquake.

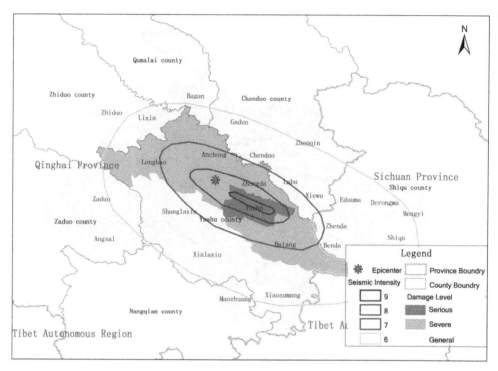

Fig. 20. Disaster scope assessment map of Yushu Earthquake

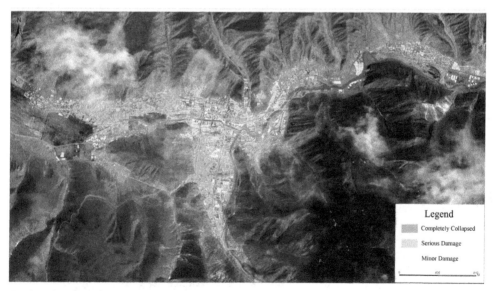

Fig. 21. Housing damage assessment results

| Collapse & damage type | Completely collapsed (ten thousand square meters, %) | | | | Serious damage (ten thousand square meters, %) | | | | Minor damage (ten thousand square meters, %) | | | |
|---|---|---|---|---|---|---|---|---|---|---|---|---|
| | Bungalow | Building | Total | Ratio | Bungalow | Building | Total | Ratio | Bungalow | Building | Total | Ratio |
| Residential | 118.5 | 1.6 | 120.1 | 27.8 | 116.6 | 31.2 | 147.8 | 34.2 | 27.7 | 8.5 | 36.2 | 8.4 |
| Office | 0.8 | 4.9 | 5.7 | 1.3 | 3.8 | 62.7 | 66.5 | 15.3 | 4.2 | 17.3 | 21.5 | 4.9 |
| School | 0.7 | 0 | 0.7 | 0.2 | 0.2 | 1.4 | 1.6 | 0.4 | 0 | 0.9 | 0.9 | 0.2 |
| Industrial and mining enterprises | 0.1 | 0.5 | 0.6 | 0.1 | 2.2 | 4.8 | 7 | 1.6 | 0.6 | 2.9 | 3.5 | 0.8 |
| Special | 4 | 0.2 | 4.2 | 1.0 | 8.9 | 1.4 | 10.3 | 2.4 | 5.4 | 0.5 | 5.9 | 1.4 |
| Total | 124.1 | 7.3 | 131.4 | 30.4 | 131.7 | 101.5 | 233.2 | 53.9 | 37.9 | 30.1 | 68 | 15.7 |

Table 2. Housing damage construction area statistic table of Jiegu town urban area.

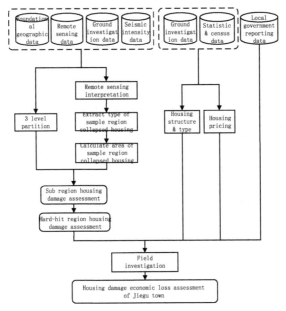

Fig. 22. Remote sensing hazard monitoring and assessment flow

### 3.3 Zhouqu debris flow

On August 8th 2010, a serious debris flow disaster occurred in Zhouqu County, Gannan Tibetan Autonomous Prefecture, Gansu Province, and China. The debris flow breaks many records since 1949, including number of death, extent of damage, and relief difficulty. The casualty consists of 1447 deaths, 318 missing, and about 21000 residents were forced to evacuate by p.m.4, August 24th. The disaster lasted for a long time, and repeatedly occurred in some regions, resulting in several barrier lakes, and part of town under water. After 8th, the disaster area had suffering from several heavy rains, which, triggered more debris flow, and interrupted the traffic. The hard-hit region was in remote mountain area with narrow road, making disaster relief a very difficult task.

### 3.3.1 Preliminary assessment

An UAV and airplane were sent to Zhouqu one day after the debris to acquire affected area images, including UAV image of 0.2m resolution and aerial remote sensing images of 1m resolution. Once the images were sent back, we compared pre-disaster and post-disaster remote sensing images, continually monitored disaster scope, damage extent of housing, residents , variation of river water level, newly happened landslide, and preliminary analyzed and estimated the damage condition and disaster trend.

Figure 23 illustrates preliminary assessment result. Collapsed area was 0.14 km², among which more than 90% were bungalows; serious damaged area was 0.15 km², among which more than 79% were buildings; minor damaged area was 1.12 km², among which more than 40% were buildings. Affected crop area was about 0.45 km², and affected forest area was about 0.27 km².

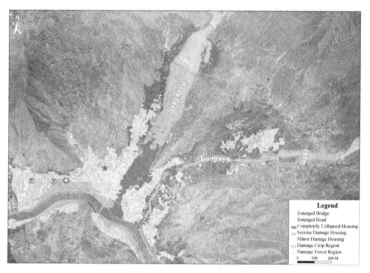

Fig. 23. Preliminary assessment results of Zhouqu Debris Flow disaster

According to debris flow and landslide monitoring result (Figure 24), Sanyanyu debris flow area was 0.66 km²; Luojiayu debris flow area was 0.2 km². The 5 landslide regions in the mountain area of Zhouqu had a total area of 17272 km².

Figure 25 and Table 3 shows refugee settlement monitoring result. Tents number increased a lot from August 15th to August 8th.

| Settlement | Area (m²) | Tents number in 8th August | Tents number in 15th August | Increasing number |
|---|---|---|---|---|
| No.1 high school | 2079 | 13 | 55 | 42 |
| No.3 high school | 3590 | 36 | 98 | 62 |
| total | 5669 | 49 | 153 | 104 |

Table 3. Comparison of tents number

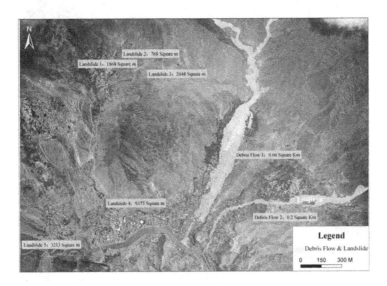

Fig. 24. Debris flow and landslide monitoring map

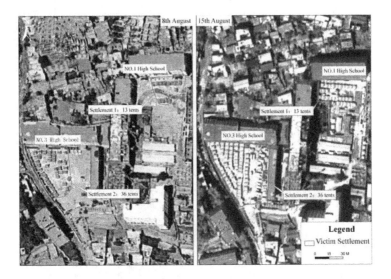

Fig. 25. Settlement monitoring map

The flooded area expanded obviously (Figure 26), with about 226,000 m² in August 15th, 23,000 m² more compared with that in August 10th, and 102,000 m² more compared with that in May 5th.

Fig. 26. Water area monitoring map

### 3.3.2 Ground investigation

Then according to the preliminary analysis results, we partitioned affection area into 2-level grids similar with previous case, using house as unit in high-resolution UAV and airborne remote sensing images. A total of 2457 girds were indexed (Figure 27). Hazard assessment experts were sent for field investigation. The damage extent, building size, architectural style, and field pictures of each grid were collected and sent back to experts in NDRCC for interpretation and verification.

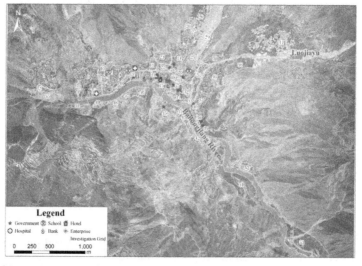

Fig. 27. Field investigation grids partition map

First level includes 3 categories of 59 grids: seriously damaged area, seriously flooded area, and minor affect area; Second level includes 2457 subgrids.

### 3.3.3 Disaster scope and housing damage assessment

The disaster scope and damage extent were defined again by integrating pre-disaster, post-disaster remote sensing information, ground investigation data and data reported by local government (Figure 28).

Disaster area was partitioned into three parts: serious affected region refers to the regions which bore main force of landslide, resulting in serious damage to houses, infrastructure and farmland; severe affected area refers the region which suffered from a long time soaked in water and mud, causing damage to houses; minor affected region refers to the region where houses and infrastructures are mildly damaged by floods and landslides.

The serious affected area was 1.2 km², severe affected area was 0.2 km² and minor affected area was 1.0 km².

Fig. 28. Disaster scope assessment map of Zhouqu debris flow disaster

### 3.3.4 Physical quantity assessment

In the following Physical Quantity Assessment process, with the help of data reported by local government, we carefully assessed the physical quantity of houses, roads, power facilities, communication facilities, water conservation facilities, municipal utilities and land resources in UAV and airborne images to estimate approximate direct economic loss. It is worth to mention that some damage quantity ignored in field investigation was detected in remote sensing images, such as submerse street trees, telegraph poles, smart street pavilions.

According to remote sensing images and ground investigation verification result (Figure 29), in rush destroyed region, collapsed residential building area was about 125,000 m², seriously damaged area was about 102,000 m², minor damaged area was about 147,000 m². In flooded region, collapsed building area was around 419,827 m².

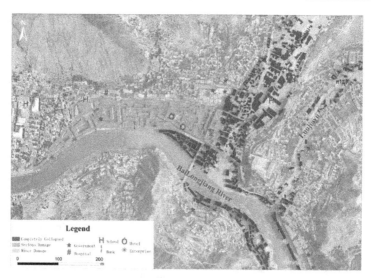

Fig. 29. Housing damage assessment map

Roads were divided into 3 classes (Figure 30): provincial road, urban road and rural road. According to remote sensing assessment result, 2 provincial roads were damaged, with a length of 2 km; 25 urban roads were damaged, with a length of 7.5 km; 5 rural roads were damaged, with a length of 4.8 km. Three bridges of a total length of 242 meters were out of function.

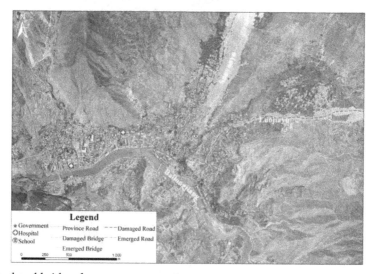

Fig. 30. Road and bridge damage assessment map

The physical quantity assessment results were used for direct economic losses estimation. It is for the first time for China that remote sensing images are integrated into the national catastrophe assessment flow as a major data source.

## 4. Conclusion

Throughout the three cases discussed in this chapter, one interesting thing is that the importance of remote sensing images is increasing in the whole process of disaster assessment: preliminary assessment step, disaster scope and damage assessment step, and physical quantity assessment. Also, the resolution of remote sensing images is improved, from 0.5m to 0.2m, providing more accurate data for assessment and decision making.

Although the whole process of integration of remote sensing images and data from ground investigation has been growing mature, the sample region choosing, physical quantity assessment still need further improvement.

Another thing to notice is that, automatic operation system is needed to connect the whole disaster assessment flow to provide efficient and accurate data for decision making. The satellite-airplane-ground disaster monitoring and assessment system is the trend we are and will realize in the near future.

## 5. Acknowledgment

This chapter is sponsored by High Resolution Earth Observation System Project of China.

## 6. References

Chen, S.; Ma, H. & Fan, Y. et al. (2008). Road Damage Assessment from High Resolution Satellite Remote Sensing Imagery in Wenchuan Earthquake. *Journal of Remote Sensing*, Vol.12, No.6, pp. 949-955, ISSN 1007-4619.

Fan, Y.; Yang, S. & Wang, L. et al. (2008a). Study on Urgent Monitoring and Assessment in Wenchuan Earthquake. *Journal of Remote Sensing*, Vol.12, No.6, pp. 858-864, ISSN 1007-4619.

Fan, Y.; Yang, S. & Wang, W. et al. (2008b). Comprehensive Assessment of Disaster Scope, In:*Comprehensive Analysis and Assessment of Wenchuan Earthquake*, Zhu, H., pp. (94-103), Science Press, ISBN 978-7-03-023855-9, Beijing, China.

Shi, F.; He, H.; Zhang, Y. (2010). Remote Sensing Interpretation of the MS7.1 Yushu Earthquake Surface Ruptures, *Technology for Earthquake Disaster Prevention*, Vol.5, No.2, pp. 220-227, ISSN 1673-5722.

The sixth editing room of Press of China Standards (2010). *Compilation of Earthquake and Related Standards*, Press of China Standards, ISBN 978-7-5066-5651-1, Beijing, China.

Wen, Q.; He H. & Wang X. et al. (2011). UAV remote sensing hazard assessment in Zhouqu Debris Flow Disaster, Proceedings of SPIE European Remote Sensing, ISBN 9780819488022, Prague, Czech Republic, September, 2011.

Xie, L.; Zhang, J. (2000). Application of Satellite Remote Sensing Technology in Earthquake Disaster Reduction, *Journal of Natural Disasters*, Vol.9, No.4, pp. 1-8, ISSN 1004-4574.

Yang, S.; Liu, S. & Wu, W. et al. (2011). Remote Sensing Applications in Qinghai Yushu Earthquake Monitoring and Assessment, *Spacecraft Engineering*, Vol.20, No.6, pp. 90-96, ISSN 1673-8748.

Zhang, D. (1993). Preliminary Study on Visual Interpretation Marks of Building Damages Caused by Earthquakes on Aero Photograph, *Earthquake*, Vol.13, No.1, pp. 26-30, ISSN 1000-3274.

# 8

# Automatic Mapping of the Lava Flows at Piton de la Fournaise Volcano, by Combining Thermal Data in Near and Visible Infrared

Z. Servadio[1,2], N. Villeneuve[1] and P. Bachèlery[3,4]

[1]*Laboratoire Géosciences Réunion, Université de la Réunion, Institut de Physique du Globe de Paris, CNRS, UMR 7154, Géologie des Systèmes Volcaniques, Saint Denis*
[2]*Institut de Recherche pour le Développement, US 140, BP172, 97492 Sainte-Clotilde Cedex*
[3]*Clermont Université, Université Blaise Pascal, Laboratoire Magmas et Volcans, CNRS, UMR 6524, Observatoire de Physique du Globe de Clermont-Ferrand, BP 10448, F-63000 Clermont-Ferrand*
[4]*IRD, R 163, LMV, F-63038 Clermont-Ferrand*
*France*

## 1. Introduction

Knowing the eruptive history of a volcano is an essential key to the understanding of its functioning, and therefore of the evolution of the character of dangerousness of its eruptions. For an essentially effusive basaltic volcano such as the Piton de la Fournaise, the spatial and temporal distribution of lava flows allows to deduct numerous parameters of its activity, on a magmatic and a structural point of view. Satellite imaging brings more advantages than the methods used in aerial pictures studies, especially by supplying bigger temporal and spectral series. The revisiting of satellites over a region can allow the generation of dynamic mappings of the implementation of the lava flow, and also bring information on the phenomenology of the eruptions: Surface, volume, flow, spatial distribution...

Furthermore, satellite images have the advantage of supplying data that grant a global visualization of the study area, and information on not easily accessible areas. The interpretation of these satellite data enables obtaining information on the surfaces and volumes of the lava field flows, but also on its nature and behavior. In a tropical environment such as La Reunion, where the climatological context presents a strong cloudiness, a satellite revisit is statistically necessary.

The optical satellite images have already been successfully used to realize mappings of lava flows. For example, in Nevado Subancaya in Peru (Legelay-Padovanie et al., 1997) or in Etna in Italy (Honda et al.,2002), the combination of spectral and morphological properties helped to elaborate surface lava flows mappings and also allowed to individualize the main structures. Lu et al. (2004) propose an association of OPTICAL and RADAR imaging in order to define more accurate outlines for the lava flows. In the later example, the downstream part of the lava flow presenting vegetation, the discrimination was realized by the infrared. The upstream part being a heavily snow covered zone, RADAR images properties become a precious source of information. Thanks to the use of LIDAR (Airborne Light Intensity Detection and Ranging) on volcanoes, Digital Ground Models of very high-resolution can be generated and various retro-reflecting properties of the lava's different textures can be studied (Favalli et al., 2009) .The comparison of the different DGM produced at different dates allow Favalli et al. (2009) to obtain mappings of the thickness and outlines of the lava flow.

The use of thermal satellite imaging to characterize the relative chronology of the implementation of lava flows was the aim of Kahle et al. (1998)'s work. Abrams et al. (1991), used it in association with optical satellite imaging to realize a chronological mapping of the lava flows in Hawaii. Other authors have realized mappings of lava flow's temperature (Hirn et al., 2005) and implementation maps by using a thermal camera (Harris et al., 2007; James et al., 2007; Lombardo et al., 2009).

At the Piton de la Fournaise, the first research that used photogrammetry to realize mappings of the outlines and thickness of lava flows took place on lava flows dated back to 1972 and 1976 (Lénat, 1987). Bonneville et al. (1989) mapped the main geological units by using SPOT1 images. After which, Despinoy (2000) realized a mapping of the lava flows above Les Grandes Pentes using a CASI hyperspectral sensor. Villeneuve (2000) realized outlines, volume calculations and a chronological follow up of the implementation thanks to stereophotogrammetry and the use of DGPS (Digital Global Positioning System). Lénat et al. (2001) associated RADAR and SPOT images to map the field of lava of the Enclos Fouqué's caldera. Recently, De Michele et Briole (2007) used a technic of correlation of images to extract lava flows which implemented between two series of aerial pictures. The study of the incoherencies in the interferograms helped realize a dynamic follow up and a mapping of the lava flows (Tinard, 2007 ; Froger et al., 2007 ; Froger et al., in press).

The aim of this article is to propose an original approach by combining thermal images with ones acquired in the visible and near infrared in order to extract independent outlines for each lava flow. The extraction is therefore independent of the operator's subjectivity. Only one automatic extraction is possible when associating thermal and optical images for the implementation of a lava flow. An automatic extraction of the outlines of lava flow is realized and then compared to a mapping of references realized by photo-interpretation. We can therefore estimate the precision of the automatic extraction.

## 2. Context

The Piton de la Fournaise (Reunion Island, Indian Ocean) is a basaltic volcano whose functioning is connected to the activity of a hot spot (Courtillot et al. 1986). The eruptions take place inside the Enclos Fouqué's caldera and have contributed to the creation of a volcanic cone whose summit is occupied by two craters: the Bory on the east side and the Dolomieu on the west side (figure 1).

During the past three decades and besides the 1986 and 1998 exceptions, the lava flows have been taking place inside de Enclos Fouqué's caldera (figure 1 and 4). They are mainly fed by

Automatic Mapping of the Lava Flows at Piton de la Fournaise Volcano, by Combining Thermal Data in Near
and Visible Infrared

203

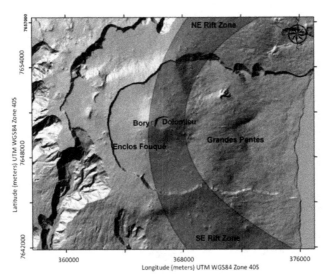

Fig. 1. Major morphological figures of the Piton de la Fournaise, the grey area is the rift zone
defined by Bachèlery (1981).

intrusions situated along the rift zones (Bachèlery, 1981). The intrusions are interconnected
at the same level as the central cone. Eruptions are of three kinds: summit zone eruptions,
proximal eruptions, and distal eruptions (Peltier et al. 2009). It is clear that in the recent
history of the Piton de La Fournaise, summit zone eruptions are the most frequent
(Villeneuve et Bachèlery, 2006; Peltier et al., 2009). Most of these eruptions take place
according to leveled cracks that progressively migrate by furthering from the central cone
(Bachèlery, 1981).

Degassing at the event during eruptions can generate lava fountains which cause pyroclastic
deposits and cones of several meters high. Two morphologies of lava flow are observed at
Piton de la Fournaise: 'a'a type lava flows and pahoehoe type lava flows.

The occidental part of the Enclos Fouqué is largely recovered by a vast field of lava that
Lénat et al. 2001 name the CLEF (Champ de Lave de l'Enclos Fouqué). This field of lava,
essentially formed by a pahoehoe lava type flow, may have been constituted from slow
emissions from the volcano's summit zone, between 1750 and 1794.

At the Piton de la Fournaise, the flow of lava emitted by eruptive cracks represent long
(several meters to several kilometers), thin, (about ten meters), shallow (one meter in the
slopes to 5 meters in flat zones (Letourneur et al., 2008)) lava flows, which shows the poor
viscosity of the emitted magma. The juxtaposition of several individual flows during a same
phase will contribute to the constitution of fields of lava, and particularly for long time
eruptions (more than one month). In this case, in regards to the initial lava flow, the new
income contributes essentially to the thickening and widening of the lava field flow. In this
case, the lava field will be considered here as a same unit.

The spectral properties of the lava flows differ according to the type of the surface (mainly
'a'a type at Piton de la Fournaise), but also according to the age, either because of a chemical

transformation of the rock on the surface, or because of the implantation (always very fast in Reunion Island) of a vegetation cover (lichen, moss, shrub...). Lava flows which implementations were separated by several years can therefore be distinguished by their spectral properties. For the summit zone of the volcano, all the more in the Dolomieu crater, where the rocks are superimposed with only a few years or a few months of interval, the spectral properties of the diverse lava flows can then be very similar.

The oldest known eruption at the Piton de la Fournaise dates back to 1644. About 200 events have been counted since that date thanks to archives, 95% of them took place in the caldera (Lacroix, 1936; Stieltjes et al., 1989; Peltier et al. 2009). Never the less, this database is incomplete, particularly in the case of short time and low scale eruptions, before 1980. The mean average magma emission at Piton de la Fournaise estimated over a century, is 0,01 $km^3.an^{-1}$ (Lénat et Bachèlery, 1987), or 0.3 $m^3.s^{-1}$. The debit estimations show a temporal estimation. For example, Stieltjes et al. (1989), calculate a mean debit of 0.3 $m^3.s^{-1}$ over 54 years (1931-1985), but obtain 0.78 $m^3.s^{-1}$ for a period of 25 years (1960-1985). These variations are partly due to the existence of long periods of inactivity. For example, no eruption took place during 1992 and 1998; witch is to say 6 years of inactivity. Also, another inactivity as long was observed between 1966 and 1972 (Villeneuve, 2000). Peltier et al. (2009) illustrate these debits variation and show a more important activity since 1998. Between January 1990 and January 2010, 61 eruptions have been registered with a total volume of emitted lava estimated at 473 $Mm^3$ (figure 2), and 33 eruptions between 1998 and 2010, with a total volume of emitted lava of 313 $Mm^3$ (Peltier et al. 2009, OVPF 2009; 2010). From these observations, we have calculated a mean debit estimated between 0.45 $m^3.s^{-1}$ and de 0.82 $m^3.s^{-1}$, from 1980 to 2010. These estimations are superior to those obtained by Stieltjes et al. (1989), on former periods.

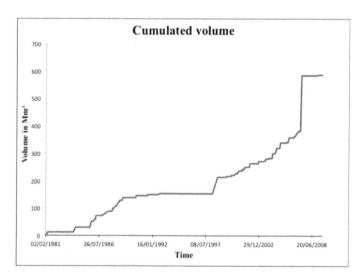

Fig. 2. Estimation of the cumulative volume of lava emitted from 1980 to 2010 by the Piton de la Fournaise.

Automatic Mapping of the Lava Flows at Piton de la Fournaise Volcano, by Combining Thermal Data in Near and Visible Infrared

205

## 3. Method

The originality of this research states in the use of thermal data as an analyze mask. In spite of its low spatial resolution (90m), thermal data brings essential information in our automatic mapping method. It allows determining with certitude the zone where the newly implemented lava flow is localized. The automatic extraction of the outline can be realized in this analyze mask. Also, its utilization enables treating the lava flows separately from one another because for one thermal image, only one lava flow is associated in this methodology. This is particularly adapted in the case of the constitution of a lava field flow.

### 3.1 Optical and thermal data

The automatic extraction method has been realized by the combination of thermal and optical data. SPOT and ASTER data have been used. SPOT data have a wavelength from the visible to mid infrared, and a spatial resolution from 2,5 to 20 meters. ASTER data have a spectrum from visible to thermal infrared and a spatial resolution that varies from 15m (VNIR) to 90m (TIR).

The ASTER TIR thermal data have to be acquired at the end of the eruption or very little time afterwards, in order for the thermal anomaly to be clearly visible on the entire zone. The maximum post eruption delay of acquisition is variable and depends on the thickness of the lava flow and therefore on its speed of cooling. In most of the cases, it is less than a month. ASTER VNIR and SPOT data can be acquired long after the lava flow's implementation. The principal is not having a new lava flow implementing on the same zone. The recent lava flows present low reflectances between the visible and mid infrared wavelength. The basalt spectrum, in the visible and short wavelength of the infrared (0.4-2.4 µm), is dominated by the presence of iron, which, at different levels of crescent oxidation, increases the reflectance (Despinoy, 2000). In the same way, the presence of lichen that grows on the lava flow increases the reflectance. In near and medium infrared, the presence of chlorophyll in the vegetation induces a strong signal (Kahle et al. 1995), permitting to discriminate precise outlines in zones with vegetation cover, especially near the Grandes Pentes area as for the Piton de la Fournaise.

The KALIDEOS project from the CNES (Centre National d'Etudes Spatiales, and GEO Grid (AIST/METI), and the NASA, grant free satellite data in the case of a research program. The data used in this article are from SPOT data from the KALIDEOS program, and from the GEO Grid program for the ASTER (Advanced Spaceborne Thermal Emission and Reflection Radiometer) data. The climatological conditions must be optimal because the presence of clouds masks, deforms or reduces the thermal anomaly, which in a tropical zone and especially in the volcano's zone is frequent. Only seven ASTER TIR data (the ASTER satellite was released in orbit in December 1999), acquired at the end of the eruption don't present these types of issues, and have therefore been used for the treatments (Table 1 and figure 3). Optical data acquired after the eruption have been associated to the former images.

### 3.2 Error calculation and outlines precision of photo-interpretation

The precision of the outlines extracted by automatic method leans entirely on their comparison with a base of outlines realized from very high-resolution satellite data' photo interpretation, and aerial pictures from IGN. This base is considerate as a reference.

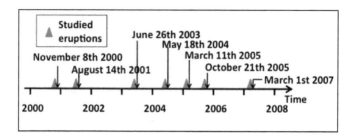

Fig. 3. Chronology of the acquisitions of the thermal data from the eruptions at the Piton de la Fournaise.

| Eruption | ASTER TIR thermal Data | | SPOT et ASTER optical data | | |
|---|---|---|---|---|---|
| | Acquisition date (AAAAMMJJ) | Resolution | Acquisition date | Angle | Resolution |
| 20070218 | 20070301 | 90m | 20070314 | R 19.1 | 10m |
| 20051014 | 20051021 | 90m | 20051222 | R 25.9 | 10m |
| 20050217 | 20050311 | 90m | 20050217 | 6.7 | 15m |
| 20040502 | 20040518 | 90m | 20040518 | −5.6 | 15m |
| 20030530 | 20030626 | 90m | 20031120 | R 25.8 | 20m |
| 20010611 | 20010814 | 90m | 20010715-20011223 | R 26.2 / R 19.2 | 20m |
| 20001012 | 20001108 | 90m | 20001213 | L 4.3 | 20m |

Table 1. SPOT and ASTER data used for automatic treatments.

The outlines realized after photo interpretation, also known as reference, are not without errors. They contain errors linked to the operator's subjectivity and from the resolution of the data used to extract the outlines. One has to take into consideration these imprecisions before interpreting the results. This will be approached in the photo interpretation chapter, where outlines extracted by satellite data' photo interpretation are tested and compared on the base of those extracted from IGN's aerial pictures.

The method chosen lies on the constitution of an error matrix, because it permits to know the precision of all or part of the classification. It's expressed in percent or area.

Automatic Mapping of the Lava Flows at Piton de la Fournaise Volcano, by Combining Thermal Data in Near and Visible Infrared

207

## 3.2.1 Error matrix

The error matrix enables comparing two thermal maps. These matrixes are constructed according to a methodology developed by the Remote sensing Center of Canada (http://ccrs.nrcan.gc.ca/glossary/index_e.php?id=3124).

In our study case, an outline is compared to another that will be considered as reference (table 2).

In the matrix, numbers contained by cells correspond to areas. The total of the areas by column represent the total area of the class obtained by automatic classification. The total of the areas of each line corresponds to what has been correctly classified. The area values present in the diagonal correspond to the ones correctly classified. The other cells represent the areas which were classified wrong, either by omission, or by commission (table 2).

| Areas (Km²) | Classification Data | | |
|---|---|---|---|
| Reference Classification | Lava Flow | No Lava Flow | Σ |
| Lava Flow | Correctly classified Lava Flow | Omission Error | **Reference Lava Flow Area** |
| No Lava Flow | Commission Error | Correctly classified No Lava Flow | **Reference No Lava Flow Area** |
| Σ | **Automatic Lava Flow Area** | **Automatic No Lava Flow Area** | **Study Area** |

Table 2. An error matrix for two classes
(http://ccrs.nrcan.gc.ca/glossary/index_e.php?id=3124 (modifiée)).

It is therefore possible to calculate a global precision (in percent) of the classification, which is to say the area of what has been classified correctly in regards to the total area of the zone that has been classified

$$Pg = \frac{\Sigma \text{Well class area}}{\text{total area of the study}} * 100$$

It is also interesting to look at the mean precision of each class of classification:

$$Pm = \frac{Pmi}{i}$$

The precision of each class is different. Some pixels can be attributed to classes that don't suit them, if the spectral properties are similar. The exactitude producer (table 2) bases itself on the area of the pixels correctly classified in regards to the area of the lava flow considered as reference. As for the exactitude user (table 2), it bases on the area of the pixel correctly classified in regard to the area of the lava flow obtained by extraction.

The exactitude producer of the "lava flow" class allows knowing the accuracy of the outline obtained by automatic extraction:

$$Pm\ (lava) = \frac{Well\ class\ area}{Real\ lava\ flow\ area} * 100$$

The mean distance separating outlines vectors is calculated from the area which was not correctly classified, divided by the smallest of the two compared lava flow outline's perimeter:

$$Error\ \beta = \frac{\sum Not\ correctly\ class\ area}{Perimeter} * 100$$

### 3.2.2 Photo interpretation

The « referenced » outlines have been realized on a period from 1980 to 2010, by vectorization on the base of IGN aerial data from 1997 and 2003. As for the eruptions post August 2003, the research was done on a base of SPOT THR, or SPOT 4 AND 5 satellite data (Figure 4). The vectorization of the lava flows by using aerial photos is one of the most precise (Paparoditis et al. 2006), but it is also one of the most binding to realize. It could ask for several worked days.

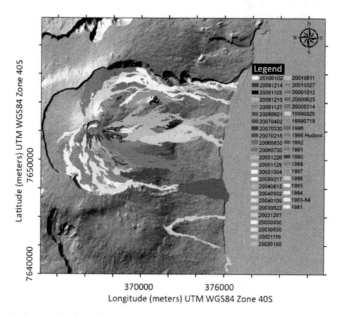

Fig. 4. Piton de la Fournaise lava flow cartography between 1980 and 2010 (Servadio et al. 2008 modified).

For the entire French territories, the temporal recurrence in the IGN's acquisitions is 5 years. It is a problem when dealing with superimposed lava flows. Satellite imaging is then a complementary tool because, even if the spatial resolution of SPOT data is lower (2,5m to 20m with optical SPOT data, or inferior or equal to 1 meter with aerial photos delivered by IGN for the BDORTHOs), the temporal resolution grants data between eruptions and defines outlines for the superimposed zones. Unfortunately, the resolution of the outlines is

Automatic Mapping of the Lava Flows at Piton de la Fournaise Volcano, by Combining Thermal Data in Near and Visible Infrared

209

then less precise. The SPOT satellite was put into function in 1986, the outline data base was completed by the mappings of the OVPF and the BRGM (Stieltjes et al., 1985 et 1989 ; Billard, 1974 ; Bussière, 1967 ; Mc Dougall, 1971).

Our methodology leans on different types of satellite data, it is important to know the influence of the spatial resolution on the extracted outlines. A comparison between the outlines obtained by photo-interpretation of different data and the referenced outlines is then done by using an error matrix and a mean distance between the outlines (table 3). According to the used satellite data, the awaited error on the outline is about the same size of the pixel (table 3). The classification's precision can tell that a mapping by satellite data' photo-interpretation is 85% more reliable for satellite data with a 10m to 20 m resolution, and 95% reliable for THR SPOT data (table 3). The other error due to the referenced outlines extraction can be the consequence of the operator's subjectivity. It presents an error from 2% to 5%. The same test was run from aerial photos, and the errors didn't exceed 2%.

### 3.2.3 Automatic process: TIR- VNIR method

The lava flows which implement in zones where spectral properties are different, such as a vegetation zone or a soil that presents high spectral reflectances, are the easiest to identify (figure 5).

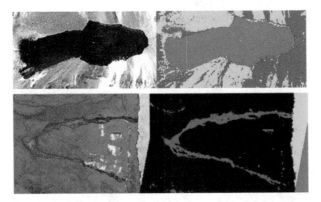

Fig. 5. Examples of automatic extraction of the lava flow outline when the reflectance of the substrate is very different from that of the lava. A and B: lava flow from the eruption of September 2009. C and D: lava flow from the eruption of 1986 off enclosures.

To automatically extract an outline, different methods can be used: classifications, threshold, or automatic detection of change (Inglada et al., 2003; Habib et al., 2007). The distance between the referenced outlines and the outlines automatically extracted is then proportional to the pixel's size, and the lava flow's classification precision is between 95% and 99%. On the other hand, lava flows which implement in low spectral reflectance zones, such as the central cone of the Piton de la Fournaise or the upperstream part of the Grandes Pentes, ask for more complex treatments. A data treatment methodology is then put together by using ENVI software (Figure 6). The visible data Principal Composant Analysis (PCA) is applied in order to maximize the data's anti-correlation. The thermal bands, near

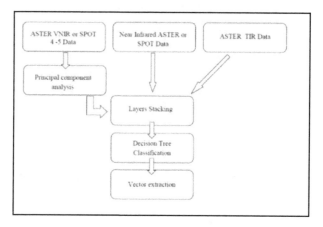

Fig. 6. TIR-VNIR Treatment sequences

infrared and the PCA are grouped in a multi-band data. A multi-level binary classification named decisional tree classification is then applied by using the following steps:

- By using thermal data, distinguish « hot » from « cold » zones.
- Extract and keep only « hot » zones.
- Discretization of the bare soil and covered vegetation zones by using the Near Infra-red band.
- Classify the pixels with low anti-correlation values.

The seven lava flows for which ASTER data were available (table 1) were tested. For half of them, the tested eruptions are on a weak slope zone, with a substrate composed of lava flows with similar spectral properties; the other half shows various substrates and slopes (figure 7).

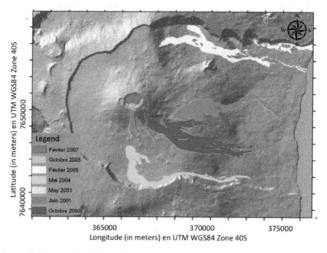

Fig. 7. Localization of the studied lava flows

**Table 3. Continued**

Aerial Pictures — Spot 4 20m

| 1991 | Lava Flow class | No lava flow class | Producer's Accuracy |
| --- | --- | --- | --- |
| Lava flow | 0.7632 | 0.0594 | 0.8226 |
| No Lava flow | 0.3848 | 3.0936 | 3.4784 |
| User's accuracy | 1.148 | 3.153 | 4.301 |
| Global Accuracy | | | 89.67% |
| Mean Accuracy | | | 90.86% |
| Producer accuracy lava flow | | | 92.78% |
| Producer accuracy no lava flow | | | 88.94% |
| Mean distance between the outlines | | | 30.5m |

Aerial Pictures — ASTER VNIR 15m

| 20070402 | Lava Flow class | No lava flow class | Producer's Accuracy |
| --- | --- | --- | --- |
| Lava flow | 3.568 | 0.041 | 3.609 |
| No Lava flow | 0.141 | 1.738 | 1.879 |
| User's accuracy | 3.709 | 1.779 | 5.488 |
| Global Accuracy | | | 96.68 % |
| Mean Accuracy | | | 95.68% |
| Producer accuracy lava flow | | | 98.86% |
| Producer accuracy no lava flow | | | 92.49% |
| Mean distance between the outlines | | | 13.7m |

| 20001012 | Lava Flow class | No lava flow class | Producer's Accuracy |
| --- | --- | --- | --- |
| Lava flow | 3.354 | 0.727 | 4.081 |
| No Lava flow | 1.273 | 9.796 | 11.069 |
| User's accuracy | 4.627 | 10.523 | 15.15 |
| Global Accuracy | | | 86.80% |
| Mean Accuracy | | | 85.34% |
| Producer accuracy lava flow | | | 82.18% |
| Producer accuracy no lava flow | | | 88.99% |
| Mean distance between the outlines | | | 38.86m |

Aerial Pictures — Spot 5 THR 5m

| 1998 | Lava Flow class | No lava flow class | Producer's Accuracy |
| --- | --- | --- | --- |
| Lava flow | 7.533 | 0.364 | 7.897 |
| No Lava flow | 1.327 | 23.636 | 24.963 |
| User's accuracy | 8.860 | 24 | 32.86 |
| Global Accuracy | | | 94.85% |
| Mean Accuracy | | | 95.04% |
| Producer accuracy lava flow | | | 95.09% |
| Producer accuracy no lava flow | | | 96.57 % |
| Mean distance between the outlines | | | 5m |

Spot 5 10m

| 20000214 | Lava Flow class | No lava flow class | Producer's Accuracy |
| --- | --- | --- | --- |
| Lava flow | 1.705 | 0.088 | 1.7893 |
| No Lava flow | 0.219 | 6.171 | 6.39 |
| User's accuracy | 1.924 | 6.259 | 8.183 |

Global Accuracy

| 1986 | Lava Flow class | No lava flow class | Producer's Accuracy |
| --- | --- | --- | --- |
| Lava flow | 1.017 | 0.165 | 1.182 |
| No Lava flow | 0.126 | 11.912 | 12.038 |
| User's accuracy | 1.143 | 12.077 | 13.22 |

|  | | |
| --- | --- | --- |
| Global Accuracy | | 97.95% |
| Mean Accuracy | | 92.5% |
| Producer accuracy lava flow | | 86.04% |
| Producer accuracy no lava flow | | 98.95% |
| Mean distance between the outlines | | 10.7m |

| 20010611 | Lava Flow class | No lava flow class | Producer's Accuracy |
| --- | --- | --- | --- |
| Lava flow | 3.119 | 0.339 | 3.458 |
| No Lava flow | 0.412 | 13.22 | 13.632 |
| User's accuracy | 3.531 | 13.559 | 17.09 |
| Global Accuracy | | | 95.61% |
| Mean Accuracy | | | 93.59% |
| Producer accuracy lava flow | | | 90.20% |
| Producer accuracy no lava flow | | | 96.98% |
| Mean distance between the outlines | | | 8.9m |

|  | | |
| --- | --- | --- |
| Global Accuracy | | 96.25% |
| Mean Accuracy | | 95.83 % |
| Producer accuracy lava flow | | 95.09% |
| Producer accuracy no lava flow | | 96.57 % |
| Mean distance between the outlines | | 6.6m |

| 20010327 | Lava Flow class | No lava flow class | Producer's Accuracy |
| --- | --- | --- | --- |
| Lava flow | 2.510 | 0.105 | 2.615 |
| No Lava flow | 0.879 | 11.476 | 12.355 |
| User's accuracy | 3.389 | 11.581 | 14.97 |
| Global Accuracy | | | 93.43% |
| Mean Accuracy | | | 94.43% |
| Producer accuracy lava flow | | | 95.98% |
| Producer accuracy no lava flow | | | 92.89% |
| Mean distance between the outlines | | | 3.2m |

| 20030530 | Lava Flow class | No lava flow class | Producer's Accuracy |
| --- | --- | --- | --- |
| Lava flow | 0.2536 | 0.0295 | 0.2831 |
| No Lava flow | 0.0908 | 0.4277 | 0.5185 |
| User's accuracy | 0.3444 | 0.4572 | 0.8016 |
| Global Accuracy | | | 84.90 % |
| Mean Accuracy | | | 86.03% |
| Producer accuracy lava flow | | | 89.58% |
| Producer accuracy no lava flow | | | 82.49% |
| Mean distance between the outlines | | | 11.7m |

| 20020105 | Lava Flow class | No lava flow class | Producer's Accuracy |
| --- | --- | --- | --- |
| Lava flow | 2.581 | 0.169 | 2.750 |
| No Lava flow | 1.057 | 8.103 | 9.16 |
| User's accuracy | 3.638 | 8.272 | 11.91 |
| Global Accuracy | | | 89.71 % |
| Mean Accuracy | | | 91.16% |
| Producer accuracy lava flow | | | 93.85% |
| Producer accuracy no lava flow | | | 88.46 % |
| Mean distance between the outlines | | | 5.3m |

Table 3. Error matrices of outlines obtained by photo-interpretation, comparing those from the photo-interpretation of aerial photographs with those of the different types of satellite data.

Automatic Mapping of the Lava Flows at Piton de la Fournaise Volcano, by Combining Thermal Data in Near
and Visible Infrared

213

The classification achieved, it is then possible to export the « lava flow » class as a vector that represents the outline of the lava flow (figure 8). The extraction is realized in less than an hour, once the data collected.

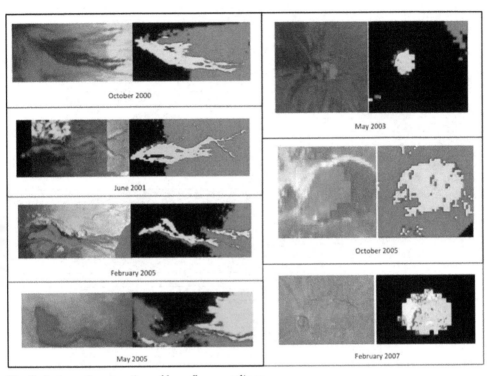

Fig. 8. Automatic extraction of lava flows outlines.

## 4. Results and interpretations

In order to validate the outlines automatic extraction method, each extraction result has been compared to the referenced outline. The error matrixes (table 4) represent the tests run on the seven tested objects localized on figure 7.

The error matrixes enable calculating global errors of classifications between 77% and 96% (table 4). We can observe a disparity between the summit zone eruptions and those implemented on sloppy substrates with various spectral properties. The first ones show a global precision between 77% and 88%, whereas the others variate from 91% to 96%. This is partly due to low reflectances observed for the substrate at the summit of the volcano. For example, the outlines of a lava flow newly implemented are hardly distinguished from the intra Dolomieu lava effusion zone that presents similar ages. The mean precisions show the same disparity, with values included between 77% and 84% at the summit zone, and included between 86% and 90% for the lava flows situated on the cone's flanks and on the slopes. The area of the lava flows also play a role in the classification precision's difference,

| 20001012 | Lava Flow class | No lava flow class | |
|---|---|---|---|
| **Lava Flow Aerial picture** | 3.422 | 0.659 | 4.081 |
| **No Lava Flow Aerial picture** | 0.667 | 11.602 | 12.269 |
| **Total Area of aerial picture** | 4.089 | 12.261 | 16.35 |
| Global Accuracy | | 91.89% | |
| Mean Accuracy | | 89.205% | |
| Producer accuracy lava flow | | 83.85% | |
| Producer accuracy no lava flow | | 94.56% | |
| Mean distance between the outlines | | 28.41m | |
| 20010611 | Lava Flow class | No lava flow class | |
| **Lava Flow Aerial picture** | 2.620 | 0.425 | 3.045 |
| **No Lava Flow Aerial picture** | 0.696 | 9.039 | 9.735 |
| **Total Area of aerial picture** | 3.316 | 9.464 | 12.78 |
| Global Accuracy | | 91.228% | |
| Mean Accuracy | | 89.445% | |
| Producer accuracy lava flow | | 86.04% | |
| Producer accuracy no lava flow | | 92.85% | |
| Mean distance between the outlines | | 28.54m | |
| 20030530 | Lava Flow class | No lava flow class | |
| **Lava Flow Aerial picture** | 0.2386 | 0.0445 | 0.2831 |
| **No Lava Flow Aerial picture** | 0.0784 | 0.3439 | 0.4223 |
| **Total Area of aerial picture** | 0.317 | 0.4223 | 0.7054 |
| Global Accuracy | | 82.58% | |
| Mean Accuracy | | 82.855% | |
| Producer accuracy lava flow | | 84.28% | |
| Producer accuracy no lava flow | | 81.43% | |
| Mean distance between the outlines | | 30.45m | |
| 20040502 | Lava Flow class | No lava flow class | |
| **Lava Flow Aerial picture** | 1.8389 | 0.6246 | 2.4635 |
| **No Lava Flow Aerial picture** | 0.6610 | 26.2154 | 26.8319 |
| **Total Area of aerial picture** | 2.4999 | 26.84 | 29.2949 |
| Global Accuracy | | 95.31% | |
| Mean Accuracy | | 86.09% | |
| Producer accuracy lava flow | | 74.64% | |
| Producer accuracy no lava flow | | 97.70% | |
| Mean distance between the outlines | | 31.55m | |
| 20050217 | Lava Flow class | No lava flow class | |
| **Lava Flow Aerial picture** | 3.291 | 0.901 | 4.192 |
| **No Lava Flow Aerial picture** | 0.816 | 17.002 | 17.818 |
| **Total Area of aerial picture** | 4.107 | 17.903 | 22.01 |
| Global Accuracy | | 92.2% | |
| Mean Accuracy | | 86.965% | |
| Producer accuracy lava flow | | 78.51% | |
| Producer accuracy no lava flow | | 95.42% | |
| Mean distance between the outlines | | 29.65m | |
| 20051004 | Lava Flow class | No lava flow class | |
| **Lava Flow Aerial picture** | 0.2727 | 0.0761 | 0.3488 |
| **No Lava Flow Aerial picture** | 0.1462 | 1.2272 | 1.3734 |
| **Total Area of aerial picture** | 0.4189 | 1.3033 | 1.7222 |
| Global Accuracy | | 87.1% | |

Automatic Mapping of the Lava Flows at Piton de la Fournaise Volcano, by Combining Thermal Data in Near and Visible Infrared

215

| Mean Accuracy | | 83.8% | |
|---|---|---|---|
| Producer accuracy lava flow | | 78.2% | |
| Producer accuracy no lava flow | | 89.4% | |
| Mean distance between the outlines | | 26m | |
| 20070218 | Lava Flow class | No lava flow class | |
| Lava Flow Aerial picture | 0.2960 | 0.0713 | 0.3673 |
| No Lava Flow Aerial picture | 0.0862 | 0.2617 | 0.3479 |
| Total Area of aerial picture | 0.3822 | 0.333 | 0.7152 |
| Global Accuracy | | 77.98% | |
| Mean Accuracy | | 77.9% | |
| Producer accuracy lava flow | | 80.588% | |
| Producer accuracy no lava flow | | 75.22% | |
| Mean distance between the outlines | | 55.14m | |

Table 4. Error matrices for the seven outlines obtained by automatic extraction, by comparing those from the photo-interpretation of aerial photographs with those from automatic extraction.

the summit zone lava flows have less superficy than the ones observed along the slopes. An error on the outline will echo as importantly as the aria of **the mapped** lava flow is low. As to say, the more the mapped lava flow's area is important, the less the error will echo on the precision of the classification. The exactitude of the producer of the "lava flow" class is the most important in the validation of an automatic lava flow outline extraction method. The later varies from 74% and 87% on the entire tested lava flow. There is no observed difference between the eruptions that occur in the Dolomieu crater and the ones that occur on the flanks and the Grandes Pentes of the volcano. On another hand, the morphology of the lava flows plays a role on this precision (figure 8). The more the lava flow's superficy is important and the more compact it is, like a cercle shape form or very large, the more precise the "lava flow" class will be.

The distances between the automatic extracted lava flow outlines, in the case of superimposed lava flows, and those considered as reference is about 30m (table 4 and figure 9). Now, the satellite data used in the automatic extraction have a visible, near and medium infrared spatial resolution from 10m to 20m. If the extraction was realized without any possible confusion, we should have mean distance of about the size of these pixels. Nevertheless, in some low reflectances zones, only the thermal imaging is able to bound the outline's localization zone (figure 8 and 9). It therefore considerably reduces the precision because this data has a 90m pixel. We notice that in zones where the lava regularly effuses, the reflectance's difference cannot allow a free from error extraction of the outline, and it is therefore deducted from the thermal mask. This mask is essential because it allows the definition of a zone of interest inside witch the outline can be determinated. There is no possible confusion: to one thermal data, only one lava flow can be associated.

In vegetalized zones, the extraction can be considered as a simple threshold of the infrared, the outline is then about the size of the pixel. In some zones, especially the Dolomieu crater or certain parts of the cone, the differences of reflectances of the lava flows don't allow to differentiate them. The outline is then obtained thanks to the thermal data. It can represent up to 40% of the summit zone lava flows outline, but only 5% to 0% of the proximal and distal eruption's outlines. The mean error observed on the outline's extraction can then reach twice the size of the optical data's pixel. Some light diffusion and wedging effects can interact on the precision.

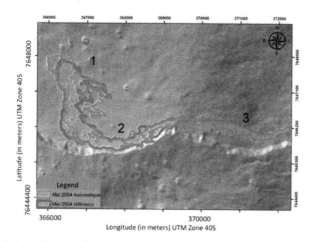

Fig. 9. Automatic lava flow outline extraction of May 2004 (blue) compared with the referenced outline (purple).

Lets considerate the example of the May 2004 lava flow (figure 9 and table 4). Three zones are sharply distinguished in the extraction: a very low reflectance and high thermal zone (1), a low reflectance with few or no thermal diffusion zone (2), and a various reflectance with few or no thermal diffusion zone (3). The weakest precision is for the first zone, for the second zone, the low reflectance is due to the substrate's nature and to luminosity issues in the zone, because the shadow projected by the rampart can interfere. A luminosity parameter is to be taken into consideration when choosing and time acquiring SPOT data. As for the extraction of the third zone, the thermal infrared band essentially obtains it, because the lava flow implemented in a vegetalized zone with a high spectral signature difference compared to the lava flow. It is the zone where the extraction is the most precise because only based on the SPOT data.

## 5. Discussion and conclusion

The vectors of the outlines obtained by automatic extraction properly match with the referenced outlines since we obtain a mean exactitude producer of 80% for the "lava flow" class. If we compare our automatic extraction results to those obtained by DGPS, they are less precise. The DGPS has a precision of about one centimeter at its antenna, now it's the operator who transports it. The error is due to the positioning of the antenna regards to the outlines. The error then is about one meter, for pluri-meter in our extractions. Nevertheless, the effusion zones are not all accessible, and cannot let realizing the outlines in their integrality, especially in the Grandes Pentes. Computer assisted drawing methods are on the other hand less reliable than our methodology for the data' distortion are not taken into account and can provoke hundred of meters error. The results of our classification are close to those obtained in other contexts' literature. (Azerzaq et al., 1997 ; Messar et Messar, 1997 ; Yüksel et al., 2008).

The vectorization time by photo-interpretation can take several days, whereas the automatic method enables obtaining this vector in less than one hour. There is therefore a considerable

Automatic Mapping of the Lava Flows at Piton de la Fournaise Volcano, by Combining Thermal Data in Near and Visible Infrared

217

gain of time. The gaps of observations due to clouds, or zone with high thermal diffusion and low reflectances, could be filled by using RADAR imaging in treatment sequences (Weisseil et al.2004). However, by adding Bi data will increase the treatment time, especially by using coherence data.

The association of thermal and optical data has already been realized in other automatic classifications with outlines extractions contexts: glaciology (Raciviteanu et al. 2008), the canopy (Joshi et al., 2006), agriculture (Kasdan, 1979; Saito et al., 2001). For similar spectral resolution data, the error matrix results are comparable to those obtained by our methodology.

This methodology was developed in order to automate lava flow outline's extraction and therefore ensure a fast update of the Piton de la Fournaise's database. The lava flow map was updated thanks to photo-interpretation and automatic extraction (figure 4). It allowed us to test the reliability of the outlines extracted according to each methodology used in this article, and to know their precision.

The errors measured by the matrixes give us the extracted surface's error, by comparing the automatic lava flow's area and the referenced one. We saw that: 1/ the difference of interpretation between two operators for the same data can be of 2% for the aerial photos and it varies between 2% and 5% with satellite data. 2/if the outlines extracted from satellite data's photo interpretation are compared to those extracted from aerial data, the exactitude producer of the "lava flow" class can reduce to 80%, but is at an average of 90%. 3/ Those obtained with the same type of data varies from 74% to 87%, which is to say an average of less than 10%. It influences the obtained area in less than 10%, exception made for low scale summit zone eruptions which represent an error of 20%, which remains modest and without major consequences on the volume and production estimation rates for the volcano.

## 6. Acknowledgements

These reasearch were financed by the "Region Reunion" and the ministry of "l'Enseignement Supérieur et de la Recherche" program. The authors would like to thank the CNES (Centre National d'Etudes Spatial) for the free access to the SPOT data via the KALIDEOS program (http://kalideos.cnes.fr).

This piece of work also used free ASTER data obtained thanks to AIST GEO Grid (Ministry of Economy, Trade and Industry) programm.

Thanks to Rebecca Roger for the translation.

## 7. Note

All the lava flow outlines' vectors will be available online on the Laboratoire de Geosciences Reunion website.

## 8. References

Abrams M., Abbott E., KahleA., 1991. Combined Use of Visible, Reflected Infrared, and Thermal Infrared Images for Mapping Hawaiian Lava Flows. Journal of Geophysical Research, n°96 (B1), pp. 475-484.

Azerzaq M., Assafi M., Fahsi A., 1997. Utilisation des images HRV de SPOT pour la détection du changement périurbain à Casablanca (Maroc). Télédétection des milieux urbains et périurbains. Éd. AUPELF-UREF, pp. 275-282.

Bachèlery P., 1981. Le Piton de la Fournaise (Ile de la Réunion). Étude volcanologique, structurale et pétrologique. Thesis manuscript, Univ. Clermont-Ferrand II, 1-215.

Billard G., 1974. Carte Géologique de l'île de la Réunion. 1 :50000.Bureau de Recherche Géologique et Minières, Paris. pp. 1-4.

Bonneville A., Lanquette A.M., Pejoux R.1989. Reconnaissance des principales unités géologiques du Piton de la Fournaise, La Réunion, à partir de SPOT 1. Bulletin de la société géologique de France, n°87, pp. 1101-1110.

Bussière P., 1967. Carte Géologique du département de La Réunion, 1 :1000000. Service de la Carte Géologique de la France.

Courtillot V., Besse J., Vandamme D., Montigny R., Jaeger J-J., Cappetta, H., 1986. Deccan flood basalts at the Cretaceous/Tertiary boundary? Earth and Planetary Sciences Letters, n°80 (3-4) , pp. 361-374.

De Michele M., Briole P., 2007. Deformation between 1989 and 1997 at Piton de la Fournaise volcano retrieved from correlation of panchromatic airborne images. Geophysical Journal International, n°169, pp. 357–364.

Despinoy M., 2000. Potentiel de la télédétection haute résolution spatiale et spectrale en milieu intertropical. Une approche transdisciplinaire à partir d'expériences aéroportées à la Réunion et en Guyane française. Thesis manuscript, pp.1-244.

Favalli M., Karátson D., Mazzarini F., Pareschi M.T., Boschi E., 2009, Morphometry of scoria cones located on a volcano flank: A case study from Mt. Etna (Italy), based on high-resolution LiDAR data. Journal of Volcanology and Geothermal Research, v. 186, p. 320-330.

Froger J.-L., Fukushima Y., Tinard P., Cayol V., Souriot T., Mora O., Staudacher T., Durand Ph., Fruneau B., Villeneuve N.,2007. Multi Sensors InSAR Monitoring of Volcanic Activity: The February & April 2007 Eruptions at Piton de la Fournaise, Reunion Island, Imaged with ENVISAT-ASAR and ALOS-PALSAR data. Proc. of FRINGE 2007 Workshop, Frascati, Italy,

Habib T., Chanussot J., Inglada J., Mercier G., 2007. Abrupt change detection on multitemporal remote sensing images: a statistical overview of methodologies applied on real cases. Proceedings IEEE IGARSS, pp. 2593-2596.

Harris A.J.L., Dehn J., James M.R., Hamilton C., Herd R., Lodato L., Steffke A., 2007, Pahoehoe flow cooling, discharge, and coverage rates from thermal image chronometry. Geophysical Research Letters, n° 34, pp. 6.

Hirn B.R., Di Bartola C., Ferrucci F.,2005. Automated, Multi-Payload, High-Resolution Temperature Mapping and Instant Lava Effusion Rate Determination at Erupting Volcanoes. Geoscience and Remote Sensing Symposium, 2005. IGARSS '05. Proceedings. IEEE International. n° 7, pp. 5056 – 5059.

Honda K., Nagai M., 2002. Real-time volcano activity mapping using ground-based digital imagery. ISPRS Journal of Photogrammetry and Remote Sensing, n°57(1-2), pp.159-168.

Inglada J., Favard J-C., Yesou H., Clandillon S., Bestault C., 2003. Lava flow mapping during the Nyiragongo January, 2002 eruption over the city of Goma (D.R. Congo) in the frame of the international charter space and major disasters. Proceedings of

Automatic Mapping of the Lava Flows at Piton de la Fournaise Volcano, by Combining Thermal Data in Near and Visible Infrared

219

the IEEE International Geoscience and Remote Sensing Symposium, IGARSS '03, n°3, pp.1540- 1542.

James M.R., Pinkerton,H., Robson S., 2007, Image-based measurement of flux variation in distal regions of active lava flows. Geochem. Geophys. Geosyst., n° 8, pp. Q03006.

Joshi C., De Leeuw J., Skidmore,A.K., Van Dure, I.C., Van Oosten, H., 2006. Remotely sensed estimation of forest canopy density: a comparison of the performance of four methods. International Journal of Applied Earth Obeservation and Géoinformation. n°8, pp.84-95.

Kahle A.B., Abrams M.J., Abbott E.A., Mouginis-Mark P.J., Realmuto V.J., 1995. Remote sensing of Mauna Loa. Mauna Loa Revealed: Structure, Composition, History, and Hazards. Geophys. Monogr. N°92, AGU, pp. 145-169.

Kahle A., Gillespie A., Abbott E., Abrams M, Walker R., Hoover G., Lockwood J.P., 1988. Mapping and relative dating of Hawaiian basalt flows using multispectral thermal infrared images. Journal of Geophysical Research, n°93( B12), pp. 239-251

Kasdan H., 1979. Feasibility of Using Optical Power Spectrum Analysis Techniques for Automatic Feature Classification from High Resolution Thermal, Radar, and Panchromatic Imagery. Recognition systems inc van nuys calif, pp. 1-194

Lacroix A., 1936. Le volcan actif de l'île de la Réunion et ses Produits. Paris, Gauthier et Villard ed., pp. 1-297

Legelay-Padovani A., Mering C., Guillande R., Huaman D., 1997. Mapping of lava flows through SPOT images - an example of the Sabancaya volcano (Peru), International Journal Remote Sensing, n°18(15), pp. 3111-3133.

Lénat J.-F., 1987. Structure et Dynamique internes d'un volcan basaltique intraplaque océanique : le Piton de la Fournaise (Ile de la Réunion). Thèse de doctorat d'Etat, Univ. Clermont II, France.

Lénat J.F., Bachèlery P. ,1987. Dynamics of Magma Transfer at Piton de La Fournaise Volcano (Réunion Island, Indian Ocean), Earth Evolution Sciences, Special Issue "Modeling of Volcanic Processes", Chi-Yu and Scarpa (Eds.), pp. 57-72, Friedr. Vieweg and Sohn, Brauschweig/Wiesbaden

Lénat J.-F., Bachèlery P. et Desmulier F., 2001 . Genèse du champ de lave de l'Enclos Fouqué : une éruption d'envergure exceptionnelle du Piton de la Fournaise (Réunion) au 18 ème siècle. Bull. Soc. Géol Fr. 2, pp.177-188.

Letourneur L., Peltier A., Staudacher T., Gudmundsson A., 2008, The effects of rock heterogeneities on dyke paths and asymmetric ground deformation: The example of Piton de la Fournaise (Réunion Island). Journal of Volcanology and Geothermal Research, n°173, pp. 289-302.

Lombardo V., Harris A.J.L., Calvari S., Buongiorno M.F., 2009, Spatial variations in lava flow field thermal structure and effusion rate derived from very high spatial resolution hyperspectral (MIVIS) data. Journal of Geophysical Research, n° 114, pp. B02208.

Lu Z., Rykhus R., Masterlark T., Dean K.G., 2004, Mapping recent lava flows at Westdahl Volcano, Alaska, using radar and optical satellite imagery. Remote Sensing of Environment, n°91(3-4), pp. 345-353.

Messar Y., Messar N., 1997. Apport de la télédétection à la cartographie de l'espace urbain au 1: 20000 en terme d'emprise au sol du bâti : cas de la ville d'Oran (Algérie). Télédétection des milieux urbains et périurbains. Éd. AUPELF-UREF., pp. 315-323.

Observatoire Volcanologique du Piton de La Fournaise internal annual report, 2009

Observatoire Volcanologique du Piton de La Fournaise internal report, janvier 2010

Paparoditis N., Souchon J-P., Martinoty G., Pierrot-Deseilligny M., 2006. High-end aerial digital cameras and their impact on the automation and quality of the production workflow. ISPRS Journal of Photogrammetry & Remote Sensing, n°60, pp. 400–412.

Peltier A., Bachèlery P., Staudacher T., 2009. Magma transport and storage at Piton de la Fournaise (La Réunion) between 1972 and 2007: A review of geophysical and geochemical data. Journal of Volcanology and Geothermal Research, n°184, pp. 93-108.

Raciviteanu A.E., Williams M.W., Barry R.G., 2008. Optical Remote Sensing of Glacier Characteristics: A Review with Focus on the Himalaya. Sensors , n°8, pp. 3355-3383

Saito G., Ishitsuka N., Matano Y., Kato M., 2001. Application on Terra/ASTER data on agriculture land mapping. Paper presented at the 22nd Asian Conference on Remote Sensing, Singapore, pp.1-6

Servadio Z., Villeneuve N., Gladys A., Staudacher T., Urai M., 2008. Preliminary results of lava flow mapping using remote sensing in Piton de la Fournaise, La Réunion island. Workshop on the "Use of Remote Sensing Techniques for Monitoring Volcanoes and Seismogenic Areas", IEEE Naples. Poster session and Proceeding.

Stieltjes L., Moutou P., 1989. A statistical and Probabilitic study of the historic activity of Piton de la Fournaise, Reunion Island, Indian Ocean. Journal of Volcanology and Geothermal Research, n°36, pp. 67-86.

Tinard P., 2007. Caractérisation et modélisation des déplacements du sol associés à l'activité volcanique du Piton de la Fournaise, île de La Réunion, à partir de données interférométriques. Août 2003 – Avril 2007. Ph.D. Thesis, Univ. Blaise Pascal, pp.1-334

Villeneuve N., 2000. Apports multi-sources à une meilleure compréhension de la mise en place des coulées de lave et des risques associés au Piton de la Fournaise : Géomorphologie quantitative en terrain volcanique. Thesis manuscript, pp.1-378.

Villeneuve N., Bachèlery P., 2006. Revue de la typologie des éruptions au Piton de La Fournaise, processus et risques volcaniques associés. Cybergéo : revue européenne de Géographie, n° 336

Yüksel A., Akay A.E., Gundogan R., 2008. Using ASTER Imagery in Land Use/cover classification of Eastern Mediterranean Landscapes According to CORINE Land Cover Project. Sensors. n° 8,pp. 1237-1251.

# Section 2

# Climate and Atmosphere

# 9

# Oceanic Evaporation: Trends and Variability

Long S. Chiu[1], Si Gao[1] and Chung-Lin Shie[2]
[1]*Department of Atmospheric, Oceanic and Earth Sciences, George Mason University*
[2]*UMBC/JCET, NASA/GSFC*
*USA*

## 1. Introduction

The global water and energy cycles are strongly coupled as two essential components of earth system. They play important roles in altering the Earth's climate.

Oceanic evaporation, or sea surface latent heat flux (LHF) divided by latent heat of vaporization $(L_v)$, is a key component of global water and energy cycle. In a bulk aerodynamic formulation, it is determined by the transfer coefficient of evaporation, $C_E$, and bulk parameters such as surface wind speed $(U)$, surface saturated and near-surface air specific humidity $(Q_s$ and $Q_a)$,

$$LHF = \rho L_v C_E U (Q_s - Q_a) \qquad (1)$$

where sea surface saturated humidity is determined by sea surface temperature (SST) and salinity, and $\rho$ is density of moist air. The transfer coefficient is dependent on the stability of the atmosphere and the sea state (Liu et al., 1979; Zeng et al., 1988). Historically, marine surface observations have provided the basis for estimating these oceanic turbulent fluxes (e.g. Bunker, 1976; Cayan, 1992; da Silva et al., 1994; Esbensen & Kushnir, 1981; Hastenrath, 1980; Hsiung, 1985; Isemer & Hasse, 1985, 1987; Josey et al., 1998; Oberhuber, 1988; Renfrew et al., 2002; Weare et al., 1981). The advent of remote sensing techniques offers means to retrieve a number of surface bulk variables. Microwave radiation interacts directly with water molecules and hence is effective in providing water vapor information. The sea surface emissivity is affected by the sea state and foam conditions, which is related to surface wind. For instance, global microwave measurement of the Special Sensor Microwave Imager (SSM/I) on board a series of Defense Meteorological Satellite Program (DMSP) satellites has been used to retrieve near-surface air humidity and winds over the ocean.

At present there are several remote sensing products of global ocean surface latent heat flux. They include the NASA/Goddard Satellite-based Surface Turbulent Flux (GSSTF) dataset version 1 (Chou et al., 1997) and version 2 (GSSTF2, Chou et al., 2003), the Japanese Ocean Flux utilizing Remote Sensing Observations (J-OFURO) dataset (Kubota et al., 2002) and the Hamburg Ocean Atmosphere Parameters and Fluxes from Satellite (HOAPS) dataset (Grassl et al., 2000). Chiu et al. (2008) examined "trends" and variations in these global oceanic evaporation products for the period 1988–2000. They found a long-term increase in global average LHF that started around 1990 in GSSTF2. They argued that the dominant patterns may be related to an enhancement of Hadley circulation and El Niño-Southern Oscillation

(ENSO), respectively. An updated version of SSM/I version 6 (V6) data released by Remote Sensing Systems (RSS) in 2006 [as used by Wentz et al. (2007), see http://www.ssmi.com] that calibrates all SSM/I sensors is available in 2008. Shie et al. (2009) have reprocessed and forward processed GSSTF2 to version 2b (GSSTF2b, Shie et al. 2010; Shie 2010) using the SSM/I V6 data (including total precipitable water, brightness temperature, and wind speed retrieval), covering the period July 1987–December 2008. We provide an assessment of these data products and examine their "trends" and variability.

The data and methodology are described in Section 2. Section 3 presents the trends of these products, compares GSSTF2 and GSSTF2b for the pre 2000 periods, assesses the post 2000 performance, and examines the GSSTF2b Set1 and Set2 differences. Summary and discussion are presented in Section 4.

## 2. Data and methodology

Earlier version of these flux products have been described elsewhere (Chiu et al., 2008). The product versions described here represent the most updated versions as of the writing of this report.

### 2.1 HOAPS

Detail descriptions of the latest version of HOAPS, (version 3, or HOAPS-3) are given in Andersson et al. (2010). Bulk variables are derived from SSM/I data except for the SST which is derived from the Advanced Very High Resolution Radiometer (AVHRR) Oceans Pathfinder SST product. A neural network algorithm is used to derive $U$. The $Q_a$ is obtained using the linear relationship of Bentamy et al. (2003). The $Q_s$ is computed from the AVHRR SST using the Magnus formula (Murray, 1967) with a constant salinity correction factor of 0.98. The near-surface air temperature ($T_a$) is estimated from the SST using the assumptions of 80% constant relative humidity and a constant surface-air temperature difference of 1 K. Latent and sensible heat fluxes are calculated using the Coupled Ocean–Atmosphere Response Experiment (COARE) 2.6a bulk algorithm (Fairall et al., 1996, 2003).

The HOAPS-3 data sets cover the time period from July 1987 to December 2005. HOAPS-G pentad and monthly data sets with 0.5-degree resolution and HOAPS-C twice daily data set with 1-degree resolution are available at the website (http://www.hoaps.zmaw.de).

### 2.2 J-OFURO

The updated version of J-OFURO, (version 2, J-OFURO2) is described in Tomita et al. (2010). Bulk variables $U$, $Q_a$ and SST ($Q_s$) are determined by multi-satellite and multiple satellite sensors. $U$ is obtained from a combination of microwave radiometers (SSM/I, AMSR-E and TMI) and scatterometers (ERS-1, ERS-2 and QuikSCAT). $Q_a$ is derived from SSM/I measurements. SST is taken from the Merged satellite and in-situ data Global Daily SST (MGDSST, Sakurai et al. 2005) analysis provided by Japanese Meteorological Agency (JMA). $T_a$ is obtained from NCEP/DOE reanalysis. COARE 3.0 bulk algorithm (Fairall et al., 2003) is used to estimate LHF and SHF. The J-OFURO2 covers the time period from January 1988 to December 2006. Daily and monthly means with 1-degree resolution are available at the website (http://dtsv.scc.u-tokai.ac.jp/j-ofuro).

## 2.3 GSSTF

The GSSTF2 product has daily and monthly fields with a 1°x1° resolution for July 1987–December 2000 (Chou et al., 2001), based on the method of Chou et al. (1997) with some improvements (Chou et al., 2003). The temporal and spatial resolutions of GSSTF2b are the same as those of GSSTF2, except that GSSTF2b product covers a longer period (July 1987–December 2008).

GSSTF2b dataset is processed using improved input datasets, namely the recently released NCEP SST analysis, and a uniform (across satellites) surface wind and microwave brightness temperature (TB) V6 dataset from the SSM/I produced by RSS. Table 1 summarizes characteristics of input data and parameters for HOAPS3, J-OFURO2, GSSTF2 and GSSTF2b, in that order. As we focus on LHF, only detailed descriptions and discussions on input parameters of LHF are presented.

A major improvement in the input parameters of GSSTF2b is the use of the newly released SSM/I V6 product (see discussions in http://www.ssmi.com). The SSM/I V6 product removes the spurious wind speed trends found in the Wentz/RSS SSM/I V4 wind speed retrievals. To be consistent, the SSM/I V4 total precipitable water ($W$) and bottom-layer precipitable water ($WB$) used in GSSTF2 are replaced by the corresponding SSM/I V6 products in the production of GSSTF2b. Moreover, the weekly 1° spatial resolution Optimum Interpolation (OI) SST version 1 (V1) dataset (Reynolds & Smith, 1994) used in GSSTF2 is replaced by the improved OI SST version 2 (V2) dataset. The OI SST V2 has a lower satellite bias, a new sea ice algorithm, and an improved OI analysis (Reynolds et al., 2002) resulting in a modest reduction of the satellite bias and global residual biases of roughly −0.03°C. The major improvement in the V2 analysis shows up at high latitudes where local differences between the old and new analysis can exceed 1 °C due to the application of a new sea ice algorithm. There are two GSSTF2b sets, Set1 and Set2 (Shie et al., 2010; Shie 2010). Set1 is developed using all available DMSP SSM/I sensor data. In a preliminary analysis, it was noted that there are large trends associated with LHF which are mostly attributed to the DMSP F13 and F15 satellites. Set2 was produced by excluding satellite retrievals that are judged to caused relatively large artificial trends in LHF (mostly post 2000) from Set1. Consequently Set2 is identical to Set1 before 1997 and shows a smaller trend than Set1 for the whole period, while Set1 has better spatial coverage (less missing data). Hilburn & Shie (2011) further found a drift in the Earth incidence angle (EIA) associated with the SSM/I sensors on the DMSP satellites that introduces artificial trends in the SSM/I TB data. These artificial trends introduce large changes in the boundary water ($WB$), which affects the $Q_a$, and thus the LHF retrievals. An improved version, GSSTF2c, incorporating the corrected SSM/I brightness temperature, has been produced as of this writing (Shie et al., 2011). The retrieved $WB$, $Q_a$ and LHF have genuinely improved, particularly in the trends post 2000 (Shie & Hilburn, 2011). An extensive study involving the GSSTF2c will be presented in a separate paper.

In this chapter, "trend" is used to indicate results from linear regression analysis and/or Empirical Mode Decomposition (EMD) for the period of study. Linear regression of the time series with time is used to detect linear trends and the significance can be estimated from the slope of the regression. EMD is based on local characteristic time scales of the data and is therefore applicable for analyzing nonlinear and non-stationary processes (Huang et al., 1998). It decomposes the time series into a finite and often small number of intrinsic mode

| Datasets | HOAPS3 | J-OFURO2 | GSSTF2 | GSSTF2b |
|---|---|---|---|---|
| $C_E$ (transfer coefficient) | Fairall et al. (1996, 2003) | Fairall et al. (2003) | Chou (1993) | Chou (1993) |
| $U$ (speed) | SSM/I V6 TB and neural network algorithm | SSM/I, AMSR-E, TMI, ERS-1, ERS-2 and QuikSCAT | Wentz V4 (1997) | Wentz V6 (2007) |
| $U$ (vector) | N/A | N/A | Atlas et al. (1996) | CCMP Level-2.5 (SSM/I, TMI, and AMSR-E) |
| $W/WB$ | N/A | N/A | Wentz V4 (1997) | Wentz V6 (2007) |
| SST | AVHRR | MGDSST Sakurai et al. (2005) | NCEP/NCAR Reanalysis (V1) Reynolds & Smith (1994) | NCEP/DOE Reanalysis (V2) Reynolds et al. (2002) |
| $Q_a$ | Bentamy et al. (2003) | Schlussel et al. (1995) | Chou et al. (1995, 1997) | Chou et al. (1995, 1997) |
| $T_a$ | Estimated from SST with assumptions of 80% humidity and 1 K surface-air temperature difference | NCEP/DOE Reanalysis (V2) | NCEP/NCAR Reanalysis (V1) | NCEP/DOE Reanalysis (V2) |
| Spatial resolution | 1°x1°, 0.5°x0.5° | 1°x1° | 1°x1° | 1°x1° |
| Spatial coverage | Global Oceans | Global Oceans | Global Oceans | Global Oceans |
| Temporal resolution | Twice daily, pentad and monthly | Daily and monthly | Daily and monthly | Daily and monthly |
| Temporal coverage | Jul. 1987 – Dec. 2005 | Jan. 1988 – Dec. 2006 | Jul. 1987 – Dec. 2000 | Jul. 1987 – Dec. 2008 |

Table 1. Characteristics of input data and parameters for HOAPS3, J-OFURO2, GSSTF2 and GSSTF2b.

functions (IMFs) of increasing time scales. The existence of a trend is dependent on the length of the dataset. If the last IMF (one with longest time scale) is monotonically increasing or decreasing, a trend is indicated. The EMD is a more stringent test for significance of "trends."

Non-seasonal variability is examined using Empirical Orthogonal Function (EOF) analysis. Non-seasonal data are obtained by subtracting the monthly climatology of the study period from the monthly data. EOF analysis decomposes a spatio-temporal dataset into a series of orthogonal spatial EOF patterns and the associated time series (also called principal components). The test proposed by North et al. (1982) is used to judge the EOFs to see if they are significant and distinct. To examine the significance of each EOF, the logarithm of the variance explained by the EOF is plotted against its EOF number. The variance explained by the $n$th EOF is given by $\lambda_n / \sum_{i=1}^{N} \lambda_i$ where $\lambda_n$ is the $n$th eigenvalue and $N$ is the number of time samples. Linear regression between the logarithms of $\lambda_n$ vs $n$ is computed. EOFs above the regression line are judged to be significant.

## 3. Results

### 3.1 Global and zonal average

Figure 1 shows the area weighted global (60°N–60°S) average LHF of HOAPS3, J-OFURO2, GSSTF2, and Set1 and Set2 of GSSTF2b. The merged OAFLUX product is included for comparison. Visual inspection of the data products does not show large missing gaps in the spatial distribution of the products. All satellite products-HOAPS3, J-OFURO2 and all GSSTF datasets show increases while there is no obvious trend in OAFLUX. All products have similar global means in the early period 1988–1991. The means of GSSTF2b (both Set1 and Set2) are generally lower than that of GSSTF2 which is the highest among all products. All products show a dip in 1991 and an increase afterwards. The dip is clearly evident in HOAPS3. GSSTF2b Set1 and Set2 are identical up to 1998 after which they diverge, but tend to come close again after 2006.

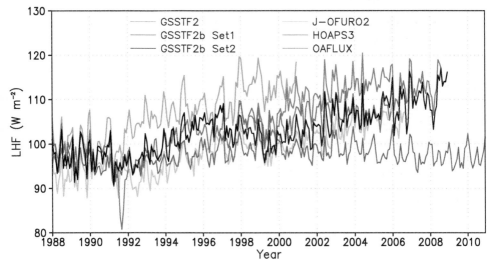

Fig. 1. Time series of global (60°N–60°S) oceanic average latent heat flux derived from HOAPS3, J-OFURO2, GSSTF2, GSSTF2b Set1 and Set2 and OAFLUX.

The zonal annual averages of LHF are depicted in Figure 2. The general features of the zonal means are quite similar among these products: they showed maxima in the subtropics, minima at the poles and relative minima at the equator. The subtropical maxima in the southern hemisphere are slighter higher than those in the northern hemisphere for the same product. While GSSTF2 is the highest among these estimates, the

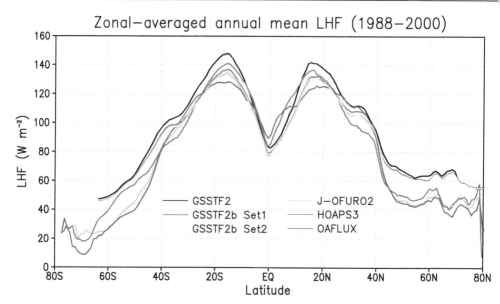

Fig. 2. Zonal annual mean of LHF for oceanic evaporation computed from HOAPS3, J-OFURO2, GSSTF2, GSSTF2b Set1 and Set2 and OAFLUX.

GSSTF2b Set2 is slightly lower than HOAPS3 but higher than J-OFURO2 and OAFLUX at their maxima. Poleward of 30°, the GSSTF zonal means are generally higher than the other products.

### 3.2 Trend analysis

Linear regression analyses of the time series with time were performed on the global mean time series. The significance of the slopes of the regression (trend) is tested using a t-test. The degree of freedom for the significance test takes into account the serial correlation of the time series (Angell, 1981; Chiu & Newell, 1983). GSSTF2 shows the largest trend. It is followed by GSSTF2b Set1 while the GSSTF2b Set2 trend is comparable to HOAPS3 and J-OFURO2 for the period of overlap (1988–2005). OAFLUX exhibits the smallest trend. Table 2 summarizes our results.

To map out the geographic differences, Figure 3 compares the spatial distribution of linear trends of GSSTF2b Set1 and Set2, HOAPS3, J-OFURO2 and OAFLUX. The linear trends are calculated for the common time period 1988–2005. While the magnitudes of the trends are different, the locations of maximum change are similar among HOAPS3, J-OFURO2 and GSSTF2b Set1 and Set2 - all show increasing trends in the storm tracks in the north Atlantic and north Pacific, the oceanic dry zones off the Inter-tropical Convergence Zone (ITCZ) in the western south Pacific and in latitude bands between 30–40°S off the coast of Australia and in the Indian Ocean. OAFLUX shows increasing trends in the storm tracks in both the North Atlantic and North Pacific, and in the eastern coastal regions off South America and Australia. There are large areas showing a decrease, notably in the south Indian Ocean, tropical eastern North Pacific and in North Atlantic.

| Period | GSSTF2 | GSSTF2b Set1 | GSSTF2b Set2 | HOAPS3 | J-OFURO2 | OAFLUX |
|--------|--------|--------------|--------------|--------|----------|--------|
| 1988–2000 | 10.44 | 9.88 | 5.98 | 7.75 | 6.51 | 3.71 |
| 1988–2005 | N/A | 11.69 | 6.34 | 7.35 | 7.62 | 2.41 |
| 1988–2008 | N/A | 10.45 | 7.08 | N/A | N/A | 1.50 |

Table 2. Linear trends of LHF products (in W m$^{-2}$ decade$^{-1}$) for the different periods. All values are significant at 99% level.

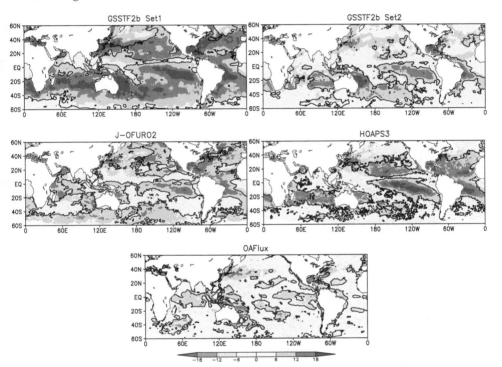

Fig. 3. Linear trends of LHF (W m$^{-2}$ decade$^{-1}$) over the period 1988–2005. Contours give the trends above 95% confidence level.

### 3.3 EMD analysis of global average LHF

We performed EMD analyses for both GSSTF2 and GSSTF2b for the periods 1988-2000 and 1988–2008, respectively. Figure 4 shows the last IMFs from EMD analysis of GSSTF2b global average (60°N-60°S) LHF for the period 1988-2008, along with those of HOAPS3, J-OFURO2 and OAFLUX. EMD analysis of GSSTF2 global average LHF for the period 1988-2000 (Chiu et al., 2008; Xing, 2006) showed an increase starting around 1990 before the Mt Pinatubo eruption event in 1991. The last IMF of GSSTF2b Set2 and HOAPS3 show a dip around 1990 while J-OFURO2 and GSSTF2b Set1 show monotonic increases. From EMD analysis, a "trend" cannot be ascertained for the whole period for GSSTF2b Set2 and HOAPS3.

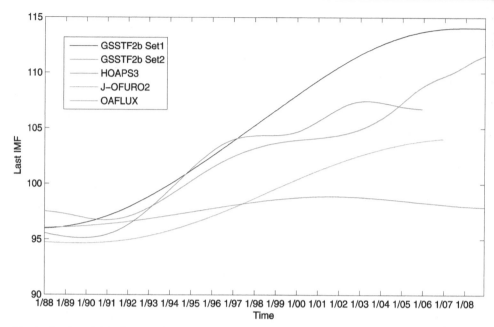

Fig. 4. The last IMFs (W m⁻²) from EMD analysis of global (60°N-60°S) average LHF for
GSSTF2b Set1 (black), GSSTF2b Set2 (red), and OAFLUX (blue) for the period 1988-2008,
HOAPS3 (purple)for the period 1988-2005 and J-OFURO2 (green) for the period 1988-2006.

### 3.4 GSSTF2b Set1 and Set2

In this subsection, we analyze the difference between GSSTF2b Set1 and Set2. The linear
trends of GSSTF2b LHF for the entire period 1988-2008 are depicted in Figure 5. The
magnitudes of the trends are reduced when compared to GSSTF2 (see Chiu et al., 2008) and
Set1 shows larger trends than Set2, as anticipated.

Trends of the zonal averages are shown in Figure 6. Both Set1 and Set2 show similar trend
patterns while the magnitude in Set2 is much reduced. Maximum increasing trends are located
in the subtropics, with maxima at 30°S and 35°N, poleward of the latitude of maximum zonal
evaporation, while the trend at the thermal equator (~5°N) shows a minimum.

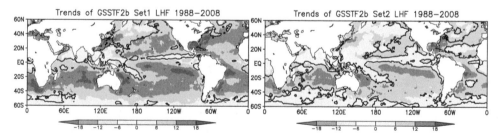

Fig. 5. Linear trends of GSSTF2b Set1 and Set2 LHF (1988-2008). Unit in W m⁻² decade⁻¹.
Contours give the trends above 95% confidence level.

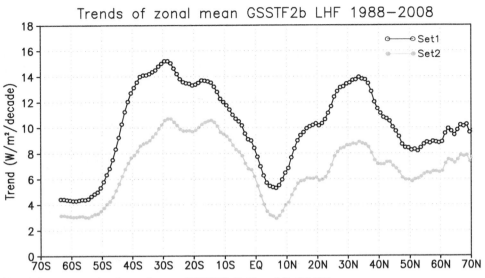

Fig. 6. Linear trends of zonal mean GSSTF2b Set1 and Set2 LHF.

Figure 7 shows the spatial distribution of trends in $U$, $DQ$ and $Q_a$ in Set 1 and Set2. It can be seen that the trend pattern in $U$ is almost identical in both Set1 and Set2. Major difference is found in the $DQ$ field, the magnitudes of the trends in Set1 are larger than that in Set2. In the equatorial eastern Pacific, in portion of the South China Sea, and the Bay of Bengal, Set2 actually shows a decreasing but non-significant trend. For the $Q_a$ trend, Set1 shows large areas of decreases. While the patterns are similar for Set2, the magnitudes are much reduced, and in some areas, such as the eastern North Pacific and the bulk of the North Atlantic, the decreasing trends actually reverse to increasing trends.

LHF is a product of the surface wind ($U$) and the humidity difference ($DQ=Q_s-Q_a$) if we assume the variation in $C_E$ is small (Equation 1). Judging from the change pattern of $U$ and $DQ$ they are essentially decoupled. Equation (1) can be integrated globally to get

$$\frac{\delta\overline{\overline{LHF}}}{\overline{\overline{LHF}}} \approx \frac{\delta\overline{\overline{U}}}{\overline{\overline{U}}} + \frac{\delta\overline{\overline{DQ}}}{\overline{\overline{DQ}}} \tag{2}$$

$$\delta\overline{\overline{DQ}} = \delta\overline{\overline{Q}}_s - \delta\overline{\overline{Q}}_a \tag{3}$$

where $\delta x$ represent the change in the quantity $x$, $\delta x/x$ represent fractional changes in $x$, and the over-bar $\overline{x}$ represents global average of $x$. For GSSTF2 (1988-2000), the terms in equation (2) is approximately 17%, 6%, and 11%, in that order (Xing, 2006). Most of the increase in $\overline{DQ}$ was attributed to increase in $\overline{Q}_s$ and decrease in $\overline{Q}_a$.

Figure 8 shows the time series of global average $U$, $DQ$ and $Q_a$ for GSSTF2b Set1 and Set2, respectively. It clearly indicates the divergence of Set1 and Set2 in late 1997. The large difference in LHF between Set1 and Set2 is mostly attributed to $DQ$, which is due to the higher $Q_a$ in Set2. The difference in $U$ between the datasets is small. There is a large decrease in $Q_a$ at the end of 2008.

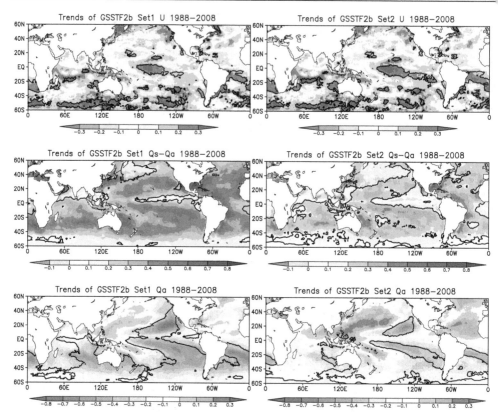

Fig. 7. Linear trends of surface wind speed ($U$, upper panel), surface humidity difference ($DQ=Q_s-Q_a$, middle panel), and surface air humidity ($Q_a$, lower panel) for GSSTF2b Set1 (left column) and Set2 (right column). Units are in m s$^{-1}$ decade$^{-1}$ for $U$ and in g kg$^{-1}$ decade$^{-1}$ for $DQ$ and $Q_a$. Contours give the trends above 95% confidence level.

Table 3 shows the changes in LHF in Set1 and Set2 of GSSTF2b (1988–2008) and the associated changes in $U$, $DQ$ and the changes in $Q_s$ and $Q_a$. The change in $DQ$ contribute most to the change in LHF for both Set1 and Set2, while the changes in $DQ$ is due both to an increase in $Q_s$ and decrease in $Q_a$. The difference in the change of LHF between Set1 (23.1%) and Set2 (15.5%) is mostly attributed to $DQ$ (20.0% vs. 12.3%) and changes in $Q_a$ (–0.51 g kg$^{-1}$ vs. –0.24 g kg$^{-1}$). It is clear that the impact of DMSP F13 is the introduction of a much lower $Q_a$, thus affecting $DQ$ and ultimately LHF.

|      | $\dfrac{\delta \overline{LHF}}{\overline{LHF}} \approx \dfrac{\delta \overline{U}}{\overline{U}} + \dfrac{\delta \overline{DQ}}{\overline{DQ}}$ | $\dfrac{\delta \overline{U}}{\overline{U}}$ | $\dfrac{\delta \overline{DQ}}{\overline{DQ}}$ | $\delta \overline{DQ}$ (g kg$^{-1}$) | $\delta \overline{Qs}$ (g kg$^{-1}$) | $\delta \overline{Qa}$ (g kg$^{-1}$) |
|------|------|------|------|------|------|------|
| Set1 | 23.1% | 3.1% | 20.0% | 0.73 | 0.22 | –0.51 |
| Set2 | 15.5% | 3.1% | 12.3% | 0.46 | 0.22 | –0.24 |

Table 3. Summary of changes in LHF, $U$ and $DQ$ and $Q_s$ and $Q_a$ for GSSTF2b Set1 and Set2.

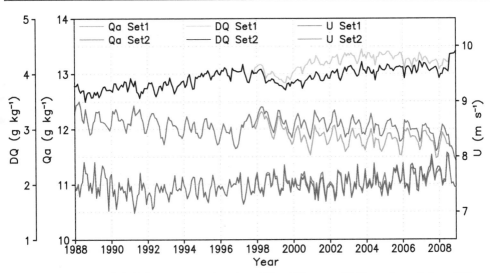

Fig. 8. Time series of global average surface humidity difference ($DQ$), surface air humidity ($Q_a$) and wind speed ($U$) for GSSTF2b Set1 and Set2.

## 3.5 EOF analyses and teleconnections

Empirical Orthogonal Function (EOF) analyses are performed on LHF of both GSSTF2b Set1 and Set2 for the period 1988-2008. Monthly means for the entire period are first removed from the data to form the non-seasonal dataset. The first, second and third EOF of Set1

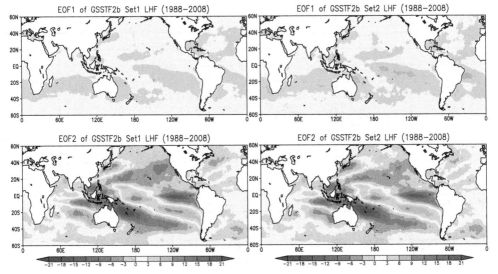

Fig. 9. Spatial patterns of the first (top) and second (bottom) EOF of GSSTF2b LHF for the period 1988-2008. The variances explained are 8.6% (11.5%) and 4.3% (4.3%) for EOF1 and EOF2 for Set2 (Set1), respectively.

explains 11.5%, 4.3%, and 3.4% of the total variance, and the corresponding variance explained for Set2 are 8.6%, 4.3% and 3.4%. Figure 9 shows the spatial patterns of the first (EOF1) and second (EOF2) EOF for Set1 and Set2, respectively. Their associated time series, accompanied by a rescaled Southern Oscillation Index (SOI), are presented in Figure 10. The general patterns of EOF1 of Set1 and Set2 are very similar, with large weights in subtropical Indian Ocean, the dry zone in the eastern tropical South Pacific and South Atlantic. They also bear striking resemblance to the first EOF pattern computed from GSSTF2 which shows

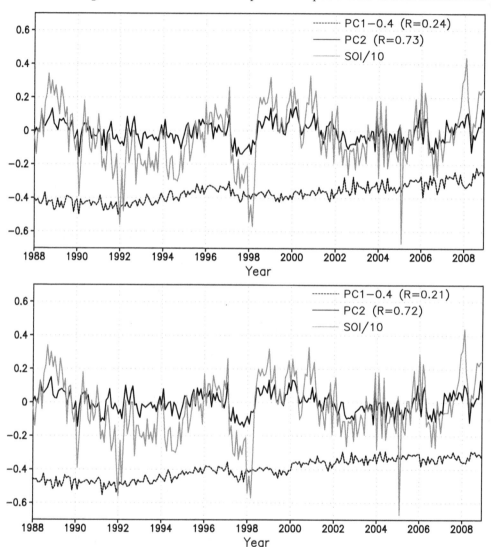

Fig. 10. Time series associated with EOF of non-seasonal GSSTF2b Set2 LHF for the period 1988–2008 and the SOI for the same period. The SOI is divided by 10 and the time series of the first EOF is shifted by –0.4 units for clarity. The lower panel shows the same for Set1.

weights of the same sign everywhere (Chiu et al., 2008). However, in GSSTF2b, there are regions of opposite sign in the South China Sea area in both Set1 and Set2, with higher weights in Set2 in the centers of highs. The EOF2 patterns for Set1 and Set2 are almost identical. Similarity is also noted for the EOF2 pattern derived from GSSTF2 (Chiu et al., 2008).

The time series of EOF1 also show a slight increasing trend, suggesting an increase of LHF over most of the global ocean. The associated time series of EOF2 have a significant correlation of 0.73 (0.72 for Set1) with SOI, reaffirming that EOF2 - ENSO events association.

The third EOF pattern (not shown) is characterized by a negative - positive - negative (- + -) zonal changes centering around 10-20°N, 20-30°N and north of 40°N in the north Pacific and a (+ - +) centering at 20°N, 30°N and 50°N in the North Atlantic. The pattern correlations for EOF1, EOF2 and EOF3 between Set1 and Set2 are 0.08, 0.49, and 0.39, and the corresponding temporal correlation for the associated time series are 0.92, 0.97, and 0.83, in that order.

EOF3 is reminiscent of the North Atlantic Oscillation (NAO) and the North Pacific Oscillation (NPO) patterns (Walker & Bliss, 1932; Wallace & Gultzer, 1981, Hurrell et al., 2003). These teleconnection patterns are further discussed in terms of an atmospheric annular mode, or Arctic Oscillation (AO), showing the opposition between subtropical highs and the polar lows (Wallace, 2000; Deser, 2000; Aubaum et al., 2001). We compute the correlations between an AO index with the time series of EOF3 and found no significant correlation with Set2, however a correlation of 0.32 is found for EOF3 Set1, significant at 95% level.

## 4. Summary and discussion

Four satellite based sea surface latent heat flux (LHF) products, HOSAPS3, J-OFURO2, GSSTF2, and GSSTF2b (Set1 and Set2) and a merged analysis OAFLUX are compared. Linear trend analysis of all satellite based products show large increasing trend, with GSSTF2 the largest, followed by GSSTF2b Set1, J-OFURO2, HOAPS3, and GSSTF2b Set2. OAFLUX exhibits the lowest linear increasing trend. Most of the satellite products used SSM/I as input. Small drifts in the SSM/I brightness temperature (TB) associated with changes in Earth incidence angle (EIA) was noted in most of the SSM/I data (Hilburn & Shie, 2011; Shie & Hilburn, 2011). Because of the sensitivity of the boundary layer water ($WB$) to the TB, these small drifts can introduce artificial trends in bulk quantities such as $Q_a$. A second data set, GSSTF2b Set2, which excludes satellite retrievals that were judged to introduce these biases, was introduced. The most-excluded satellite data are SSM/I onboard the DMSP F13 and F15 satellites. The new set, GSSTF2b Set2, was found to have a much reduced increasing trend, the magnitude of which is comparable to HOAPS3 and J-OFURO2 for the period of overlap. To account for the drift in the EIA, a new version of GSSTF, GSSTF2c, that takes account of the correction in EIA, has been completed as of this writing, and will be officially released to the public via NASA/GES DISC by the end of October 2011 (Shie et al., 2011). These trend issues will be revisited after its release.

Empirical Mode Decomposition (EMD) analyses, which are designed for examining non-stationary non-homogeneous time series, are performed on the global LHF. The last IMF of GSSTF2b Set1 shows a monotonic increase indicating the existence of a trend in this period.

The corresponding IMF of the other data products do not show monotonic increases, hence trends cannot be ascertained.

To examine the attribution of the increase in GSSTF2b Set1 and Set2, the linear trends in both the surface wind ($U$) and surface humidity difference ($DQ$) are computed. There is no significant difference between the trend patterns in the wind field. However, large difference in the $DQ$ trend is noted. The $DQ$ trend difference is attributed to a reduction in the negative trend in surface air humidity ($Q_a$) in Set2.

A major difference between GSSTF2 and GSSTF2b is the use of RSS SSM/I V4 for GSSTF2 and RSS SSM/I V6 for GSSTF2b. The changes in LHF, $U$, and $DQ$ are 16%, 6%, and 11% for GSSTF2 (Xing, 2006). The corresponding changes are 23%, 3%, and 20% for GSSTF2b Set1 and 16%, 3%, and 12% for Set2 (Table 3). The use of RSS SSM/I V6 products reduces the wind trend from 6% to 3% but increases the $DQ$ trend from 11% to 20% for Set1. The exclusion of F13 and F15 data reduces the LHF trend to 16%, mostly due to a reduction in the $DQ$ trend to 12% (from 20%) with $U$ changes remain at 3%.

Interannual variability is examined using EOF analyses. The first three significant non-seasonal EOF patterns are similar, and they explaining 10.5%, 4.3% and 3.4% for Set1 and 8.6%, 4.3% and 3.4% for Set2, respectively. The first EOF pattern of GSSTF2 for 1998-2000, with opposite changes between the equatorial eastern Pacific and the subtropics in the Pacific and Indian ocean, may be indicative of an enhance Hadley circulation (Chiu & Xing, 2004; Chiu et al., 2008). Observations also indicate large decadal variability in the Hadley Circulation (Wielicki et al., 2002; Cess & Udelhofen, 2003; Chen et al., 2002; Mitas and Clements, 2005).

This seesaw pattern is much reduced in the EOF1 pattern in both GSSTF2b Set1 and Set2 of 1998-2008, which may indicate a reduction, change of phase, or mixing of the signal with the trend in GSSTF2b. The contribution to the total variance is smaller for Set2, which excluded DMSP datasets that contains large long-term trends introduced by drifts in the Earth incidence angle in the SSM/I sensors. The difference in the fraction of variance explained in Set1 and Set2 is attributed to the artificial trend in F13 and F15 and the EIA drift effect.

Examination of the trends of the zonal means show that the latitude of maximum increase, situated in the subtropics, is found poleward of the LHF maximum in the tropic. This pattern is consistent with the expansion of the Hadley Circulation associated with global warming as predicted in climate models (Lu et al., 2007).

The EOF2 patterns of Set1 and Set2 are almost identical, both contributed to 4.3% of the variance of the dataset. The association with the El Nino/Southern Oscillation phenomena is corroborated by a high correlation between their time series and an index of the Southern Oscillation (SOI). The patterns for EOF3 and their associated time series are also similar, indicating that both GSSTF2b Set1 and Set2 are useful for examining interannual variability.

## 5. Acknowledgment

This study is supported by the MEaSUREs Program of NASA Science Mission Directorate-Earth Science Division. The authors are especially grateful to their program manager M. Maiden and program scientist J. Entin for their valuable supports of this research.

# 6. References

Andersson, A.; Fennig, K.; Klepp, C.; Bakan, S.; Graßl, H. & Schulz, J. (2010). The Hamburg Ocean Atmosphere Parameters and Fluxes from Satellite Data—HOAPS-3. *Earth System Science Data*, Vol. 2, No. 1, pp. 215-234, ISSN 1866-3508

Angell, J. K. (1981). Comparison of Variations in Atmospheric Quantities with Sea Surface Temperature Variations in the Equatorial Eastern Pacific. *Monthly Weather Review*, Vol. 109, No. 4 , pp. 230-243, ISSN 0027-0644

Aubaum, M. H. P.; Hoskins, B. J. & Stephenson, D. B. (2001). Arctic Oscillation or North Atlantic Oscillation. *Journal of Climate*, Vol. 14, No. 16, pp. 3495-3507, ISSN 0894-8755

Atlas, R.; Hoffman, R. N.; Bloom, S. C.; Jusem, J. C. & Ardizzone, J. (1996). A multiyear global surface wind velocity dataset using SSM/I wind observations. *Bulletin of the American Meteorological Society*, Vol. 77, No. 5, pp. 869-882, ISSN 0003-0007

Bentamy, A.; Katsaros, K. B.; Mestas-Nuñez, A. M; Drennan, W. M.; Forder, E. B. & Roquet, H. (2003). Satellite estimates of wind speed and latent heat flux over the global oceans. *Journal of Climate*, Vol. 16, No. 4, pp. 637-656, ISSN 0894-8755

Bunker, A. F. (1976). Computations of surface energy flux and annual air-sea interaction cycle of the north Atlantic Ocean, *Monthly Weather Review*, Vol. 104, No. 9, pp. 1122-1140, ISSN 0027-0644

Cayan, D. R. (1992). Latent and sensible heat flux anomalies over the Northern Oceans: The connection to monthly atmospheric circulation. *Journal of Climate*, Vol. 5, No. 4, pp. 354-369, ISSN 0894-8755

Cess, R. D. & Udelhofen, P. M. (2003). Climate change during 1985–1999: Cloud interactions determined from satellite measurements. *Geophysical Research Letters*, Vol. 30, No. 1, 1019, ISSN 0094-8276

Chen, J., Carlson B. E., & Del Genio A. D. (2002). Evidence for strengthening of the tropical general circulation in the 1990s. *Science*, Vol. 295, No. 5556, pp. 838-841, ISSN 0036-8075

Chiu, L. S. & Xing, Y. (2004). Modes of interannual variability of oceanic evaporation observed from GSSTF2. *Gayana: International Journal of Biodiversity, Oceanology and Conservation*, Vol. 68, No. 2, pp. 115-120. ISSN 0717-652X

Chiu, L. S. & Newell, R. E. (1983). Variations of zonal mean sea surface temperature and large-scale air-sea interaction. *Quarterly Journal of the Royal Meteorological Society*, Vol. 109, No. 459, pp. 153-168, ISSN 0035-9009

Chiu, L. S.; Chokngamwong, R.; Xing, Y.; Yang, R. & Shie, C.-L. (2008). Trends and Variations of Global Oceanic Evaporation Datasets from Remote Sensing. *Acta Oceanologica Sinica*, Vol. 27, No. 3, pp. 1-12, ISSN 0253-505X

Chou, S.-H. (1993). A comparison of airborne eddy correlation and bulk aerodynamic methods for ocean-air turbulent fluxes during cold-air outbreaks. *Boundary-Layer Meteorology*, Vol. 64, No. 1-2, pp. 75-100, ISSN 0006-8314

Chou, S.-H.; Shie, C.-L.; Atlas R. M. & Ardizzone, J. (1997). Air-sea fluxes retrieved from special sensor microwave imager data. *Journal of Geophysical Research*, Vol. 102, No. C6, pp. 12706-12726, ISSN 0148-0227

Chou, S.-H., Atlas R. M., Shie C.-L. & Ardizzone J. (1995). Estimates of surface humidity and latent heat fluxes over oceans from SSM/I data. *Monthly Weather Review*, Vol. 123, No. 8, pp. 2405-2425, ISSN 0027-0644

Chou, S.-H.; Nelkin, E.; Ardizzone, J.; Atlas, R. & Shie, C.-L. (2001). The Goddard Satellite-Based Surface Turbulent Fluxes Dataset Version 2 (GSSTF2.0) [global (grid of 1° x 1°) daily air-sea surface fluxes from July 1987 to December 2000]. Available from <http://disc.gsfc.nasa.gov/precipitation/gsstf2.0.shtml>

Chou, S.-H.; Nelkin, E.; Ardizzone, J.; Atlas, R. M. & Shie, C.-L. (2003). Surface turbulent heat and momentum fluxes over global oceans based on the Goddard satellite retrieval, version 2 (GSSTF2). *Journal of Climate*, Vol. 16, No. 20, pp. 3256-3273, ISSN 0894-8755

da Silva, A. M.; Young, C. C. & Levitus, S.(1994). *Atlas of Surface Marine Data. Vol. 3: Anomalies of Heat and Momentum Fluxes*. NOAA/NESDIS, Washington, D.C., USA

Deser, C. (2000). On the teleconnectivity of the "Arctic oscillation". *Geophysical Research Letters*, Vol. 27, No. 6, pp. 779-782, ISSN 0094-8276

Esbensen, S. K. & Kushnir,V. (1981). *The heat budget of the global oceans: An atlas based on estimates from marine surface observations*. Climatic Research Institute, Report No. 29, Oregon State University, 27 pp

Fairall, C. W.; Bradley, E. F.; Rogers, D. P.; Edson, J. B. & Young, G. S. (1996). Bulk parameterization of air-sea fluxes for Tropical Ocean Global Atmosphere Coupled Ocean-Atmosphere Response Experiment. *Journal of Geophysical Research*, Vol. 101, No. C2, pp. 3747-3764, ISSN 0148-0227

Fairall, C. W.; Bradley, E. F.; Hare, J. E.; Grachev, A. A. & Edson, J. B. (2003). Bulk parameterization of air-sea fluxes: Updates and verification for the COARE algorithm. *Journal of Climate*, Vol. 16, No. 4, pp. 571-591 , ISSN 0894-8755

Grassl, H.; Jost, V.; Kumar, R.; Schulz, J.; Bauer P. & Schluessel, P. (2000). *The Hamburg Ocean-Atmosphere Parameters and Fluxes from Satellite Data (HOAPS): A climatological atlas of satellite-derived air-sea interaction parameters over oceans*. Max Planck Institute for Meteorology, Report No. 312, ISSN 0937-1060, Hamburg, Germany

Hastenrath, S. (1980). Heat budget of tropical ocean and atmosphere. *Journal of Physical Oceanography*, Vol. 10, No. 2, pp. 159-170, ISSN 0022-3670

Hilburn, K. A. & Shie, C.-L. (2011). Decadal trends and variability in Special Sensor Microwave Imager (SSM/I) brightness temperatures and Earth incidence angle. Report No. 092811, Remote Sensing Systems, 53 pp

Huang, N. E.; Shen, Z.; Long, S. R.; Wu, M. C.; Shih, H. H.; Zheng, Q.; Yen, N.-C.; Tung, C. C. & Liu, H. H. (1998). The empirical mode decomposition and the Hilbert spectrum for nonlinear and non-stationary time series analysis. *Proceedings of the Royal Society of London. Series A*, Vol. 454, No. 1971, pp. 903-995, ISSN 1364-5021

Hsiung, J. (1985). Estimates of global oceanic meridional heat transport. *Journal of Physical Oceanography*, Vol. 15, No. 11, pp. 1405-1413, ISSN 0022-3670

Hurrell, J. W.; Kushnir, Y.; Ottersen, G. & Visbeck, M. (Eds.). (2003). *The North Atlantic Oscillation-Climatic Significance and Environmental Impact*, Geophysical Monograph Vol. 134, American Geophysical Union, ISBN 0-875-90994-9, Washington D.C., USA

Isemer, H.-J. & Hasse, L. (1985). *The Bunker Climate Atlas of the North Atlantic Ocean: 1. Observations*. Springer-Verlag, ISBN 0-387-15568-6, New York, USA

Isemer, H.-J. & Hasse, L. (1987). *The Bunker climate atlas of the North Atlantic Ocean: 2. Air-sea interactions*. Springer-Verlag, ISBN 0-387-17594-6, New York, USA

Josey, S. A.; Kent,E. C. & Taylor,P. K. (1998). *The Southampton Oceanography Centre (SOC) Ocean-Atmosphere Heat, Momentum and Freshwater Flux Atlas*. Southampton Oceanography Centre, Report No.6, Southampton, United Kingdom

Kubota, M.; Ichikawa, K.; Iwasaka, N.; Kizu, S.; Konda, M. & Kutsuwada, K. (2002). Japanese Ocean Flux Data Sets with Use of Remote Sensing Observations (J-OFURO). *Journal of Oceanography*, Vol. 58, No. 1, pp. 213-225, ISSN 0916-8370

Liu, W. T.; Katsaros, K. B. & Businger,J. A. (1979). Bulk Parameterizations of Air-Sea Exchanges of Heat and Water Vapor Including Molecular Constraints at the Interface. *Journal of Atmospheric Science*, Vol. 36, No. 9, pp. 1722-1735, ISSN 1520-0469

Lu, J.; Vecchi, G. & Reichler, T. (2007). The expansion of the Hadley Cell under global warming. *Geophysical Research Letters*, Vol. 34, L06805, ISSN 0094-8276

Murray, F. W. (1967). On the computation of saturation vapor pressure. *Journal of Applied Meteorology*, Vol. 6, No. 1, pp. 203-204, ISSN 0021-8952

Mitas, C. M. & Clement, A. (2005). Has the Hadley cell been strengthening in recent decades?. *Geophysical Research Letters*, Vol. 32, L03809, ISSN 0094-8276

North, G. R.; Bell, T. L.; Cahalan, R. F. & Moeng, F. J. (1982). Sampling Errors in the Estimation of Empirical Orthogonal Functions. *Monthly Weather Review*, Vol. 110, No. 7, pp. 699-706, ISSN 0027-0644

Oberhuber, J. M. (1988). *An Atlas Based on the 'COADS' Data Set: The Budgets of Heat, Buoyancy and Turbulent Kinetic Energy at the Surface of the Global Ocean*. Max-Planck-Institut für Meteorologie, Report No. 15, Hamburg, Germany

Renfrew, I. A.; Moore, G. W. K.; Guest, P. S. & Bumke, K. (2002). A comparison of surface layer and surface turbulent flux observations over the Labrador Sea with ECMWF analyses and NCEP reanalyses. *Journal of Physical Oceanography*, Vol. 32, No. 2, pp. 383-400, ISSN 0022-3670

Reynolds, R. W. & Smith, T. S. (1994). Improved global sea surface temperature analyses using optimum interpolation. *Journal of Climate*, Vol. 7, No. 6, pp. 929-948, ISSN 0894-8755

Reynolds, W. R.; Rayner, N. A.; Smoth, T. M.; Stokes, D. C. & Wang, W. (2002). An improved in situ and satellite SST analysis for climate. *Journal of Climate*, Vol. 15, No. 13, pp. 1609-1625, ISSN 0894-8755

Sakurai, T.; Kurihara, Y. & Kuragano, T. (2005). Merged satellite and in-situ data global daily SST. *Procceeding of International Geoscience and Remote Sensing Symposium*, pp. 2606-2608, ISBN 0-7803-9050-4, Seoul, South Korea, November 14, 2005

Schlussel, P.; Schanz, L. & Englisch, G. (1995). Retrieval of latent-heat flux and longwave irradiance at the sea-surface from SSM/I and AVHRR measurements. *Advances in Space Research*, Vol. 16, No. 10, pp. 107-116, ISSN 0273-1177

Shie, C.-L.; Chiu, L. S.; Adler, R.; Nelkin, E.; Lin, I-I; Xie, P.; Wang, F-C; Chokngamwong, R.; Olson, W. & Chu, A. D. (2009). A note on reviving the Goddard Satellite-based Surface Turbulent Fluxes (GSSTF) dataset. *Advances in Atmospheric Sciences*, Vol. 26, No. 6, pp. 1071-1080, ISSN 0256-1530

Shie, C.-L.; Chiu, L. S.; Adler, R.; Lin, I-I; Nelkin, E. & Ardizzone, J. (2010). The Goddard Satellite-Based Surface Turbulent Fluxes Dataset --- Version 2b (GSSTF2b), In: *NASA Goddard Earth Sciences (GES) Data and Information Services Center (DISC)*, October 2010, Available from
<ftp://auraparlu.ecs.nasa.gov/data/s4pa/GSSTF/> or
<http://disc.sci.gsfc.nasa.gov/daac-bin/DataHoldingsMEASURES.pl?PROGRAM_List=ChungLinShie>

Shie, C.-L. (2010). Science background for the reprocessing and Goddard Satellite-based Surface Turbulent Fluxes (GSSTF2b) Data Set for Global Water and Energy Cycle Research, In: *NASA GES DISC*, 18 pp, October 12, 2010, Available from < http://disc.sci.gsfc.nasa.gov/measures/documentation/Science-of-the-data.pdf>

Shie, C.-L. & Hilburn, K. (2011). A satellite-based global ocean surface turbulent fluxes dataset and the impact of the associated SSM/I brightness temperature, *Proceedings of The 2011 EUMETSAT Meteorological Satellite Conference*, Oslo, Norway, September 5-9, 2011

Shie, C.-L.; Hilburn, K. A.; Chiu, L. S.; Adler, R.; Lin, I-I; Nelkin, E. & Ardizzone, J. (2011). The Goddard Satellite-Based Surface Turbulent Fluxes Dataset --- Version 2c (GSSTF2c). In: *NASA GES DISC*, by the end of October 2011.

Tomita, H.; Kubota, M.; Cronin, M. F.; Iwasaki, S.; Konda, M. & Ichikawa, H. (2010). An assessment of surface heat fluxes from J-OFURO2 at the KEO and JKEO sites. *Journal of Geophysical Reserach*, Vol. 115, C03018, ISSN 0148-0227

Walker, G. T. & Bliss, E. W. (1932). World weather V. *Memoirs of the Royal Meteorological Society*, Vol. 4, No. 36, pp. 53-84, LCCN 7021-1603

Wallace, J. M. (2000). North Atlantic Oscillation/annular mode: Two paradigms—one phenomenon. *Quarterly Journal of the Royal Meteorological Society*, Vol. 126, No. 564, pp. 791-805, ISSN 0035-9009

Wallace, M. & Gutzler, D. S. (1981). Teleconnections in the geopotential height field during the Northern Hemisphere winter. *Monthly Weather Review*, Vol. 109, No. 4, pp. 784-812, ISSN 0027-0644

Weare, B. C.; Strub, P. T. & Samuel, M. D. (1981). Annual Mean Surface Heat Fluxes in the Tropical Pacific Ocean. *Journal of Physical Oceanography*, Vol. 11, No. 5, pp. 705-717, ISSN 0022-3670

Wentz, F. J. (1997). A well calibrated ocean algorithm for Special Senor Microwave/Imager. *Journal of Geophysical Research*, Vol. 102, No. C4, pp. 8703-8718, ISSN 0148-0227

Wentz, F. J.; Ricciardulli, L.; Hilburn K. A. & Mears C. A. (2007). How Much More Rain Will Global Warming Bring? *Science*, Vol. 317, No. 5835, pp. 233-235, Supporting material, ISSN 0036-8075

Wielicki B.; Wong, T.; Allen, R. P.; Slingo, A.; Kiehl, J. T.; Soden, B. J.; Gordon, C. T.; Miller, A. J.; Yang, S.-K.; Randall, D. A.; Robertson, F.; Susskind, J. & Jacobowitz, H. (2002). Evidence for large decadal variability in the tropical mean radiative energy budget. *Science*, Vol. 295, No. 5556, pp. 841-844, ISSN 0036-8075

Xing, Y. (2006). *Recent changes in oceanic latent heat flux from remote sensing*. Ph.D. dissertation, George Mason University, 119 pp

Zeng, X.; Zhao, M. & Dickinson, R. E., (1998). Intercomparison of bulk aerodynamic algorithms for the computation of sea surface fluxes using TOGA COARE and TAO data. *Journal of Climate*, Vol. 11, No. 10, pp. 2628-2644, ISSN 0894-8755

# 10

# Coupled Terrestrial Carbon and Water Dynamics in Terrestrial Ecosystems: Contributions of Remote Sensing

Baozhang Chen

*Institute of Geographic Sciences and Nature Resources Research,*
*Chinese Academy of Sciences, Beijing*
*P.R. China*

## 1. Introduction

The Earth climate is a complex, interactive system, determined by a number of complex connected physical, chemical and biological processes occurring in the atmosphere, land and ocean. The terrestrial biosphere plays many pivotal roles in the coupled Earth system providing both positive and negative feedbacks to climate change (Treut et al., 2007). Terrestrial vegetation via photosynthesis converts solar energy into carbon that would otherwise reside in the atmosphere as a greenhouse gas, thereby regulating climate. Vegetation also transfers water between belowground reservoirs and the atmosphere to maintain precipitation and surface water flows.The terrestrial carbon (C) cycle is closely linked to hydrological and nutrient controls on vegetation (Betts et al., 2000; Cox et al., 2000). Understanding the coupled terrestrial C and water cycle is required to gain a comprehensive understanding of the role that terrestrial ecosystems play in the global climate change. Much progress has been made in gaining insight of the coupling processes between C and water cycles across a range of time and spatial scales (Pielke Sr, 2001; Friedlingstein et al., 2003; Seneviratne et al., 2006; Betts et al, 2007a,b; Baldocchi, 2008). Since the early 1990s, there has been an increased interest in monitoring of the $CO_2$, water vapor and energy exchange between the atmosphere and terrestrial ecosystems by a variety of methods, such as the eddy-covariance techniques (EC), satellite and other airborne remote sensing, $CO_2$ concentration and isotope measurements. Meanwhile, there are various kinds of models have been developed to better understanding of these processes and for large-scale C and water budgeting.

Remote sensing (RS) from satellite and airborne platforms, along with many other sources of land ground-based measurements (e.g., eddy covariance flux tower network, biometric plots, radar network, etc.) is playing and will continue to play a vital role in better understanding the coupled C and water cycle. Satellite RS allows the study of ecosystems from a completely new vantage point, facilitating a holistic perspective like viewing the Earth does for astronauts. Satellite-borne RS offers unique opportunities to parameterize land surface characteristics over large spatial extents at variable spatial and temporal resolutions. While there are challenges relating RS data recorded in radiance or backscatter to variables of interest, and RS has poor temporal resolution compared to ground-based measurement devices, RS and spatial analytical techniques and distributed biogeochemical modeling embedded in Geographical Information Systems (GIS) have allowed us to better understand the coupled C and hydrological dynamics across a large range of temporal and spatial scales.

The large number of papers published since the 1980s on the terrestrial and C/water cycles have resulted in the publication of several major reviews from different perspectives. For example, Running et al. (2004) described a blueprint for more comprehensive coordination of the various flux measurement and modeling activities into a global terrestrial monitoring network by reviewing the literature published before the middle of 1990s. Baldocchi (2008) recently provided a comprehensive review of research results associated with a global network of C flux measurements systems. The topics discussed by this review include history of the network, errors and issues related with the EC method, and a synopsis of how these data are being used by ecosystem and climate modellers and the remote-sensing community (Baldocchi, 2008). Kalma et al. (2008) reviewed satellite-based algorithms for estimating evepotranspiration (ET) and land surface temperatures at local, regional and continental scales, with particular emphasis on studies published since the early 1990s; while Verstraeten et al. (2008) provided a comprehensive review of remote sensing methods for assessing ET and soil moisture content across different scales based on the literature published after 1990s. Marquis and Tans (2008) reviewed satellite-based instruments on $CO_2$ concentration measurements.

In this chapter, I distil and synthesise the rapidly growing literature on C and water cycles using remote sensing in direct or indirect ways across local to global spatial scales and over a range of time scales. To give the reader a perspective of the growth of this literature, a search of Web of Science produced over 1500 papers with the key words 'ecosystem carbon, water cycles and remote sensing' published since 1990 which is indicative of the large amount of research recently being undertaken on these topics. In order to filter through this large body of literature, I concentrate on papers discussing on the coupling processes between C and water and I extract information from a database of published results that I have collated during the past decade (available on request). In terms of content, the report covers the state of knowledge, monitoring and modeling of the coupled terrestrial C and water cycles. My aim is to highlight the recent advances in this field, and propose areas of future research based on perceived current gaps in the literature.

This is a synthesis of state-of-the-art research on how RS has informed the study of coupled C and hydrology cycles. The review is divided into several inter-connected sections. First, I review the scientific background of the linkage between terrestrial ecosystems and climate, and revise the state of knowledge on terrestrial C cycling, coupling of the C and water cycles. Second, I discuss the ground-based and satellite-based monitoring methods and observation networks associated with measuring C and water fluxes, $CO_2$ concentration and C isotopes. Third, I report on the recent advances in modeling approaches associated with the terrestrial biochemical and hydrological studies. Fourth, I discuss research gaps in C sinks/sources estimates and finally, I discuss the current research trends and the near-future directions in this field and propose an upscaling framework for landscape and regional C and water fluxes estimates.

## 2. Scientific background and state of knowledge

### 2.1 Overview of terrestrial ecosystems and climate

The climate system is controlled by a number of complex coupled physical, chemical and biological processes (Figure 1). The terrestrial biosphere plays a crucial role in the climate system, providing both positive and negative feedbacks to climate change through biogeophysical and biogeochemical processes (Treut et al., 2007). Couplings between the climate system and biogeochemistry are mainly through tightly linked dynamics of C and water cycles. The importance of coupled C and water dynamics for the climate system has been increasingly recognized (Cox et al., 2000; Pielke Sr, 2001; Friedlingstein et al., 2003; Seneviratne et al., 2006; Betts et al, 2000, 2007a,b); however the mechanisms behind these coupled cycles are still far from well understood.

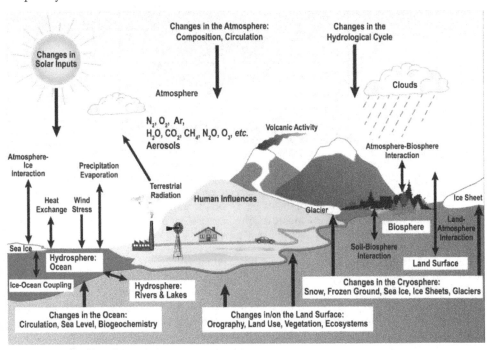

Fig. 1. Schematic view of the components of the climate system, their processes and interactions (Treut et al., 2007).

### 2.2 Terrestrial C cycling

One of the crucial issues in the prognosis of future climate change is the global budget of atmospheric $CO_2$. The growth rate of atmospheric $CO_2$ is increasing rapidly. Three processes contribute to this rapid increase: fossil fuel emission, land use change (deforestation), and ocean and terrestrial uptake. As shown in Figure 2, terrestrial C budgets have large uncertainties and interannual variability.

Terrestrial ecosystems mediate a large part of $CO_2$ flux between the Earth's surface and the atmosphere, with ~120 Pg C yr[-1] taken up by photosynthesis and roughly the same amount

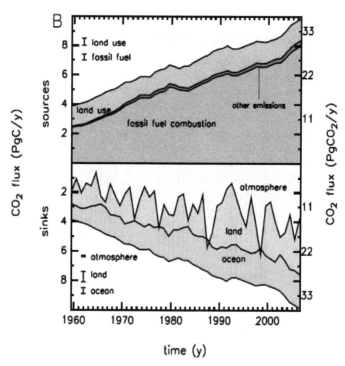

Fig. 2. Global $CO_2$ budget from 1959 to 2006. *Upper* panel: $CO_2$ emissions to the atmosphere (sources) as the sum of fossil fuel combustion, land-use change, and other emissions. *Lower* panel: The fate of the emitted $CO_2$, including the increase in atmospheric $CO_2$ plus the sinks of $CO_2$ on land and in the ocean (Canadell et al., 2007).

released back to the atmosphere by respiration annually (Treut et al., 2007; Prentice et al., 2001). Imbalances between gross ecosystem photosynthesis or gross primary productivity (GPP) and ecosystem respiration ($R_e$) lead to land surfaces being either $CO_2$ sinks or sources. The magnitudes of sinks and sources have fluctuated on annual and longer time scales due to variable climate, land use change, disturbance, and changes in the age distribution and species composition of ecosystems (Battle et al., 2000; Arain et al., 2002; Law et al., 2002; Morgenstern et al., 2004; Humphreys et al., 2005, 2006; Urbanski et al., 2007). Terrestrial ecosystems modify atmospheric C balance through many mechanisms. A detailed understanding of the interactive relationships in atmosphere–biosphere exchange is relevant to ecosystem-scale analysis and is needed to improve our knowledge of the global C cycle (Falk et al., 2008).

In recent years, scientists have learnt that terrestrial ecosystems' vegetation, soil (Melillo et al., 1989; Knapp et al., 1993) and animals (Naeem et al., 1995; Hattenschwiler and Bretscher, 2001) play key roles in mediating the terrestrial C cycle. Plants being the primary producers, it is from them that mass and energy gets transformed to other living organisms (Engel and Odum, 1999) within an ecosystem. The process of photosynthesis fixes atmospheric C into the biosphere. Atmospheric $CO_2$ enters the plant through stomatal opening that is controlled by a variety of environmental factors (Jarvis, 1976; Griffis et al., 2003). These factors include ambient temperature, atmospheric $CO_2$ concentration, nutrient availability, soil water availability and forest age (Schimel, 1995; Prentice et al., 2001). Changes in the atmospheric $CO_2$ concentration and the corresponding changes in the climate have altered the magnitudes of terrestrial C cycling. For example, a climate change induced increases in vegetation growth due to earlier springs and lengthened growing seasons were detected by the phase shift of seasonal atmospheric $CO_2$ cycle by Keeling et al. (1996) and satellite-based vegetation index analysis by Myneni et al. (1997). Studies indicate that an increase in atmospheric $CO_2$ enhances photosynthesis (e.g. Woodward and Friend, 1988) and hence increases assimilation of atmospheric $CO_2$ by the terrestrial vegetation. Nitrogen (N) availability to plants is another factor that can affect photosynthesis. This is because N is a primary nutrient for plant growth. In the recent years, variations in plant N availability have also altered the trends in the terrestrial C cycles. Variations in plant N availability occur mainly due to natural and anthropogenic N-deposition. Based on modeling studies, e.g., researchers (Townsend et al., 1996; Asner et al., 1997; and Holland et al., 1997) have demonstrated that N deposition is responsible for about 0.1-2.3 PgC yr$^{-1}$ fixed by terrestrial vegetation which is almost half of the magnitude of C flux due to fossil fuel emission. Another factor that determines the nature of terrestrial C balance of an ecosystem is the age of the vegetation. Schimel et al. (1995) have demonstrated that forest re-growth can account for part of terrestrial C uptake as much as 0.5 $\pm$ 0.5 PgC yr$^{-1}$, especially in northern mid and high latitudes. This is because younger vegetation actively grows and hence sequesters more atmospheric $CO_2$ as opposed to mature forest stands. There are many other processes that directly and indirectly affect photosynthesis and thus, the C cycle. They include land use and land cover change (Caspersen et al., 2000; Houghton and Hackler, 2006; Easter et al., 2007), reforestation (House et al., 2002; Paul et al., 2002), agricultural and grazing activities (Cerri et al., 2005), insect attack (Chapman et al., 2003; Throop et al., 2004) and invasive species (Szlavecz et al., 2006). Respiration is a process by which C is added to the atmosphere from the biosphere. There are studies that indicate that total ecosystem respiration is a major determinant of terrestrial C balances (Valentini et al., 2000). Total ecosystem respiration includes respiration by aboveground plant parts (boles, branches, twigs, and leaves) and soil respiration, which is the sum of the heterotrophic respiration, and root respiration including respiration of symbiotic microorganisms. The temporal variability of respiratory metabolism is influenced mostly by temperature and humidity conditions (Davidson and Janssens, 2006). Although ecosystem respiration has received considerable attention in recent decades, much less is known about the relative contributions of its sub- components (Jassal et al., 2007), and our understanding of how they will respond to global warming is poor. Soil respiration (root + heterotrophic respiration) is a dominant component of C exchange in terrestrial ecosystems which accounts for more than half of the total ecosystem respiration (Black et al. 2005). This is because soils of terrestrial ecosystems contain more C than the atmosphere and live biomass together (Eswaran et al., 1993). Components of respiration can have different responses to temperature and soil water content (Boone et al., 1998; Lavigne et al., 2004), thus the effects of these environmental controls needs to be understood in order to fully comprehend the soil C cycling mechanism. There are many other mechanisms that can release terrestrial C to atmosphere. This includes both natural and anthropogenic reasons. Emission of large amounts of C to the atmosphere from vegetation can occur during forest fires (Amiro et al., 2002; Soja et al., 2004; Amiro et al., 2004) or biomass burning (Fernandez et al., 1999; Tanaka et al., 2001). These C emissions are of very high magnitudes although their duration is very short. Forest fires and biomass burning also affect the nutrient status of the soil which could have positive effects on the succeeding vegetation (Prietofernandez et al., 1993; Deluca and Sala, 2006). Another form of C flux in almost all terrestrial ecosystems is the import and export of dissolved organic carbon (DOC) (Neff and Asner 2001; Hornberger et al. 1994). DOC fluxes include C in the form of simple amino acids to large molecules that are transported through water flows. Fluxes of DOC into the ocean via runoff from terrestrial ecosystems are estimated to be 0.2 (Harrison et al., 2005) to 0.4 Pg C per year (IPCC, 2001). Since these fluxes are very small compared to the C fluxes due ecosystem is a net C sink due to the presence of soil C-pools having much longer residence times (Thompson et al., 1996; Chen et al., 2003; Canadell et al., 2007; Schulze, 2006). The strength of the terrestrial C sink was estimated to be 0.5-2.0 Pg C yr$^{-1}$ (Schimel et al., 1995). By sequestering atmospheric C, the terrestrial ecosystems help decrease the rate of accumulation of anthropogenic $CO_2$ in the atmosphere, and its associated climate change (Cihlar, 2007). Terrestrial C sinks may be responsible for taking up about one-third of all the $CO_2$ that is released into the atmosphere (Canadell et al., 2007). The terrestrial C sink, inferred based on our current understanding, may not be permanent (Luo et al., 2003; Cox et al., 2000 ; Friedlingstein et al., 2003). Over the last few years there have been several studies suggesting that the size of this terrestrial C sink is vulnerable to global warming (Martin et al., 1998; Nemani et al., 2002; Canadell et al., 2007). The metabolism of terrestrial ecosystems is complex and highly dynamic because ecosystems consist of coupled, non-linear processes that possess many positive and negative feedbacks (Levin, 2002; Ma et al. 2007). How the C budget of major ecosystems will respond to changes in climate is not quantitatively well understood (Baldocchi & Meyers 1998, Goulden et al., 1998; Black et al., 2000; Baldocchi et al., 2001a; Baldocchi & Wilson, 2001; Law et al., 2002; Barr et al., 2004, 2007). A detailed understanding of the interactive relationships in atmosphere-biosphere exchange is relevant to ecosystem-scale analysis and is needed to improve our knowledge of the global C cycle (Falk,M et al., 2008). The metabolism of terrestrial ecosystems is complex and highly dynamic because ecosystems consist of coupled, non-linear processes that possess many positive and negative feedbacks (Levin et al., 2002; Ma et al., 2007). Complex features of ecosystem metabolism are relatively unknown and how C budget of major ecosystems will respond to changes in climate is not quantitatively well understood (Black et al., 2000; Baldocchi et al., 2001; Baldocchi et al.,2001; Barr et al., 2004; Law et al., 2002).

## 2.3 Terrestrial water cycling

Most of the Earth is covered by water, amounting to more than one billion km$^3$. The vast majority of that water, however, is in forms unavailable to land-based or freshwater ecosystems. Less than 3 percent is fresh enough to drink or to irrigate crops, and of that total, more than two-thirds is locked up in glaciers and ice caps. Freshwater lakes and rivers hold 100,000 km$^3$ globally, less than one ten-thousandth of all water on earth (Jackson et al, 2001).

Water vapor in the atmosphere exerts an important influence on climate and on the water cycle, even though only 15,000 km$^3$ of water is typically held in the atmosphere at any time. This tiny fraction, however, is vital for the biosphere. Water vapor is the most important of the so-called greenhouse gases (others include $CO_2$, $CH_4$ and $N_2O$) that warm the Earth by trapping heat in the atmosphere. Water vapor contributes approximately two-thirds of the total warming that greenhouse gases supply. Without these gases, the mean surface temperature of the earth would be well below freezing, and liquid water would be absent over much of the planet. Equally important for life, atmospheric water turns over every ten days or so as water vapor condenses and falls as rain to the Earth and the heat of the Sun evaporates new supplies of vapor from the liquid reservoirs on earth. Solar energy typically evaporates about 425,000 km³ of ocean water each year. Most of this water returns back directly to the oceans as precipitation, but approximately 10% falls on land. If this were the only source of rainfall, average precipitation across the earth's land surfaces would be only 25 cm a year, a value typical for deserts or semi-arid regions. Instead, a second, larger source of water is recycled from plants and the soil through evapotranspiration. The water vapor from this source creates a direct feedback between the land surface and regional climate. This second source of recycled water contributes two-thirds of the 70 cm of precipitation that falls over land each year. Taken together, these two sources account for the 110,000 km³ of renewable freshwater available each year for terrestrial, freshwater, and estuarine ecosystems. Because the amount of rain that falls on land is greater than the amount of water that evaporates from it, the extra 40,000 km³ of water returns to the oceans, primarily via rivers and underground aquifers. A number of factors affect how much of this water is available for human use on its journey to the oceans. These factors include whether the precipitation falls as rain or snow, the timing of precipitation relative to patterns of seasonal temperature and sunlight, and the regional topography. For example, in many mountain regions, most precipitation falls as snow during winter, and spring snowmelt causes peak flows that flood major river systems. In other regions, excess precipitation percolates into the soil to recharge ground water or is stored in wetlands.

## 2.4 Coupling of the C and water cycles

The cycling of other materials such as C and N is strongly coupled to this water flux through the patterns of plant growth and microbial decomposition, and this coupling creates additional feedbacks between vegetation and climate. Thermodynamically, a terrestrial ecosystem is an open system. Therefore, hydrological and C cycles are closely coupled at various temporal and spatial scales (Betts, 2007; Ball et al., 1987; Levis et al., 1999; Rodriguez-Iturbe, 2000; Joos, 2001; Arain et al., 2006; Blanken and Black, 2004; Snyder 2004). C uptake for example, is closely coupled to water loss by ecosystems mainly through leaf stomatal pathway governed principally through leaf conductance (Jarvis, 1976; Harris et al., 2004; Rodriguez-Iturbe, 2001). Soil organic C decomposition is very sensitive to soil moisture content via microbial activity and other processes (Betts, 2007; Levis 1999; Snyder et al., 2004; Parton et al., 1993; D'odorico 2004). The flux of terrestrial organic C by river runoff to the ocean and wetland discharge is an important component of the global organic C cycle (Hedges, 1992; Wang et al., 2004). It is estimated that $0.25 \times 10^{15}$ g dissolved organic carbon (DOC) is discharged to the ocean by the world rivers each year (Meybeck, 1982). The land surface hydrological processes (in particular the terrestrial river systems) play an important role in transport of dissolved and particulate organic C from terrestrial to marine ecosystems (Wang et al., 2004). However, the interactions between C and water cycles and the mechanisms how these interactions will shape future climatic and biosphere conditions are far from well understood.

## 3. The array of airborne and satellite sensors developed for monitoring of the coupled C and water cycles

### 3.1 Satellite monitoring

RS is the observation of a phenomenon from a distance, using devices that detect electromagnetic radiation. Satellite-borne remote sensing offers unique opportunities to parameterize land surface characteristics over large spatial extents at variable spatial and temporal resolutions. There has been a substantial increase in the number of satellite sensors for Earth observations that cover a large range of the electromagnetic radiation spectrum (Tables 1 and 2) since 1960s when the earlier Landsat satellites were launched into orbit, such as the Television Infrared Observation Satellite (TIROS-1) launched in 1960. None of these sensors have been designed exclusively for C, water or vegetation applications. For example, the TIROS-1 was focused on weather analysis and forecasting (Natl. Res. Counc., 2008). However, scientists were applying these observations to vegetation studies by the next decade (Rouse et al., 1974; Tucker et al., 1979). Tuker et al. (1986) exploited the properties of chlorophyll pigments to absorb wavelengths in the red spectral region and structural properties of leaves to reflect near-infrared spectra based on the imagery data obtained by the Advanced Very High Resolution Radiometer (AVHRR) sensor onboard TIROS. This pioneer study that synoptic view of the coupled atmosphere-biosphere as C sequestration by photosynthesis from the atmosphere in the Northern Hemisphere (Tuker et al., 1986) opened possibilities for global perspectives in ecology. The first Landsat satellite launched in 1972 carried the Multispectral Scanner System (MSS) sensors which were specifically designed to map land resources with finer spatial resolution (68 m × 82 m) than the AVHRR. The program was the first civil, non-weather satellite program and Landsat provided observations for any place on Earth once every 18 days, offering a wide range of studies on terrestrial vegetation and C and water cycles. The Landsat Thematic Mapper sensors carried onboard the Landsat series of satellites, acquire images at a 30-m spatial resolution with a 16-day interval. The acquired data have been the backbone for land-cover, vegetation and C cycle studies. NASA's Earth Observing System (EOS), launched in 1999 (Tilford S. 1984), brought new capabilities for monitoring vegetation productivity and other properties with near-daily and global coverage. The multispectral sensors---Moderate Resolution Imaging Spectroradiometer (MODIS), onboard the EOS platform, have built invaluable global observation dataset for C and water cycles research since the early 2000s. MODIS provides a global coverage every 1-2 days with 36 bands. The spatial resolution of MODIS (pixel size at nadir) is 250 m for channels 1 and 2 (0.6μm - 0.9μm), 500 m for channels 3 to 7 (0.4μm - 2.1μm) and 1000 m for channels 8 to 36 (0.4μm - 14.4μm), respectively. Data from the satellite-borne MODIS are currently used in the calculation of global weekly GPP and ET at 1-km spatial resolution (Running et al., 2004).

Sensors that have potential applications in C and hydrology studies fall into two groups---optical (Table 1) and microwave (Table 2). Optical sensors cannot penetrate vegetation or clouds. In contrast, microwave sensors are able to penetrate vegetation and can collect data independently of cloud cover and solar illumination. This is important because it is difficult to acquire cloud-free imagery using optical sensors. There are two types of microwave sensors: active sensors and passive sensors. The former send and receive their own energy; while the latter detect the microwaves emitted by the Earth's surface. The microwave bands, being useful for vegetation and carbon and water cycles, are K, X, C, and L, ranked in increasing wavelengths. K- and X-bands are useful for detecting surface temperature, snow density, and rainfall rates, whereas C- and L-bands are sensitive to soil moisture (Sass and Greed, 2011).

## 3.2 Other airborne measurements

Besides satellite monitoring, other airborne observation techniques (*e.g.* aircraft, airplane and land surface remote sensing) have been developed rapidly since the latest decade.

| | Sensor | Bands (nm) | | | | Spatial Resolution | Spatial Coverage | Temporal Resolution | Passive /Active | Operational years |
|---|---|---|---|---|---|---|---|---|---|---|
| | | Vsible | NIR | SWIR | Thermal | | | | | |
| Very high resolution | LiDAR | | - | | | 0.15-1m | Global | No regular repeat cycle | Active | - |
| | GeoEye | 450–510 510–580 655–690 | 780–920 | - | - | Pan:0.41m Multi:1.65m | Global | Less than 3 days | Passive | 2008-09-06 to now |
| | Worldview-2 | 400–450 585-625 705-745 | 860-1040 | - | - | 0. 5m | Global | 1.1day | Passive | 2009-10 to now |
| | Quikbird | 450-520 520-660 630-690 | 760-900 | - | - | Pan:0.61m Multi:2.44m | Global | 1-3,5days depending on latitude | Passive | 2001-10-18 to now |
| | Ikonos | 450-530 520-610 640-720 | 770-880 | - | - | Pan:1m Multi:4m | Global | 1.5-2.9days | Passive | 1999-09-24 to now |
| | Orbview-3 | 450-900 450-520 520-600 625-695 | 760-900 | - | - | Pan:1m Multi:4m | Global | >3days | Passive | 2003-06-26 to now |
| | KOMPSAT-2 | 500-900 450-520 520-600 630-690 | 760-900 | - | - | Pan:1m Multi:4m | Global | 3 days | Passive | 2006-07-28 to now |
| | Resurs-DK1 | 580-800 500-600 600-700 700-800 | - | - | - | Pan:0.9-1.7m Multi:1.5-2m | Global | 5 days | Passive | 2006-06-15 to now |
| | TopSat | 500-700 450-500 500-600 600-700 | - | - | - | Pan:2.5m Multi:5m | Global | 4 days | Passive | 2005-10-27 to now |
| | MTI | 450-520 520-600 620-680 | 760-860 860-900 910-970 | 990-1040 1360-1390 | 1550-1750 3500-4100 4870-5070 8000-8400 8400-8850 10200-11500 20800-23500 | | Global | 5-8 min | Passive | 2000-03-12 to now |
| | RapidEye EOC | 440-510 20-590 630-685 | 760-850 | - | - | 5m | Global | Daily | Passive | 2008-8-29 to now |
| | Formosat-2 | 450-900 450-520 520-600 630-690 | 760-900 | - | - | Pan:2m Multi:8m | Global | Daily | Passive | 2004-04-21 to now |

| | Sensor | Bands (nm) | | | | Spatial Resolution | Spatial Coverage | Temporal Resolution | Passive /Active | Operational years |
|---|---|---|---|---|---|---|---|---|---|---|
| | | Vsible | NIR | SWIR | Thermal | | | | | |
| **Fine** | Spot4 | 510-730 500 – 590 610 – 680 | 780–890 | 1580-1750 | - | Pan:10m Multi:20m | Global | 2-3days | Passive | 1998-03 to now |
| | Spot5 | 490-690 490-610 610-680 | 780– 890 | 1580-1750 | - | Pan:2.5m,5m Multi:10m | Global | 2-3days | Passive | 2002-05-04 to now |
| | ALOS AVNIR-2 | 420-500 520-600 610-690 | 760-890 | - | - | 10m | Global | 2days | Passive | 2006-01-24 to now |
| | Terra ASTER | 520-600 630-690 | 760-860 | 1600-1700 | - | 15m 30m 90m | Global | 4-16days | Passive | 1999-12-18 to now |
| | JERS-1 OPS | 520-600 630-690 | 760-876 | 1600-1710 2010-2120 2030-2250 2270-2400 | - | 18m | Global | 44days | Passive | 1992-02-11 to now |
| | SPOT1-3HRV | 500-730 500-590 610-680 | 780-890 | - | - | Pan:10m Multi:20m | Global | 26days | Passive | 1986-02 to now |
| | CBERS IRMSS | 500-900 | - | 1550-1750 2080-2350 1040-1250 | 10400-12500 | 78m, 156m | Global | 26days | Passive | 1994-10-14 to now |
| | Deimos-1/ UK DMC-2 | 520– 600 630-690 | 770-900 | - | - | 22m | Global | 3 days | Passive | 2009-07-29 to now |
| | IRS LISS3 | 20-590 620-680 | 770-860 | 1550-1700 2200-25000 | - | 23m | Global | 24days | Passive | 2003-10-17 to now |
| | Landsat7 ETM | 520-900 450-520 520-600 620-690 | 760-960 | 1550-1750 2080-3350 | 1040-1250 | Pan:15m Muti:30m,60m | Global | 16days | Passive | 1999-04 to now |
| | Landsat4-5TM | 450-520 520-600 630-690 | 760-900 | 1550-1750 20800-23500 | 104000-125000 | 30m,120m | Global | 16days | Passive | 1984-03-16 to now |
| | Eo-1 ALI | 520-900 450-520 520-600 620-690 | 760-960 | 1550-1750 2080-3350 | 1040-1250 | Pan:10m Muti:30m,60m | Global | 16days | Passive | 2000-11-21 to now |

| Sensor | Bands (nm) | | | | Spatial Resolution | Spatial Coverage | Temporal Resolution | Passive/Active | Operational years |
|---|---|---|---|---|---|---|---|---|---|
| | Vsible | NIR | SWIR | Thermal | | | | | |
| Eo-1 Hyperion | 400-2500 | | | | 30m | Global | 16days | Passive | 2000-11-21 to now |
| Meteor 3M-1 | 450nm-1000 10500-12500 5700-7100 | | | | 1.4km,3km, | Global | daily | Passive | 2001-12-10 to now |
| Mos-1,2 MESSR | 510-590 610-690 | 720-800 800-1100 - | - | | 50m | global | 17days | Passive | 1987-02-19 to now |
| Okean MSU-SK | 530-590 610-690 | 700-800 900-1000 | - | 104000-126000nm | Visible:200m Ir:200m Thermal:600m | Global | 17days | Passive | 1983-7-10 to now |
| landsat1-5 MSS | 500-600 600-700 | 700-800 800-1100 - | - | | 80m | Global | 1,2,3:18 days; 4,5:16days | Passive | 1972 -07 to now |
| CBERS IRMSS | 500-1100 | 1550-1750 800-2350 - | | 104000-125000nm | Visible, Ir:78m Thermal :56m | Global | 26days | Passive | 1999-10-14 to now |
| SAC-C MMRS | 480-500 540-560 630-690 | 795-835 | 1550 - 1700 | - | 175m | Global | 16days | Passive | 2000-06-08 to now |
| Terra MODIS | B1-B36:400-144000 B1-B2:250m B3-B7:500m B8-B36:1000m | | | | | Global | 1/4day | Passive | 1999-12-18 to now |
| Fengyun-3a MERSI | 410-125000 | | | | B1-B5:250m B6-B20:1000m | Global | 1/2 day | Passive | 2008-05-27 to now |
| ENVISAT MERSI | 410-125000 | | | | B1-B5:250m B6-B20:1000m | Global | 35days | Passive | 2002-03-1 to now |
| MOS-1,2 VTIR | 520-590 620-680 | - | | 115000-125000 | Visibl:900m Thermal:2700m | Global | 17days | Passive | 1987-02-19 to now |
| GEOS Imager | Shortwave:3800-4000 Vsible:550-750 Moisture:6500-7000 IR:10200-11200 IR:11500-12500 | | | | Shortwave:4km Vsible:1km Moisture:8km IR:4km | Global | - | Active | 1985-04 to now |
| Meteosat SEVIRI | 12spectral channels in visible and near infrad red region | | | | 1.25km,5km | Global | daily | Passive | 2002-12-28 to now |
| NOAA AVHRR | 550-680 | 725-1100 - | 3550-3930 10500-11300 | 11500-12500 | 1.1km | Global | 1/4 day | Passive | 1978-10 to now |
| GMS5 | Visible:550-900 Water vapour:6500-7000 IR:10500-11500 IR: 11500-12500 | | | | Visible:1.25km Water vapour:5km IR:5km | Global | daily | Passive | 1995-03-18~2003-05 |

Medium

COARSE

| Sensor | Bands (nm) | | | | Spatial Resolution | Spatial Coverage | Temporal Resolution | Passive /Active | Operational years |
|---|---|---|---|---|---|---|---|---|---|
| | Vsible | NIR | SWIR | Thermal | | | | | |
| GMS1-4 | Visible:550-900 Water vapour:6500-7000 IR:10500-11500 IR: 11500-12500 | | | | Visible:1.25km Water vapour:5km IR:5km | Global | daily | Passive | 1997-7 to now |
| Fengyun-2CD | Visible:550-1050 Water vapour:6200-7600 IR:10500-12500 | | | | Visible:1.25km IR:5km | Global | 1 hour | Passive | C:2004-10-19 D: 2006-12--08 |
| fengYun-2AB | Visible:550-1050 Water vapour:6200-7600 IR:10500-12500 | | | | Visible:1.25km Water vapour:5km IR:5km | Global | 1 hour | Passive | A:1997-06-10 B:1997-07-21 |
| INSAT-2E | 550-750 | - | 10500-12500 | - | Visible:2km IR:8km | Global | daily | Passive | 1992-04-02 to now |
| INSAT-2 VHRR | 550-750 | - | 10500-12500 | - | Visible:2km IR:8km | Global | daily | Passive | 1995-12-06 to now |
| Meteosat MVIRI | Visible:450-1000 Water vapour:10500-12500 IR: 5700-7100 | | | | Visible:1.25km Water vapour:5km IR:5km | Global | daily | Passive | Meteosat—7:1993-11 |

Abbreviations: Ali, Advanced land Imager; ALOS AVNIR-2, Advanced Visible and Near Infrared Radiometer type 2; ALOS, Advanced land observing Satellite; ASTER, Japanese Earth Resources Satellite 1; AVHRR, The Advanced Very High Resolution Radiometer; CBERS, The *China–Brazil Earth Resources Satellite;* Deimos-1, Spanish Earth imaging satellite; DMC, Disaster Monitoring Constellation; Envisat, Environmental Satellite; EOS, Earth Observing System; Etm, Enhanced Thematic Mapper; Formosat-2, the first and only high-resolution satellite; GMS, Geosynchronous Meteorological Satellite; *HRV, High Resolution Visible* ; INSAT-2E, Indian geostationary communications and weather satellite; KOMPSAT, Korea Multi-Purpose Satellite; IRMSS, Infra-Red Multispectral Scanner; IRMSS, Infrared Multispectral Scanner Camera; IRS, Indian Remote Sensing; LISS-3, Linear Imaging Self-Scanning Sensor - 3. Satellites; Lidar, Light Detection And Ranging; MTI, moving target indication radar; MOS -1, Marine Observation Satellite 1; MESSR, Multi Spectral Electronic Self Scanning Radiometer; MSS, Multi-spectral Scanner; Meteor 3M-1, Meteorological Satellite; 3M, Monitoring of ocean and land surfaces, Meteorological observations, and Measurement of vertical profiles of aerosol, ozone and other constituents in the atmosphere; MMRS, Multispectral Medium Resolution Scanner; MODIS, The Moderate Resolution Imaging Spectroradiometer; MERSI, Medium Resolution Spectral Imager; *MVIRI, METEOSAT* Visible and Infrared Imager; NOAA, National Oceanic and Atmospheric Administration; Orbview, the satellite of Orbitally company; OPS, Optical System.
Okean MSU-SK: Multispectral Scanner - Conical Scanning; RapidEyeEOC, Electro-Optical Camera; SEVIRI, Spinning Enhanced Visible Infra-Red Imager; Spot, systeme probatoire d'observation de laterre, TM, Thematic Mapper; Topsat, Tactical Operational Satellite; UK-DMC 2, British Earth imaging satellite, operated by DMC International Imaging; VHRR, Very High Resolution Radiometer; VTIR, Visible and ThermalInfrared Radiometer.

Table 1. Optical Remote Sensing Systems

| Spatial resolution | Sensor | Bands | Spatial Resolution | Spatial Coverage | Temporal Resolution | Passive/ Active | Operational years |
|---|---|---|---|---|---|---|---|
| **Very high resolution** | Radarsat-2 | C | 3m | Global | 24days | Active | 2007-12-14tonow |
| | Radarsat-1 | C | 10m,25m,30m,35m,50m, 100m | Global | 24days | Active | 1995-11-04to now |
| | COSMOS Skymed | X | 1m | Global | several times a day | Passive | COSMO1:08.06.2007to now |
| | | | | | | | COSMO2:09.12.2007to now |
| | | | | | | | COSMO3:25.10.2008to now |
| | | | | | | | COSMO4:06.11.2010to now |
| | ALOS PALSAR | L | 7-100m | Global | 2days | Active | 03.01.2002to now |
| **FINE** | ERS-1,2 | C | 30m,50km | Global | 2days | Active | ERS-1:03.17.1991to now |
| | | | | | | | ERS-2:04.1995to now |
| | TerraSAR-X | X | 1m,3m,16m | Global | 11days | Active | 06.15.2007to now |
| | 7JERS-1 | L | 18mX18m | Global | 44days | | 02.11.1992~10.12.1998 |
| | ENVISAT ASAR | X | 9mX6m;30X30m;150X15 0m;450mX450m;1800mX 1800m | Global | 35days | Active | 03.01.2002to now |
| | SIR-C | X,C,L | 50m 100m | Global | - | Active | 04.091994~04.12.1999; 30.091994~11.10.1999; |
| **medium** | SRTM | X,C | 30mX30m | 60°N~56°S | 16days | Active | 02.11.2000~02.10.2000 |
| | Nimbus-7 SMMR | Ka,K,Ku, X,C | 30km,60km, 97.5km,156km | Global | 5~6days | Passive | 26.10.1978~08.21.1987 |
| **COARSE** | TRMM PR | Ku, | 4.3~5.0km | 50°S~50°N 180°W~180°E | 3h,1d, 3d,7d | Active | 12.281997 to now |
| | TRMM TMI | Ka,K,C | 6~50km | 50°S~50°N 180°W~180°E | 3h,1d, 3d,7d | Active | 12.28.1997 to now |
| | FengYun 3a MWRI | Ka,K,Ku, X | 250m | Global | 5d | Passive | 05.27.2008 to now |
| | Aqua AMSR-E | Ka,k,x | 25km | Global | 16day | Passive | 05.04.2002 To now |
| | DMSP SSMI | Ka,k | | Global | 4h | Passive | 01.19.1965to now |
| | DMSP SSMT | Ka k | 174km | Global | 4h | Passive | 01.19.1965to mow |
| | MOS-1,2 MSR | Ka,K | 32km,23km | Global | 17days | Passive | 02.19.1987~04.19.1996 |
| | ADEOS-1 NSCAT | Ku, | 25km | 90%global sea | 2days | Active | 08.17.1996~06.30.1997 12.2002~10.24.2003 |
| | SMOS MIRAS | L | 50km,200km | Global | 16days | Passive | 02.11.2009 to now |
| | CRACE | Ka,K | | Global | 16days | Passive | 03.17.2002to now |

Abbreviations: ADEOS, Advanced Earth Observing Satellite; ALOS, Advanced land observing Satellite; *AMSR-E*, The Advanced Microwave Scanning Radiometer for EOS; ASAR, An Advanced Synthetic Aperture Radar; *COSMO*, COnstellation of small Satellites for the Mediterranean basin Observation; DMSP, Defense Meteorological Satellite Program; Envisat, Environmental Satellite; ERS, European Remote-Sensing Satellites; GRACE, Gravity Recovery And Climate Experiment; JERS-1, Japanese Earth

Resources Satellite 1; MIRAS, Instrument. Synthetic. Apperture Radiometer; MSR, MicrowaveScanning Radiometer; MWR, The Microwave Radiation Imager; **NSCAT, NASA SCATterometer;** PR, Precipitation Radar; SAR, Synthetic Aperture Rada; SIR-C, Space Imaging Radar; *SRTM*, Shuttle Radar Topography Mission; *SMMR*, Scanning Multichannel Microwave Radiometer; SMOS, Soil Moisture and Ocean Salinity; SSMI, Special Sensor Microwave Imager; SSMT, Special Sensor Microwave Temperature; TRMM, Tropical Rainfall Mearsuring Mission; *TMI,TRMM*'s Microwave Imager.

Table 2. Microwave Remote Sensing Systems

Microwave wavelengths penetrate greater depths into plant canopies than optical sensors (Kasischke et al. 1997). The potential for using RADAR (RAdio Detection And Ranging) for studying terrestrial carbon and water cycles, particularly for assessing standing woody biomass is promising. The sensitivity of RADAR to vegetation biomass strongly depends on wavelength: the longer wavelengths, the greater vegetation volumes and biomass levels. Single-band RADAR is able to detect aboveground biomass up to approximately 100 Mg per hectare (Dobson et al. 1992, Luckman et al. 1998). In addition, multiband RADAR enables to separate biomass into component fractions (e.g., stem and canopy) (Saatchi and Moghaddam 2000). Synthetic aperture RADAR (SAR) is also sensitive to vegetation structure and to the amount of biomass, including both photosynthetic (green) and nonphotosynthetic vegetation components (Turner et al., 2004). LiDAR (Light Detection and Ranging) is a remote sensing technology that determines distances to an object or surface using laser pulses, which is a relatively new technology compared to optical sensors, and has the added capability of characterizing the distribution of foliage with height in the canopy (Lefsky et al. 2002, Treuhaft et al. 2002, 2004; Turner et al., 2004). LiDAR data have proved to be highly effective for the determination of three dimensional forest attributes. The suitability of airborne LiDAR for the determination of forest stand attributes including LAI and the probability of canopy gaps within different layers of canopy has been widely acknowledged by various studies (Coops et al., 2004; Coops et al., 2007). The interpreted LiDAR data have been further used for landscape C modeling and scaling (Hilker et al. 2008; Chen et al., 2009). The number and types of sensors used for research on C and water cycles have multiplied many times over since the first sensor launched into orbit. Remote sensing provides consistency of coverage and repeat measurements through time are now indispensable in the C and hydrological scientist's toolbox.

### 3.3 Remote sensing of GPP

Satellite-based studies have used the light-use efficiency ($\varepsilon$) approach to estimate GPP (Prince & Goward, 1995; Running et al., 2000, 2004; Behrenfeld et al., 2001) or net primary production (NPP) (Field et al., 1995; Ruimy et al., 1999). Significant effort and progress have been made in developing the satellite-based GPP algorithms (Running et al., 2004; Xiao et al., 2004; 2005).The algorithm relies on $\varepsilon$ approach relating GPP to the amount of absorbed photosynthetically active radiation (APAR) (Monteith, 1966, 1972), such that,

$$GPP = \varepsilon \times fPAR_{chl} \times PAR , \tag{1}$$

where PAR is the photosynetically active radiation (in $\mu$mol photosynthetic photon flux density, PPFD), $fPAR_{chl}$ is the fraction of PAR absorbed by leaf chlorophyll in the canopy, and $\varepsilon$ is the light use efficiency ($\mu$mol $CO_2$/$\mu$mol PPFD). Light use efficiency ($\varepsilon$) is affected by leaf phenology, temperature, and water:

$$\varepsilon = \varepsilon_0 \times P_m \times W_m \times T_m , \tag{2}$$

where $\varepsilon_0$ is the apparent quantum yield or maximum light use efficiency ($\mu$mol $CO_2$/$\mu$mol PPFD) for a given land cover type or vegetation function type, and $P_m$, $W_m$ and $T_m$ are the modifiers for the effects of leaf phenology, water and temperature on light use efficiency of vegetation, respectively.

Different parameters and inputs for the satellite-based algorithm were estimated in different ways: (i) the fraction of PAR absorbed by leaf chlorophyll in the canopy ($f$PAR$_{chl}$) and the modifiers ($P_m$, $W_m$); (ii) PAR and temperature modifier ($T_m$) were calculated using climate data (either from tower measurements or climate models); and (iii) the maximum light use efficiency ($\varepsilon_0$) was referred to the land-cover-related look-up table and then modified/optimized using EC tower C measurements and footprint climatology.

To accurately estimate $f$PAR$_{chl}$ in forests is a challenge to both radiative transfer modeling and field measurements. Significant efforts and progress have been made in developing advanced vegetation indices that are optimized for retrieval of $f$PAR from individual optical sensors (Gobron et al., 1999; Govaerts et al., 1999). The $f$PAR$_{chl}$ within the photosynthetically active period of vegetation was estimated as a linear function of the the Enhanced Vegetation Index (EVI),

$$f\text{PAR} = f(\text{EVI}) . \tag{3}$$

EVI is similar in design to NDVI but uses spectral information from the blue band ($\rho_{blue}$). Following Huete et al. (1997) it was computed,

$$EVI = G \times (\rho_{nir} - \rho_{red}) / (\rho_{nir} + C_1 \times \rho_{red} - C_2 \times \rho_{blue} + L), \tag{4}$$

where $G = 2.5$, $C_1 = 6$, $C_2 = 7.5$, and $L = 1$. EVI is found to be significantly correlated with the fraction of the photosynthetically active radiation absorbed by leaf chlorophyll in the canopy providing a good surrogate of the spatial variability index for photosynthesis rate.

The parameter $P_m$ was estimated using the Normalized Difference Vegetation Index (NDVI) and the Land Surface Water Index (LSWI) and was calculated at two different phases, depending upon life expectancy of leaves (deciduous versus evergreen):

$$P_m = \begin{cases} \frac{1+LSWI}{2} & \text{Duirng bud burst to leaf full expansion} \\ 1 & \text{After leaf full expansion} \end{cases} \tag{5}$$

NDVI (Tucker 1979; Field et al., 1995) was calculated as,

$$NDVI = (\rho_{nir} - \rho_{red}) / (\rho_{nir} + \rho_{red}), \tag{6}$$

where $\rho_{nir}$, and $\rho_{red}$ are the reflectance in the near infrared and red bands, respectively. NDVI is generally related to green vegetation cover or vegetation canopy density and has been shown to be well correlated with green LAI and biomass (e.g., Sellers, 1985; Myneni et al., 1995).

LSWI (Xiao et al. 2002) is a useful water index and was calculated as the normalized difference between the NIR (0.78-0.89 $\mu$m) and AWIR (1.58-1.75 $\mu$m) spectral bands:

$$LSWI = (\rho_{nir} - \rho_{swir}) / (\rho_{nir} + \rho_{swir}), \tag{7}$$

where $\rho_{nir}$ and $\rho_{swir}$ are the reflectance of near infrared bands, red bands and short infrared bands, respectively.

The timings of bud burst and leaf full expansion can be identified using NDVI. The effect of water on plant photosynthesis ($W_m$) has been estimated as a function of available soil content in plant root zone and water vapor pressure deficit (VPD) in a number of process-based ecosystem models (e.g. Chen et al., 2007) and remote-sensing based models (e.g. Running et al., 2000). Soil moisture represents water supply to the leaves and canopy, and VPD represents evaporative demand in the atmosphere. Leaf and canopy water content is largely determined by the dynamics of both soil moisture and VPD. As the first order of approximation, here following the alternative and simple approach that uses a satellite-derived water index (Xiao et al., 2004), the seasonal dynamics of $W_m$ was estimated,

$$W_m = \alpha \times (1 + \text{LSWI}) / (1 + \text{LSWI}_{\max}), \tag{8}$$

where $\alpha$ is a magnifier (its default value equals 1.0) and $\text{LSWI}_{\max}$ is the maximum LSWI within the plant growing season for individual pixels. The temperature modifier $T_m$ was estimated at each time step, using the equation developed for the terrestrial ecosystem model (Raich et al., 1991),

$$T_m = \frac{(T - T_{\min})(T - T_{\max})}{[(T - T_{\min})(T - T_{\max})] - (T - T_{opt})^2} \tag{9}$$

where $T_{min}$, $T_{max}$ and $T_{opt}$ are the minimum, maximum and optimal temperature for photosynthetic activities, respectively. Their default values are respectively set to be 0, 35 and 20 °C in this study. If air temperature falls below $T_{min}$, $T_m$ is set to be zero.

The $\varepsilon_0$ values vary with vegetation types, and the information about $\varepsilon_0$ for individual vegetation types can be obtained from a survey of the literature (Ruimy et al., 1995) and optimized using EC tower measurements. According to the work (Zhang et al. 2006), the default $\varepsilon_0$ value was estimated to be 0.032 µmol $CO_2$/µmol PPFD in this study stand in 2004.

## 3.4 Remote sensing of ET

We follow a drop of water traveling through a watershed from input, storage, and finally output and assess how RS can be used to track water fluxes and reservoirs. Table 3 summarizes the potential application of RS to study of hydrology. ET, the largest component of water loss from ecosystems, plays an important role in affecting soil moisture, vegetation productivity, C cycle, and water budgets in terrestrial ecosystems (Dirmeyer, 1994; Hilker et al. 2008; Chen et al., 2009). In this section, I mainly discuss application of RS to ET.

Verstraeten (Verstraeten et al., 2008) provided a comprehensive review of remote sensing methods for assessing ET and soil moisture content across different scales and Kalma (Kalma et al., 2008) reviewed satellite-based algorithms for estimating ET and land surface temperatures at local, regional and continental scales, with particular emphasis on studies published since the early 1990s.

In general, water evapotranspired from ecosystems into the atmosphere will reduce the land surface temperature $(T_a)$. Reduction in soil moisture will decrease plant transpiration and evaporation from soil and plant surfaces. Reduction in ET will increase $T_a$. $T_a$ can be derived from remotely-sensed thermal-infrared (TIR) band (8-14 microns) from various operational satellites. Based on the relationship between $T_a$ and ET, remotely sensed $T_a$ has been used to estimate regional ET (Gillies et al,. 1997; Kite et al., 2000; Su et al., 2000., Coops et al., 2002). The existing thermal imaging sensors provide adequate coverage of thermal dynamics that are useful for operational monitoring applications of ET. For example, thermal images at 15 minutes intervals and at a spatial resolution of 5 kilometers can be obtained from the NOAA Geostationary Operational Environmental Satellites (GOES), and TIR data at a fine spatial resolution (60 m or 120 m) with a much longer time interval (16 days) have been provided by the Thematic Mapper (TM) and ETM+ instruments on Landsat 5 and Landsat 7.

## 4. Modeling of C and water dynamics in terrestrial ecosystems based on remote sensing

The land surface of the Earth represents significant sources, sinks, and reservoirs of C, heat and moisture to the atmosphere. C and energy fluxes and water cycles at soil-atmosphere and plant-atmosphere interfaces are therefore important land surface processes. Due to the complexity and non-linearity of C, N and water dynamics in terrestrial ecosystems, various modeling tools are needed for better understanding of these biogeochemical and hydrological processes and their feedback mechanisms with the land surface climate system (Rannik et al., 2006). The rapidly proliferating volume of spatial data generated by RS has created a significant challenge in terms of designing model algorithms. A spatially distributed process-based model uses spatial data for computing ecohydrological and biophysical processes. The model algorithms represent hypotheses that can be assessed and potentially revised after confrontation with RS and land surface-based observations. It is well known that realistic simulations of C and water dynamics in terrestrial ecosystems is of critical importance, not only for the surface microclimate, but also for the large-scale physics of the atmosphere (Cox et al., 1999; Gedney et al., 2006; Dickinson et al., 2002). Depending on the scientific objectives or applications, C and water cycle models have been designed with varying degrees of aggregation with respect to ecosystem processes, components, and RS data as model inputs. Such models can be flagged by land surface, ecosystem and hydrological models based on their objectives and emphases. The former focus on ecosystem processes and the interactions between ecosystems and the atmosphere; while the latter place emphasis on the land surface hydrology processes, including lateral flow resulting from catchment topography.

### 4.1 Land surface and ecosystem modeling

Global climate and the global carbon cycle are controlled by exchanges of water, carbon, and energy between the terrestrial biosphere and atmosphere. Thus land surface models (LSMs) are essential for the purpose of developing predictive capability for the Earth's climate on all time scales (Matthews et al., 1998). Most current LSMs can be associated with three broad types (Seth et al., 1994): soil-vegetation-atmosphere transfer schemes (SVATS), potential vegetation models (PVMs), and terrestrial biogeochemistry models (TBMs).

The first generation of SVATS evolved from simple bucket schemes focusing on soil water availability（Manabe et al., 1969）, through the schemes of Deardorff (Deardorff et al., 1978).

Marked improvements of the second generation (*e.g.*, BATS (Seth et al., 1994), SiB (Sellers et al., 1997; Sellers et al., 1986), and CLASS (Verseghy et al., 1999; Verseghy et al., 1993) from the first generation are the separation of vegetation from soil and the inclusion of multiple soil layers for dynamic heat and moisture-flow simulations (Chen et al., 2007). The second generation SVATS firstly modeled plant physiology in an explicit manner in GCMs (General Circulation Model or Global Climate Model) (Henderson et al., 1993). For most second-generation SVATS, land cover was fixed, with seasonally-varying prescriptions of parameters such as reflectance, leaf area index or rooting depth (Wang et al., 2002; Kickert et al., 1999; Kley et al., 1999; Schwalm et al., 2001). Some SVATS incorporated satellite data to characterize more realistically the seasonal dynamics in vegetation function (Kickert et al., 1999; Bonan et al., 1994). The latest (third generation) SVATS used more recent theories relating photosynthesis and plant water relations to provide a consistent description of energy exchange, ET, and C exchange by plants (Chen et al., 2007; Sellers et al., 1996). In our effort in understanding the impact of climate change on terrestrial ecosystems, energy, water, and C cycles need to be modelled simultaneously (Sellers et al., 1996; Williams et al., 2001). Recently, most of SVATS have thus been enhanced to include the $CO_2$ flux between the land surface and the atmosphere, such as SiB2 (Sellers, P.J et al, 1996), IBIS (Foley et al., 1996), NCAR-LSM (Bonan et al., 1995), BATS (Dickinson et al., 2002), CLASS-C (Wang et al., 2002) and EASS (Chen et al., 2007).

The earlier generation of PVMs comprised a suite of schemes that focus on modeling distributions of vegetation as a function of climate (Holdridge et al., 1947; Prentice et al., 1990) without influences of anthropogenic or natural disturbance. The second generation of PVMs included more sophisticated modules to account for factors controlling vegetation distributions, such as competition, varying combinations of plant functional types, and physiological and ecological constraints (Prentice et al., 1992).

TBMs developed from scaling up local ecological models, are process-based models that simulate dynamics of energy, water, and carbon and nitrogen exchange among biospheric pools and the atmosphere (Seth et al., 1994). Few of the existing TBMs incorporate PVMs. These models are not applicable to transient climate change experiments without coupling with PVMs.

In recent decades, the interactions among soil, vegetation and climate have been studied intensively and modeled successfully on the basis of water and energy transfer in the soil-vegetation-atmosphere system (Seth et al., 1994; Sellers et al., 1986; Verseghy et al., 1999; Verseghy et al., 1993; Zhang et al., 2003). Also the construction and refinement of LSMs have received increasing attention (Sellers et al., 1996; Viterbo et al., 1995; Christopher et al., 2004). Combination of these three different LSMs and utilization of remotely sensed land surface parameters are critical in the future LSM development, because of (1) the tight coupling of exchanges of water, energy and carbon between the land surface and the atmosphere; (2) the sophisticated impact/feedback mechanisms between climate change and terrestrial ecosystems; and (3) increasingly strong anthropogenic alterations to land cover. On-line coupling of a LSM with a GCM is needed for studying interannual to multi-decadal climate variations.

Several model intercomparisons have focused on evaluating SVATS and TBMs with particular objectives. For instance, the Project for Intercomparison of Land-surface Parameterization Schemes (PILPS) was initiated to evaluate an array of LSMs existing in GCMs (General Circulation Model or Global Climate Model) (Henderson et al., 1993); while the AMMA (African Monsoon Multidisciplinary Analysis) Land Surface Model Intercomparison Project (ALMIP) is being conducted to get a better understanding of the role of soil moisture in land surface processes in West Africa (de Rosnay et al., 2009). Coordinated land surface modeling activities have improved our understanding of land surface processes (de Rosnay et al., 2009).

## 4.2 Spatially-distributed hydrological processes modeling

Hydrology and ecosystem have, for the most part, been studied independently. Most LSMs and ecosystem models make an assumption of "flat Earth" with the absence of lateral redistribution of soil moisture. On the other hand, hydrological models have mostly been concerned with runoff production. Spatially-distributed models are needed, especially for hydrological simulation objective, because of heterogeneity of land surface and non-linearity of hydrological processes. Spatially-distributed hydrological models are not only able to account for spatial variability of hydrological processes, but enable computation of internal fluxes and state variables. Such kinds of models are increasingly applied to simulate spatial variability of forcing variables (*e.g.* precipitation), physiographic characteristics, detailed processes and internal fluxes within a catchment (Liang et al., 1994; Liang et al., 2004; Beldring et al., 2003; Brath et al., 2004; Christensen et al., 2007; Reed et al., 2004).

## 4.3 Modeling dynamics of stable C isotopic exchange between ecosystem and the atmosphere

It is recognized that the atmospheric measurements are still too sparse, relative to its spatial variability, to be used for inferring the surface flux at high spatial resolution (Ciais et al., 1995). The use of the isotope ratio as an additional constraint to identify various C sources and sinks can contribute to a significant reduction in the uncertainty. Though available isotopic datasets are being accumulated quickly (Griffis et al., 2005; Ponton et al., 2006; Lai et al., 2006; Lai et al., 2005) isotope measurements are still lacking considering land surface diversity and heterogeneity. This shortage of long-term measurements and of sampling frequency still limits C isotopic studies.

Mechanistic ecosystem models that couple micrometeorological and eco-physiological theories have the potential to shed light on how to extend efforts and applications of stable isotopes of $CO_2$ to global C budgeting, because biophysical models have the capacities of simulating isotope discrimination in response to environmental perturbations and can produce information on its diurnal, seasonal and interannual dynamics. Few biophysical models, however, have been developed to assess stable C discrimination between a plant canopy and the atmosphere (Suits et al., 2005; Oge'e et al., 2003; Baldocchi et al., 2003). Most existing biophysical models are based on individual leaf level discrimination equations given by Farquhar *et al.* (Farquhar et al., 1989; Farquhar et al., 1982) and only focus on the land surface layer (ignoring vertical and horizontal advection effects beyond 50~100 m above the ground (Baldocchi et al., 2003). However, in nature, the convective boundary layer (CBL) integrates the effects of photosynthesis, respiration, and turbulent transport of $CO_2$ over the landscape (Lloyd et al., 1996; Pataki et al., 2003). The influence of the CBL cannot be ignored when using isotope composition of $CO_2$ to investigate biological processes (Bowling et al., 1999), because the effect of atmospheric stability on turbulent mixing/diffusion has an important impact on scalar fluxes and concentration fields within and above canopies (Baldocchi et al., 1995; Leuning et al., 2000). Few such models considering the CBL effects on isotope fractionation have been developed to date (Lloyd et al.,1996; Lloyd et al., 2001; Chen et al,. 2006; Chen et al., 2006; Chen et al., 2007).

## 4.4 Modeling coupled C and water dynamics – An ecohydrological approach

C and N dynamics and hydrological processes are closely linked. The stomatal conductance ($g_s$) is the key linkage between C assimilation (photosynthesis) and transpiration. An empirical equation is used in the second-generation LSMs to calculate $g_s$, which is hypothesized to be controlled by the environmental conditions (Jarvis et al., 1976). While field and laboratory studies have documented that leaf photosynthesis also affects $g_s$. Therefore, Ball *et al.* (Ball et al., 1987) proposed a semi-empirical stomatal conductance formulation (Ball-Woodrow-Berry model), in which $g_s$ is controlled by both photosynthesis and the environmental conditions. Most of third-generation LSMs (Ecological models, *e.g.* SiB2 (Sellers et al., 1997; Sellers et al., 1996); CN-CLASS (Arain et al., 2006); *Ecosys* (Grant et al., 2007; Grant et al., 1999; Chen et al., 2007) fully couple photosynthesis and transpiration processes by employing the Ball-Woodrow-Berry stamatal conductance formulation.

In addition to the coupling of hydrological condition and C assimilation through the linkage of $g_s$, C assimilation is also coupling with N dynamics through another biochemical parameter, $V_{cmax}$ --- maximum carboxylation rate at 25 °C. In the photosynthesis model proposed by Farquhar *et al.* (1980), the net photosynthetic rate $A_{net}$ at leaf level is a function of two tightly-correlated parameters $V_{cmax}^{25}$ and $J_{cmax}^{25}$ (the maximum electron transport rate at 25 °C), and is calculated as,

$$A_{net} = \min(A_c, A_j) - R_d \tag{10}$$

where $A_c$ and $A_j$ are Rubiso-limited and light-limited gross photosynthesis rates, respectively, and $R_d$ is the daytime leaf dark respiration and computed as $R_d = 0.015\ V_{c\ max}$. $A_c$ and $A_j$ are expressed as,

$$A_c = V_{c\max} \frac{C_c - \Gamma^*}{C_c + K_c(1 + O_c/K_o)} \tag{11a}$$

and,

$$A_j = J_{\max} \frac{C_c - \Gamma^*}{4(C_c + 2\Gamma^*)} \tag{11b}$$

where $C_c$ and $O_c$ are the intercellular $CO_2$ and $O_2$ mole fractions (mol mol$^{-1}$), respectively; $\Gamma^*$ is the $CO_2$ compensation point without dark respiration (mol mol$^{-1}$); $K_c$ and $K_o$ are Michaelis-Menten constants for $CO_2$ and $O_2$ (mol mol$^{-1}$), respectively. In the nutrient-limited stands, $A_{net}$ is generally limited by $A_c$, while $A_c$ is dominantly controlled by a parameter $V_{cmax}$ (see Eq. 11a). Many research results showed $V_{cmax}^{25}$ is very sensitive to leaf N status (more specifically leaf Rubisco-N) (Dickinson et al., 2002; Wilson et al., 2000; Wilson et al., 2001; Warren et al., 2001). As a result in some ecosystem models (*i.e.* C&N-CLASS (Arain et al., 2006)), $V_{cmax}^{25}$ is calculated as a nonlinear function of Rubisco-N following observations made by Warren and Adams (Dickinson et al., 2002):

$$V_{c\max}^{25}(N) = \alpha\left[1 - \exp(-1.8N_{r0})\right] \tag{12}$$

where α is the maximum value of $V_{c\max}^{25}$ and $N_{r0}$ is the leaf Rubisco-N (g N m$^{-2}$ leaf area) in the top canopy.

The coupled C, N and water processes have been carefully considered in most of the third-generation LSMs (*e.g.* SiB2 (Sellers et al., 1997; Sellers et al., 1986; Sellers et al., 1996); CN-CLASS (Arain et al., 2006) and *Ecosys* (Grant et al., 2007; Grant et al., 1999), the models' grids, however, are isolated from their neighboring grids mainly due to the availability of input data. Vertical soil hydrological processes are hard to be realistically simulated if the lateral flows are ignored by assuming that the Earth is "flat". However, Simulations of the topographically-driven lateral water flows are important components in most of spatially-distributed models, while the detailed ecophsiological processes are weakly represented (Govind et al., 2009). Much effort to bridge these two different models has been increasingly made (Rodriguez et al., 2001; D'odorico et al., 2004;Govind et al., 2009; Creed et al., 1998; Band et al., 2001; Porporato et al., 2002; Porporato et al., 2003; Daly et al., 2004; Chen et al., 2005). However, a model coupling approach --- a full combination of ecosystem model and hydrological model, *i.e.* ecohydrological modeling, is still lacking.

### 4.5 Applications of remotely-sensed data in ecohydrological modeling

Remote sensing techniques, which inherently have the ability to provide spatially comprehensive and temporally repeatable information of the land surface, may be the only feasible way to obtaining data needed for land surface and ecological modeling (Sellers et al., 1986; Gurney et al., 2003; Kite et al., 1996; Engman et al., 1996; Melesse et al., 2008). The most common rationale for interfacing remote sensing and land surface-ecosystem models is using remotely sensed data as model inputs (Plummer et al., 2000). These input data, corre-sponding to forcing functions or state variables in ecological modeling, include LC, LAI, normalized difference vegetation index (NDVI), and the fraction of photosynthetically active radiation ($f_{PAR}$) (Sellers et al., 1986; Running et al., 1998; Chiesi et al., 2002; Loiselle et al., 2001). Another effort is the direct estimation of GPP and net primary productivity (NPP) (Goetz et al., 1999; Seaquist et al., 2003) of biomass (Seaquist et al., 2003; Bergen et al., 1999) and of plant growth (Maas et al., 1988; Kurth et al., 1994), by making use of $f_{PAR}$ and NDVI. It has been shown that the direct estimation has lower accuracy than the integration of remotely sensed data with process based models (Goetz et al., 1999).

Remote sensing data have also been used to parameterize hydrological models (Chen et al., 2005; Kite et al., 1996; Boegh et al., 2004). For instance, a hydrological model (TerrainLab) was further developed using remote sensing as inputs (Chen et al., 2005). TerrainLab is a spatially distributed, process-oriented hydrological model using the explicit routing scheme of Wigmosta et al. (Wigmosta et al., 2004). This model has been applied to flat areas such as boreal and wet land region, (Govind et al., 2009; Chen et al., 2005; Govind et al., 2009), but it has not yet been applied to mountainous areas.

Different from traditional hydrological models, which have coarse spatial resolutions, the grid-based-distributed ecohydrological models have a high demand for spatial data (Kite et al., 1996; Montzka et al., 2008). Some researchers highlight that the main obstacles in current distributed ecohydrological modeling is the lack of sufficient spatially distributed data for input and model validation (Stisen et al., 2008). Remote sensing can potentially fill in some of the gaps in data availability and produce means of spatial calibration and validation of distributed hydrological models. As a result the application of remote sensing techniques in hydrological studies and water resources management has progressed in the past decades (see review by (Kite et al., 1996)).

In general, the applications of remotely sensed data in ecohydrological modeling can be in the two ways (Kite et al., 1996; Chen et al., 2005; Boegh et al., 2004; Montzka et al., 2008; Stisen et al., 2008; Ritchie et al., 1996; Schultz et al., 1996; Melesse et al., 2007; Schmugge et al., 2002; Jain et al., 2004; Pietroniro et al., 2005; French et al., 2006): (i) multispectral remote sensing data are used to quantify surface parameters, such as vegetation types and density. Although the usefulness of remote sensing data is widely recognized, there remain few cases where remote sensing data have been actually used in ecohydrological simulations. Difficulties still exist in choosing the most suitable spectral data for studying hydrological processes as well as in interpreting such data to extract useful in formation (Chen et al., 2005; Kite et al., 1996; Engman et al., 1996); and (ii) processed remote sensing data are used to provide fields of hydrological parameters for calibration and validation of ecohydrological models, such as precipitation (Kite et al., 1996; Wang et al., 2001), and soil moisture (Jackson et al., 1993; Hollenbeck et al., 1996., 1996; Kim et al., 2002; Koster et al., 2006). Koster et al. (Koster et al., 2006) pointed out that remote sensing data take the form of emitted and reflected radiances and thus are not the type of data traditionally used to run and calibrate models. Hence, it is important to understand and develop relationships between the electromagnetic signals and hydrological parameters of interest (Chen et al., 2005). Kite and Pietroniro (Kite et al., 1996) stated that the use of remote sensing in hydrological modeling was limited. Even though a number of new sensors have been launched since then and research has documented that remote sensing data have promising perspectives, operational uses of satellite data in hydrological modeling still appear to be in its infancy (Stisen et al., 2008).

## 5. Research gaps in C and water flux estimates and scaling approaches

A variety of methods are being used in the C and water cycles studies. As shown in Figure 3, different approaches have different temporal and spatial scales. The most direct measurements of the terrestrial C flux are made either at the plot scale ($10^{-2}$-$10^1$ m$^2$), $e.g.$ using biometric methods and various forms of chamber, or at the ecosystem (patch) scale ($10^4$ - $10^6$ m$^2$), using the EC technique. Ecohydrological / ecosystem modeling and remote sensing estimations are generally available across variable spatiotemporal scales. These estimates are normally available within a nested framework that permits a progressive comparison of measurements made by surface instrumentation (scale: 1 to 10 m), surface flux equipment (10 m to 1 km), airborne remote sensing equipment (100 m to several km), satellite remote sensing (30 m to global scale) and EC tower (1-3 km),

The atmosphere integrates surface fluxes over many temporal and spatial scales and links scalar sources and sinks with concentrations and fluxes. This principle has been successfully used to develop inverse models to estimate annual C budgets (Tans et al., 1990; Enting et al., 1995; Fan et al., 1998; Bousquet et al., 1999; Gurney et al., 2002; Gurney et al., 2003). However, due to model limitations and paucity of continental $CO_2$ observations these studies have yielded C fluxes only at coarse resolution, over large spatial regions (Gurney et al., 2004; Gurney et al., 2005; Gurney et al., 2008).

Progress in C balance studies has been achieved at both ends of the spatial scale spectrum, either large continents (larger than $10^6$ km², e.g. global inverse modeling) or small vegetation stands (less than 1-3 km², e.g. EC-measurements). Methods to estimate $CO_2$ sources and sinks at the intermediate scale (i.e. landscape to regional scales) between continental and local scales are less well advanced. Moreover, the C cycle in different regions can vary markedly in response to changing climate (Friedlingstein et al., 2003). Reliable estimates of terrestrial C sources and sinks at landscape to regional spatial scales (finer than those used in global inversions and larger than local EC flux measurements and roughly defined as the range between $10^2$ and $10^6$ km²) are required to quantitatively account for the large spatial variability in sources and sinks in the near-field of a measurement location (Gerbig et al., 2003), as well as fundamental to improving our understanding of the C cycle (Crevoisier et al., 2006).

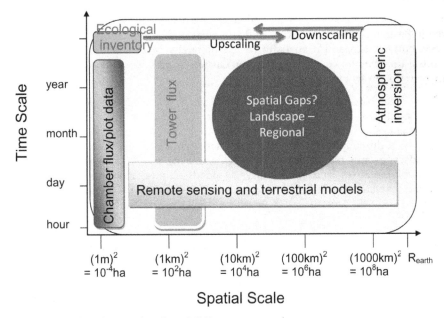

Fig. 3. Temporal and spatial scales of different approaches

It is generally considered unreliable to upscale stand-level fluxes (i.e. EC measurements) to a region by simple spatial extrapolation and interpolation because of the heterogeneity of the land surface and the nonlinearity inherent in ecophysiological processes (Levy et al., 1999). It is also challenging to apply atmospheric inversion technique to regional scales for quantifying annual C budgets because at such intermediate scales the atmosphere is often poorly constrained (Matross et al., 2006; Gloor et al., 1999). Moreover, aggregation errors and errors in atmospheric transport, both within the PBL and between the PBL and free troposphere, can also be obstacles to using these approaches to obtain quantitative estimates of regional C fluxes (Lin et al., 2004). Hence, there is a strong motivation to develop methods to quantify and validate estimates of the C balance at these intermediate scales (Lin et al., 2004; Chen et al., 2008; Bakwin et al., 2004; Matross et al., 2006). Observations of $CO_2$ over the continent within the PBL reflect exchange processes occurring at the surface at a regional scale ($10^2 - 10^5$ $km^2$). The flux information contained in $CO_2$ concentration data represents footprints of up to $10^5$ $km^2$ (Gloor et al., 2001; Lin et al., 2004), which are several orders of magnitude larger than the direct EC-flux footprint. This information is therefore needed in our effort to upscale from site to region. Moreover, the number of $CO_2$ mixing ratio measurements above the land surface, made by either tower or aircraft, is steadily increasing. Previous efforts to interpret the signal of regional $CO_2$ exchange making use of tower concentration data have focused on simple one-dimensional PBL budgets that rely on gradients in $CO_2$ concentrations between the PBL and the free troposphere (Bakwin et al., 2004; Helliker et al., 2004). These methods are limited to monthly resolution because of the need to smooth and average over several synoptic events (Matross et al., 2006).

## 6. Future research directions

A synthetic research framework is needed to strength the less well researched areas as reviewed in Section 5: bottom-up and top-down approaches integrating scalable (footprint and ecosystem) models and a spatially nested hierarchy of observations which include multispectral RS, inventories, existing regional clusters of eddy-covariance flux towers and $CO_2$ mixing ratio towers and chambers.

The current research trends and the future directions in this field include: (i) A synthesis aggregation method --- integrating ecohydrological and isotopic models, remote sensing and component flux data, is becoming a pragmatic approach towards a better understanding of the coupled C, N and water dynamics at landscape/watershed scales; and (ii) The landscape- and regional-scale C fluxes are being estimated using an integrated approach involving direct land surface measurements, RS measurements, and ecosystem-, footprint- and inversion- modeling.

## 7. Summary

After comprehensive reviewing of a variety of approaches being used in research on the C/water cycles, the concluding remarks are summed the following:

Research gaps in this field are (i) The coupled terrestrial C and hydrological dynamics are far from well understood, especially at landscape (watershed) and regional scales; (2) Much progresses have been achieved at the extreme ends of the spatial-scale spectrum, either large regions/continents or small vegetation stands. Because of the heterogeneity of the land surface and the nonlinearity inherent in ecophysiological and ecohydrological processes in response to their driving forces, it is difficult to upscale stand level results to regions and the globe by extrapolation. Budgets of C and water at landscape intermediate regional scales ($10^2$–$10^5$ $km^2$) have large uncertainties.

A coupled spatially-explicit ecohydrological model is a powerful tool for quantitative and predictive understanding of the coupled C and water mechanism. This modeling framework can be used to infer aspects of the land surface system that are difficult to measure, and will be critical to improving the accuracy of forecasts of landscape change and C dynamics in the real world.

Combining and mutually constraining the bottom-up and top-down methods to reduce their uncertainties using data assimilation techniques is a practical and effective means to derive regional C and water fluxes with reasonably high accuracy. In the future upscaling framework, spatially nested hierarchy of observations, including multispectral RS, inventories, existing regional clusters of EC flux towers and $CO_2$ mixing ratio towers and chambers, are able to integrated using scalable (footprint and ecosystem and ecohydrological) models and data-model fusion techniques.

## 8. Acknowledgements

This research is financially supported by the National Science Foundation of China (Grant No. 41071059), "One hundred talents" program funded by Chinese Academy of Sciences and Alexander Graham Bell Canada Scholarship (CGS) funded by Natural Sciences and Engineering Research Council of Canada.

## 9. References

Amiro, B.D.; Barr, A.G.; Black, T.A. Carbon, energy and water fluxes at mature disturbed forest sites, Saskatchewan, Canada, *Agric. For. Meteor.* 2006, 136, 237-251.

Arain, M.A.; Yuan, F.M.; Black, T.A. Soil-Plant Nitrogen Cycling Modulated Carbon Exchanges in a Western Temperate Conifer Forest in Canada. *Agric. For. Meteor.* 2006, *140*, 171-192.

Bakwin, P.S.; Davis, K.J.; Yi, C.; Wofsy, S. C.; Munger, J. W.; Haszpra, L.; Barcza, Z. Regional carbon dioxide fluxes from mixing ratio data. *Tellus B* 2004, *56*, 301-311.

Baldocchi ,D.D.; Fálge, E.; Gu, L. et al. FLUXNET: a new tool to study the temporal and spatial variability of ecosystem-scale carbon dioxide, water vapor, and energy flux densities. *Bull Amer. Meteorl. Soc.* 2001, *82*, 2415-2434.

Baldocchi, D.D. Breathing of the terrestrial biosphere: lessons learned from a global network of carbon dioxide flux measurement systems. *Aust. J. Bot.* 2008, *56*, 1-26.

Baldocchi, D.D., Bowling, D.R. Modeling the discrimination of [13]C above and within a temperate broad-leaved forest canopy on hourly to seasonal time scales. *Plant Cell Environ.* 2003, *26*, 231-244.

Baldocchi, D.D.; Harley, P.C. Scaling carbon dioxide and water vapour exchange from leaf to canopy in a deciduous forest: model testing and application. *Plant, Cell and Environment* 1995, *18*, 1157–1173.

Baldocchi, D.D.; Wilson, K.B. Modeling $CO_2$ fluxes and water vapor exchange of a temperate broadleaved forest across hourly to decadal time scales. *Ecol. Mod.* 2001, *142*, 155-184.

Ball, J.T.; Woodrow , I.E.; Berry J.A. A model predicting stomatal conductnance amd its contribution to the control of photosynthesis under different environmental conditions, In Progress in Photosynthesis Research, Martinus Nijhoff Publishers: *Dordrecht, Netherland*, 1987, pp 221-224.

Band, L.E.; Tague, C.L.; Groffman, P.; Belt, K. Forest Ecosystem Processes at the Watershed Scale: Hydrological and Ecological Controls of Nitrogen Export. *Hydro. Processes* 2001, *15*, 2013-2028.

Barford, C. C., Wofsy, S. C., and Goulden, M. L.: Factors controlling long- and short-term sequestration of atmospheric CO2 in a mid-latitude forest. *Science*, 2001, 294, 1688-1691.

Barr, A.G.; Black, T.A.; Hogg, E.H. *et al.* Inter-annual variability in the leaf area index of a boreal aspen-hazelnut forest in relation to net ecosystem production. *Agric. For. Meteor.* 2004, *126*, 237–255.

Behrenfeld, M. J., Randerson, J. T., McClain, C. R., Feldman, G. C., Los, S. O., and Tucker, C. J.: Biospheric primary production during an ENSO transition, *Science*, 2001, 291, 2594–2597.

Beldring, S.; Engeland, K.; Roald, L.A.; Sælthun, N.R.; Voksø, A. Estimation of parameters in a distributed precipitation runoff model for Norway. *Hydro. Earth System Sci.*, 2003, *7*, 304–316.

Bergen, K.M.; Dobson, M.C. Integration of remotely sensed radar imagery in modeling and mapping forest biomass and net primary production. *Ecol. Model.* 1999, *122*, 257–274.

Betts, A.K.; Desjardins, R.L.; Worth D. Impact of agriculture, forest and cloud feedback on the surface energy budget in BOREAS. *Agric. For. Meteor.* 2007a, *142*, 156–169.

Betts, R.A.; Boucher, O.; Collins, M.; Cox, P.M.; Falloon, P.D.; Gedney, N.; Hemming, D.L.; Huntingford, C.; Jones, C.D.; Sexton, D.M.H.; Webb, M.J. Projected Increase in Continental Runoff Due to Plant Responses to Increasing Carbon Dioxide, *Nature* 2007b, *448*, 1037-U5.

Betts, R.A.; Cox, P.M.; Woodward, F.I. Simulated responses of potential vegetation to doubled-$CO_2$ climate change and feedbacks on near-surface temperature, *Global Ecol. Biogeogr.* 2000, *9*, 171-180.

Black, T.A.; Chen, W.J.; Barr, A.G. *et al.* Increased carbon sequestration by a boreal deciduous forest in years with a warm spring. *Geophys. Res. Lett.* 2000, *27*, 1271-1274.

Blanken, P.D.; Black T.A. The canopy conductance of a boreal aspen forest, Prince Albert national park, Canada. *Hydro. Proces.* 2004, *18*, 1561–1578.

Boegh, E.; Thorsen, M.; Butts, M.B.; Hansen, S.; Christiansen, J.S.; Abrahamsen, P.; Hasager, C.B.; Jensen, N.O.; van der Keur, P.; Refsgaard, J.C.; Schelde, K.; Soegaard, H.; Thomsen, A. Incorporating remote sensing data in physically based distributed agro-hydrological modelling. *J. Hydro.* 2004, *287*, 279–299.

Bonan, G.B. Comparison of two land surface process models using prescribed forcings, *J. Geophys. Res.* 1994, 99, 25,803-25,818.

Bonan, G.B. Land-atmospheric interactions for climate system models: Coupling biophysical, biogeochemical and ecosystem dynamical processes. *Remote Sens. Environ.* 1995, *51*, 57-73.

Bousquet, P.; Ciais, P.; Peylin, P.; Ramonet, M.; Monfray, P. Inverse modeling of annual atmospheric $CO_2$ sources and sinks 1. method and control inversion. *J. Geophys. Res.* 1999, *104*, 26161-26178.

Bowling, D.R.; Baldocchi, D.D.; Monson, R.K. Dynamics of isotope exchange of carbon dioxide in a Tennessee deciduous forest. *Global Biogeochem. Cycles* 1999, *13*, 903–921.

Brath, A.; Montanari, A.; Toth, E. Analysis of the effects of different scenarios of historical data availability on the calibration of a spatially-distributed hydrological model. *J. Hydro.* 2004, *291*, 232–253.

Buermann, W.; Dong, J.; Zeng, X.; Myneni, R.B.; Dickinson, R.E. Evaluation of the utility of satellite-based vegetation leaf area index data for climate simulations, *J Clim*, 2001, 14, 3536-3550.

Canadell, J. G.; Que´ re´, C. Le; Raupacha, M. R.; Fielde, C. B.; Buitenhuisc, E.T.; Ciais, P.; Conwayg, T.J.; Gillettc, N. P.; Houghtonh, R. A.; Marlandi, G. Contributions to accelerating atmospheric $CO_2$ growth from economic activity, carbon intensity, and efficiency of natural sinks. *PNAS* 2007, *104*, 18866-18870.

Chen, B.; Black, A.; Coops, N.C.; Hilker, T.; Trofymow, T.; Nesic, Z.; Morgenstern, K. Assessing tower flux footprint climatology and scaling between remotely sensed and eddy covariance measurements. *Boundary-Layer Meteorology*, 2009a , 130, 137-167. DOI: 10.1007/s10546-008-9339-1.

Chen, B.; Black, A.; Coops, N.C.; Jassal, R.; Nesic Z. Seasonal controls on interannual variability in carbon dioxide exchange of a Pacific Northwest Douglas-fir forest, 1997 – 2006, *Global Change Biology*, 2009b , 15, 1962-1981, doi: 10.1111/j.1365-2486.2008.01832.x.

Chen, B.; Chen, J. M.; Mo, G.; Yuen, C-W.; Margolis, H.; Higuchi, K.; Chan D. Modeling and scaling coupled energy, water, and carbon fluxes based on remote sensing: An application to Canada's landmass, *J. Hydrometeorology*, 2007, 8, 123-143,.

Chen, B.; Chen, J.M.; Mo, G.; Black, T.A.; Worthy, D.E.J. Comparison of regional carbon flux estimates from CO2 concentration measurements and remote sensing based footprint integration. *Global Biogeochem Cycles*, 2008, 22, GB2012, doi:10.1029/2007GB003024.

Chen, B.; Chen, J.M. Diurnal, seasonal and inter-annual variability of carbon isotope discrimination at the canopy level in response to environmental factors in a boreal forest ecosystem. *Plant, Cell and Environment* 2007, *30*, 1223-1239, doi: 10.1111/j.1365-3040.2007.01707.

Chen, B.; Chen, J.M.; Huang, L.; Tans, P.P. Simulating dynamics of $\delta^{13}C$ of $CO_2$ in the planetary boundary layer over a boreal forest region: Covariation between surface fluxes and atmospheric mixing. *Tellus B* 2006, *58*,537-549.

Chen, B.; Chen, J.M.; Ju, W. Remote sensing based ecosystem-atmosphere simulation Scheme (EASS) --- model formulation and test with multiple-year data. *Ecol. Model.* 2007, *209*, 277-300.

Chen, B.; Chen, J.M.; Mo, G.; Black, T.A.; Worthy, D.E.J. Comparison of regional carbon flux estimates from $CO_2$ concentration measurements and remote sensing based footprint integration. *Global Biogeochem.Cycles*, 2008, 22, GB2012, doi:10.1029/2007GB003024

Chen, B.; Chen, J.M.; Tans, P.P.; Huang, L. Modeling dynamics of stable carbon isotopic exchange between a boreal ecosystem and the Atmosphere. *Global Change Biol.* 2006, *12*, 1842-1867.

Chen, J.M.; Chen, X.Y.; Ju, W.M.; Geng, X.Y Distributed Hydrological Model for Mapping Evapotranspiration Using Remote Sensing Inputs. *J. Hydrol.* 2005, *305*, 15-39.

Chiesi, M.; Maselli, F.; Bindi, M.; Fibbi, L.; Bonora, L.; Raschi, A.; Tognetti, R.; Cermak, J.; Nadezhdina, N. Calibration and application of FOREST-BGC in a Mediterranean area by the use of conventional and remote sensing data. *Ecol. Model.* 2002, *154*, 251-262.

Christensen, N.; Lettenmaier, D.P. A multimodel ensemble approach to assessment of climate change impacts on the hydrology and water resources of the Colorado River basin. *Hydro. Earth System Sci.* 2007, *11*, 1417–1434.

Christopher, S.R.; Ek, A.R., A process-based model of forest ecosystems driven by meteorology. *Ecol. Model.* 2004, *179*, 317-348.

Ciais, P; Tans P.P.; Trolier, M; White, J.W.C.; Francey, R.J. A large Northern Hemisphere terrestrial $CO_2$ sink indicated by the $^{13}C/^{12}C$ ratio of atmospheric $CO_2$. *Science* 1995, *269*, 1098-1102

Coops, N.; Wulder, M.; Culvenor, D.; St-Onge, B. Comparison of forest attributes extracted from fine spatial resolution multispectral and lidar data. *Can. J. Remot. Sens.* 2004, *30*, 855-866.

Coops, N.C.; Black, T.A.; Jassal, R.S.; Trofymow, J.A.; Morgenstern, K. Comparison of MODIS, eddy covariance determined and physiologically modelled gross primary production (GPP) in a Douglas-fir forest stand, *Remote Sens Environ*, 2007, 107, 385-401.

Coops, N.C.; Hilker, T; Wulder, M.A.; St-Onge, B.; Newnham G.; Siggins, A.; Trofymow, J.A. Estimating canopy structure of Douglas-fir forest stands from discrete-return lidar. *Trees*, 2007, *21*, 295-310.

Cosh, M. H.; Brutsaert, W. Microscale structural aspects of vegetation density variability, *J. Hydrol.*, 2003, 276,128-136.

Cox, P.M.; Betts, R.A.; Bunton, C.B.; Essery, R.L.H.; Rowntree, P.R.; Smith, J. The impact of new land surface physics on the GCM simulation of climate and climate sensitivity. *Climate Dyn.* 1999, *15*, 183-203.

Cox, P.M.; Betts, R.A.; Jones, C.D.; Spall, S.A.; Totterdell, I.J. Acceleration of Global Warming Due to Carbon-Cycle Feedbacks in a Coupled Climate Model. *Nature* 2000, *408*, 184.

Creed, I.F.; Band, L.E. Exploring Functional Similarity in the Export of Nitrate-N From Forested Catchments: a Mechanistic Modeling Approach. *Water Resour. Res.*, 1998, *34*, 3079- 3093.

Crevoisier, C.; Gloor, M.; Gloaguen, E.; Horowitz, L.W.; Sarmiento, J.; Sweeney, C.; Tans, P. A direct carbon budgeting approach to infer carbon sources and sinks: Design and synthetic application to complement the NACP observation network. *Tellus B*, 2006, *58*, 366-375.

Daly, E.; Porporato, A.; Rodriguez-Iturbe, I. Coupled Dynamics of Photosynthesis, Transpiration, and Soil Water Balance. Part I: Upscaling From Hourly to Daily Level. *J. Hydrometeo.* 2004, *5*, 546-558.

De Rosnay, P.; Drusch, M.; Boone, A.; Balsamo, G.; Decharme, B.; Harris, P.; Kerr, Y.; Pellarin, T.; Polcher, J.; Wigneron J.-P. AMMA Land Surface Model Intercomparison Experiment coupled to the Community Microwave Emission Model: ALMIP-MEM, *J. Geophys. Res.*, 2009, *114*, D05108, doi:10.1029/2008JD010724.

Deardorff, J.W. Efficient prediction of ground surface temperature and moisture, with inclusion of a layer of vegetation. *J. Geophys. Res.*, 1978, 83, 1889-1903.

Dickinson, R.E.; Berry, J.A.; Bonan, G.B. et al. Nitrogen controls on climate model evapotranspiration. *J. Climat.* 2002, *15*, 278–295.

Diego, S.; Peylin, P.; Viovy, N.; Ciais, P. Optimizing a process-based model with eddy-covariance flux measurements: A pine forest in southern France. *Global Biogeochemical cycles*, 2007, 21, GB2013, doi:1029/2006GB002834.

Dirmeyer, P.A. Vegetation stress as a feedback mechanism in mid-latitude drought. *J. Climate*, 1994, 7, 1463-1483.

Dobson, M.C.; Ulaby, F.T.; Le Toan, T.; Beaudoin, A.; Kasischke, E.S.; Christensen, N.: Dependence of radar backscatter on coniferous forest biomass. IEEE Transactions on Geoscience and Remote Sensing, 1992, 30, 412–415.

D'odorico, P.; Porporato, A.; Laio, F.; Ridolfi, L.; Rodriguez-Iturbe, I. Probabilistic Modeling of Nitrogen and Carbon Dynamics in Water-Limited Ecosystems. *Ecol. Model.* 2004, *179*, 205-219.

Drolet, G.G.; Middleton, E.M.; Huemmrich, K.F.; Hall, F.G.; Amiro, B.D.; Barr, A.G. Regional mapping of gross light-use efficiency using MODIS spectral indices, *Remote., Sens. Environ.*, 2008, 112, 3064-3078.

Efron, B.; Tibshirani, R. J. An Introduction to the Bootstrap. *Chapman & Hall/CRC, Boca Raton*, 1993.

Engman, E.T. Remote sensing applications to hydrology: future impact. *Hydro. Sci. J.* 1996, 41, 637-647.

Enting, I.G.; Trudinger, C.M.; Francey, R.J. A synthesis inversion of the concentration and $\delta^{13}C$ of atmospheric $CO_2$. *Tellus B* 1995, 47, 35-52.

Falge, E.; Baldocchi, D.; Olson, R. et al. Gap filling strategies for defensible annual sums of net ecosystem exchange, *Agric. For. Meteorol.*, 2001, 107, 43–69.

Falk, M.; Wharton, S.; Schroeder, M. et al. Flux partitioning in an old-growth forest: seasonal and interannual dynamics. *Tree Physiol.* 2008, 28, 509-520.

Fan, S.; Gloor, M.; Mahlman, J.; Pacala, S.; Sarmiento, J. et al. A large terrestrial carbon sink in North America implied by atmospheric and oceanic carbon dioxide data and models. *Science* 1998, 282, 442-446.

Farquhar, G.D.; Ehleringer, J.R.; Hubick, K.T. Carbon isotope discrimination and photosynthesis. *Annual Review of Plant Physiology and Plant Molecular Biol.* 1989, 40, 503–537.

Farquhar, G.D.; O'Leary, M.H.; Berry, J.A. On the relationship between carbon isotope discrimination and the intercellular carbon dioxide concentration in leaves. *Aust. J. Plant Physi.* 1982, 9, 121–137.

Field, C. B.; Randerson, J. T.; Malmstrom, C. M. Global net primary production-combining ecology and remote sensing. *Remote Sensing of Environment*, 1995, 51, 74–88,.

Finnigan, J. The footprint concept in complex terrain, *Agric For Meteorol.*, 2004,127,117-129,.

Foken, T., Leclerc, M.Y.: Methods and limitations in validation of footprint models. *Agricultural and Forest Meteorology*, 2004, 127, 223-234.

Foley, J.A.; Prentice, I.C.; Ramankutty, N.; Levis, S.; Pollard, D.; Sitch, S.; Haxeltine, A.; An integrated biosphere model of land surface processes, terrestrial carbon balance, and vegetation dynamics. *Global Biogeochem. Cycles* 1996, 10, 603-628.

French, R.H.; Miller, J.J.; Dettling, C.; Carr, J. Use of remotely sensed data to estimate the flow of water to a playa lake. *J. Hydro.*2006, 325, 67–81.

Friedlingstein, P.; Dufresne, J.L.; Cox, P.M.; Rayner, P. How Positive Is the Feedback Between Climate Change and the Carbon Cycle? *Tellus B* 2003, 55, 692-700.

Gao, F., Masek, J., Schwaller,M., and Hall, H: On the blending of the Landsat andMODIS surface reflectance: Predicting daily Landsat surface reflectance. *IEEE Transactions on Geosciences and Remote Sensing*, 2006, 44, 2207–2218,.

Gedney, N.; Cox, P.M.; Betts, R.A.; Boucher, O.; Huntingford, C.; Stott, P.A. Detection of a Direct Carbon Dioxide Effect in Continental River Runoff Records. *Nature*, 2006, 439, 835-838.

Gerbig, C.; Lin, J. ; Wofsy, S.C.; Daube, B.C.; Andrews, A.E., *et al.* Towards constraining regional scale fluxes of $CO_2$ with atmospheric observations over a continent: 1. Observed spatial variability from airborne platforms. *J. Geophys. Res.* 2003, 108, D4756,10.1029/2002JD003018.

Gillies, R.R.; Cui, J.; Carlson, T.N.; Kustas, W.P.; Humes, K.S. Verification of a method for obtaining surface soil water content and energy fluxes from remote measurements of NDVI and surface radiant temperature. *Inter. J. Remot. Sens.*, 1997, 18, 3145-3166.

Gloor, M.; Bakwin, P.; Hurst, D.; Lock, L.; Draxler, R.; Tans, P. What is the concentration footprint of a tall tower? *J. Geophys. Res.* 2001, *106*, 17,831-17,840.

Gloor, M.; Fan, S.M.; Pacala, S.; Sarmiento, J.; Ramonet, M. A model-based evaluation of inversions of atmospheric transport, using annual mean mixing ratios, as a tool to monitor fluxes of nonreactive trace substances like $CO_2$ on a continental scale. *J. Geophys. Res.* 1999, *104*, 14245-14260.

Gobron, N.; Pinty, B.; Verstraete, M.; Govaerts, Y. The MERIS Global Vegetation Index (MGVI): Description and preliminary application, *International Journal of Remote Sensing*, 1999, 20, 1917- 1927.

Goetz, S.J.; Prince, S.D.; Goward, S.N.; Thawley, M.M.; Small, J. Satellite remote sensing of primary production: an improved production efficiency modelling approach. *Ecol. Model.* 1999, *122*, 239-255.

Gökede, M.; Rebmann, C.; Foken, T. A combination of quality assessment tools for eddy co-variance measurements with footprint modelling for the characterisation of complex sites, *Agric. For. Meteorol.*, 2004,127, 175-188.

Govaerts, Y. M.; Verstraete, M. M.; Pinty, B.; Gobron, N. Designing optimal spectral indices: A feasibility and proof of concept study, *International Journal of Remote Sensing*, 1999, 20, 1853- 1873.

Govind, A.; Chen, J. M.; Ju, W. Spatially explicit simulation of hydrologically controlled carbon and nitrogen cycles and associated feedback mechanisms in a boreal ecosystem, *J. Geophys. Res.*, 2009, *114*, G02006, doi:10.1029/2008JG000728.

Govind, A.; Chen, J.M.; Margolis, H.; Ju, W.; Sonnentag, O.; Giasson, M.-A. A spatially explicit hydro-ecological modeling framework (BEPS-TerrainLab V2.0): Model description and test in a boreal ecosystem in Eastern North America. *J. Hydro..* 2009, *367*, 200-216.

Grant, R.F.; Black, T.A.; den Hartog, G. et al. Diurnal and annual exchanges of mass and energy between an aspen-hazelnut forest and the atmosphere: testing the mathematical model Ecosys with data from the BOREAS experiment. *J. Geophys. Res.* 1999, *27*, 27699-27718.

Grant, R.F; Flanagan, L.B. Modeling stomatal and nonstomatal effects of water deficits on CO2 fixation in a semiarid grassland. *J. Geophys. Res.* 2007, *112*, G03011, doi:10.1029/2006JG000302.

Griffis, T. J.; Black, T. A.; Morgenstern , K. Ecophysiological controls on the carbon balances of three southern boreal forests. *Agricultural and Forest Meteorology*, 2003, 117, 53-71.

Griffis, T.J.; Baker, J.M.; Zhang, J. Seasonal dynamics and partitioning of isotopic $CO_2$ exchange in a $C_3/C_4$ managed ecosystem. *Agric. For. Meteor.* 2005, *132*, 1-19.

Gurney, K. R. et al. TransCom 3 $CO_2$ Inversion Intercomparison: 1. Annual mean control results and sensitivity to transport and prior flux information. *Tellus B* 2003, *55*, 555- 579.

Gurney, K.R.; Law, R.M.; Denning, A.S.; Rayner, P.J.; Baker, D. et al. Towards robust regional estimates of $CO_2$ sources and sinks using atmospheric transport models. *Nature* 2002, *415*, 626-630.

Hall, F. G.; Hilker, T.; Coops, N. C.; Lyapustin, A.; Huemmrich, F.; Middleton, E.; Margolis, H.; Drolet, G.; Black, T. Multi-angle remote sensing of forest light use efficiency by observing PRI variation with canopy shadow fraction. *Remote Sensing of Environment.* , 2008, 112:3201-3211.

Hall, F.; Sellers, P.; Strebel, D.; Kanemasu, E.; Kelly, R.; Blad, B.; Markham, B.; Wang, J.; Huemmrich, F. Satellite remote sensing of surface energy and mass balance results: Results from FIFE. *Remote Sensing of Environment*, 1991, 35, 187–199.

Harris, P. P.; Huntingford, C.; Cox, P. M.; Gash, J. H.C.; Malhi, Y. Effect of soil moisture on canopy conductance of Amazonian rainforest. *Agric. For. Meteor.* 2004, *122*, 215-227.

Hedges, J.I. Global biogeochemical cycles: progress and problems. *Mar. Chem.*, 1992. 39, 67–93.

Helliker, B.R.; Berry, J.A.; Betts, A.K.; Bakwin, P.S.; Davis, K.J.; Denning, A.S.; Ehleringer, J. R.; Miller, J.B.; Butler, M.P.; Ricciuto D.M. Estimates of net $CO_2$ flux by application of equilibrium boundary layer concepts to $CO_2$ and water vapor measurements from a tall tower. *J. Geophys. Res.* 2004, *109*, D20106, doi:10.1029/2004JD004532.

Henderson-Sellers, A.; Yang, Z.-L.; Dickinson, R.E. The Project for Intercomparison of Land-surface Parameterization Schemes, *Bull. Amer. Meteorol. Soc.*, 1993, 74, 1335.

Hilker, T.; Coops, N.C.; Hall, F.G.; Black, T.A.; Chen, B.; Krishnan, P.; Wulder, M.A.; Sellers, P.J.; Middleton, E.M.; Huemmrich, K.F. A modeling approach for upscaling gross ecosystem production to the landscape scale using remote sensing data. *J. Geophys Res.*,2008 113, G03006, doi:10.1029/2007JG000666.

Holdridge, L. Determination of world plant formations from simple climatic data, *Science* 1947,105, 367-368.

Hollenbeck, K.J.; Schmugge, T.J.; Homberger, G.M.; Wang, J.R. Identifying soil hydraulic heterogeneity by detection of relative change in passive microwave remote sensing observations. *Water Resour. Res.* 1996, *32*, 139–148.

Hollinger, D. Y.; Goltz, S.M.; Davidson, E. A. Seasonal patterns and environmental control of carbon dioxide and water vapour exchange in an ecotonal boreal forest. *Global Change Biology*, 1999, 5, 891–902.

Hollinger, D. Y.; Richardson, A. D. Uncertainty in eddy covariance measurements and its application to physiological models. *Tree Physiology*, 2005, 25, 873-885.

Horst, T.W.; Weil, J.C. Footprint estimation for scalar flux measurements in the atmospheric surface layer. *Boundary-Layer Meteorol.*, 1992, 59, 279-296.

Huang, M.; Ji, J.; LI, K.; Liu, Y.; Yang, F.; Tao, B. The ecosystem carbon accumulation after conversion of grasslands to pine plantations in subtropical red soil of South China, *Tellus*, 2007, 59B, 439-448.

Huete, A. R.; Liu, H. Q.; Batchily, K.; vanLeeuwen, W. A comparison of vegetation indices over a global set of TM images for EOS-MODIS, *Remote Sensing of Environment*, 1997, 59, 440–451.

Huete, A.; Didan, K.; Miura, T.; Rodriguez, E. P.; Gao, X.; Ferreira, L. G. Overview of the radiometric and biophysical performance of the MODIS vegetation indices, *Remote Sensing of Environment*, 2002, 83, 195– 213.

Jackson, T.J. Measuring surface soil moisture using passive microwave remote sensing. *Hydro. Processes* 1993, 7, 139–152.

Jain, M.K.; Kothyari, U.C.; Raju K.G. A GIS based distributed rainfall–runoff model. *J. Hydro.*2004, *299*, 107–135.

Jarvis, P.G. Interpretation of variations in leaf water potential and stomatal conductance found in canopies in field. *Philos Trans R Soc Lond B Biol Sci*, 1976, 273, 593-610.

Joos, F.; Prentice, I.C.; Sitch, S.; Meyer, R.; Hooss, G.; Plattner, G.K.; Gerber, S.; Hasselmann, K. Global Warming Feedbacks on Terrestrial Carbon Uptake under the

Intergovernmental Panel on Climate Change (Ipcc) Emission Scenarios, *Global Biogeochem. Cycles*, 2001, *15*, 891-907.

Ju, J.C.; Roy, D.P. The availability of cloud-free Landsat ETM+ data over the conterminous United States and globally. *Remote Sensing of the Environment*, 2007 112, 1196–1211.

Kalma, J. D.; McVicar, T. R.; McCabe, M. F. Estimating Land Surface Evaporation: A Review of Methods Using Remotely Sensed Surface Temperature Data, *Surv. Geophys.*, 2008, *29*, 421–469.

Kalvelage, T.; Willems, J. Supporting users through integrated retrieval, processing, and distribution systems at the Land Processes Distributed Active Archive Center. *Acta Astronautica*, 2005, 56, 681–687.

Kasischke, E.S.; Melack, J.M.; Dobson, M.C. The use of imaging radars for ecological applications—a review. *Remote Sensing of Environment*, 1997, 59, 141–156

Kickert, R.N.; Tonella, G.; Simonov, A.; Krupa, S. Predictive modeling of effects under global change. *Environ. Pollut.* 1999, *100*, 87-132.

Kim, G.; Barros, A.P. Space-time characterization of soil moisture from passive microwave remotely sensed imagery and ancillary data. *Remote Sens. Environ.* 2002, *81*, 393-403.

Kite, G.W.; Pietroniro, A. Remote sensing applications in hydrological modelling. *Hydro. Sci. J.* 1996, *41*, 563-591.

Kite, G.W.; Droogers, P. Comparing evapotranspiration estimates from satellites, hydrological models and field data. *J. Hydrol.* 2000, *229*, 3-18.

Kley, D.; Kleinmann, M.; Sandermann, H.; Krupa, S. Photochemical oxidants: state of the science. *Environ. Pollut.* 1999, *100*, 19-42.

Kljun, N.; Kormann, R.; Rotach, M.W. Comparison of the Langrangian footprint model LPDM-B with an analytical footprint model. *Boundary-Layer Meteorology* , 2003, 106, 349-355.

Koster, R.D.; Guo, Z.C.; Dirmeyer, P.A.; Bonan, G.;Chan, E.; Cox, P.; Davies, H.; Gordon, C.T.; Kanae, S.; Kowalczyk, E., *et al.* Glace: the Global Land-Atmosphere Coupling Experiment. Part I: Overview. *J. Hydrometeor.* 2006, *7*, 590-610.

Kurth, W. Morphological models of plant growth: possibilities and ecological relevance. *Ecol. Model.* 1994, *75*, 299–308

Lai C.T.; Ehleringer, J.; Schauer, A.; Tans , P.P.; Hollinger, D.; Paw, K.T.U.; Munger, J.; Wofsy, S. Canopy-scale $\delta^{13}C$ of photosynthetic and respiratory $CO_2$ fluxes: observations in forest biomes across the United States. *Global Change Biol.* 2005, *11*, 633-643, doi: 10.1111/j.1365-2486.2005.00931.x.

Lai, C.-T.; Schauer, A.J.; Owensby, C. et al. Regional $CO_2$ fluxes inferred from mixing ratio measurements: estimates from flask air samples in central Kansas, USA. *Tellus B* 2006, *58*, 523-536.

Law, B. E.; Waring, R. H.; Anthoni, P. M. Measurements of gross and net ecosystem, productivity and water vapour exchange of a Pinus ponderosa ecosystem, and evaluation of two generalized models, *Global Change Biol.*, 2000, 6, 155-168.

Law, B.E.; Falge, E; Gu, L. *et al.* Environmental controls over carbon dioxide and water vapor exchange of terrestrial vegetation. *Agric. For. Meteor.* 2002, *113*, 97-120.

Lee, X.; Fuentes, J. D.; Staebler, R. M. Long-term observation of the atmospheric exchange of CO2 with a temperature deciduous forest. *Journal of Geophysical Research*, 1999, 104, 15975–15984.

Lefsky, M.A.; Cohen, W.B.; Parker, G.G.; Harding, D.J. Lidar remote sensing for ecosystem studies. *BioScience*, 2002, 52, 19–30.

Leuning, R. Estimation of scalar source/sink distributions in plant canopies using Lagrangian dispersion analysis: Corrections for atmospheric stability and comparison with a multilayer canopy model. *Boundary-Layer Meteor.* 2000, *96*, 293-314.

Levin, S.A. Complex adaptive systems: exploring the known, unknown and unknowable. *Bull. Amer. Meteor. Soc.* 2002, *40*, 1-19.

Levis, S.; Foley, J.A.; Pollard D. Potential high-latitude vegetation feedbacks on $CO_2$-induced climate change, *Geophys. Res. Lett*, 1999, *26*, 747–750.

Levy, P. E.; Grelle, A.; Lindroth, A. et al. Regional-scale CO2 fluxes over central Sweden by a boundary layer budget method. *Agric. For. Meteor.* 1999, 98-99, 169-180.

Li, Z.; Yu, G.; Xiao, X. et al. Modeling gross primary production of alpine ecosystems in the Tibetan Plateau using MODIS images and climate data, *Remote Sensing of Environment*, 2007, 107, 510–519.

Liang, X.; Guo, J.; Leung, L.R. Assessment of the effects of spatial resolutions on daily water flux simulations. *J. Hydrol.* 2004, *298*, 287-310.

Liang, X.; Lettenmaier, D.P.; Wood, E.F.; Burges, S.J. A simple hydrologically based model of land surface water and energy fluxes for GSMs. *J. Geophys. Res.*, 1994, *99*, 14415-14428.

Lin, J.C.; Gerbig, C.; Wofsy, S.C.; Andrews, A.E.; Daube, B.C.; Grainger, C.A. Stephens, B.B., Bakwin, P.S.; Hollinger D.Y. Measuring fluxes of trace gases at regional scales by Lagrangian observations: Application to the $CO_2$ Budget and Rectification Airborne (COBRA) study, *J. Geophys. Res.* 2004, *109*, D15304, doi:10.1029/2004JD004754.

Liu, Y., Yu, G., Wen, X., Wang, Y., Song, X, Li, J., Sun, X., Yang, F., Chen, Y., Liu Q.: Seasonal dynamics of CO2 fluxes from subtropical plantation coniferous ecosystem. *Science in China Series D*, 2006, 49 Supp.II, 99-109.

Lloyd, J.; Francey, R.J.; Mollicone, D. et al. Vertical profiles, boundary layer budgets, and regional flux estimates for $CO_2$ and its $^{13}C/^{12}C$ ratio and for water vapor above a forest/bog mosaic in central Siberia. *Global Biogeochem. Cycles* 2001, *15*, 267-284.

Lloyd, J.; Kruijt, B.; Hollinger, D.Y., *et al.* Vegetation effects on the isotopic composition of atmospheric $CO_2$ at local and regional scales: theoretical aspects and a comparison between rain forest in Amazonia and a boreal forest in Siberia. *Austr. J. Plant Physi.* 1996, *23*, 371-399.

Loiselle, S.; Bracchini, L.; Bonechi, C.; Rossi, C. Modelling energy fluxes in remote wetland ecosystems with the help of remote sensing. *Ecol. Model.* 2001, *145*, 243–261.

Luckman, A.; Baker, J.; Honzak, M.; Lucas, R. Tropical forest biomass density estimation using JERS-1 SAR: Seasonal variation, confidence limits and application to image mosaics. *Remote Sensing of Environment*, 1998, *63*, 126–139.

Ma, S.; Baldocchi, D.D.; Xu, L. *et al.* Interannual variability in carbon dioxide exchage of an oak/grass sacanna and open grassland in California. *Agric. For. Meteor.* 2007, *147*,157-171

Maas, S.J. Use of remotely sensed information in agricultural crop growth models. *Ecol. Model.* 1988, *41*, 247–268.

Marquis, M.; Tans, P. Carbon Crucible, *Science*, 2008, *320*, 460-470.

Matthews, E.; Hansen, J. Eds. Land Surface Modeling: A Mini-Workshop, May 1998: http://www.giss.nasa.gov/meetings/landsurface1998/section3.html

Melesse A.M.; Vijay N.; Jon H. Analysis of Energy Fluxes and Land Surface Parameters in Grassland Ecosystem: Remote Sensing Perspective. *Int. J. Remot. Sens.* 2008, *29*, 3325-3341.

Melesse, A.M.; Weng, Q.; Thenkabail, P.; Senay, G. Remote sensing sensors and applications in environmental resources mapping and modeling. *Sensors* 2007, *7*, 3209-3241

Meybeck, M. Carbon, nitrogen, and phosphorous transport by world rivers. *Am. J. Sci.* 1982, *282*, 401-450.

Monteith, J. L. Solar radiation and productivity in tropical ecosystems, *J. of Applied Ecology*, 1972, 9, 747-766.

Monteith, J. L. The photosynthesis and transpiration of crops, *Experimental Agriculture*, 1966, 2, 1-14.

Montzka, C.; Canty, M.; Kunkel, R.; Menz, G.; Vereecken, H.; Wendland F. Modelling the water balance of a mesoscale catchment basin using remotely sensed land cover data. *J. Hydro.* 2008, *353*, 322-334.

Natl. Res. Counc. *Earth Observations from Space: The First 50 Years of Scientific Achievements.* 2008, Washington, DC: Natl. Acad. Press. 142 pp.

Oge´e, J.; Brunet, Y.; Loustau, D.; Berbigier, P.; Delzon, S. MuSICA, a $CO_2$, water and energy multi-layer, multi-leaf pine forest model: Evaluation from hourly to yearly time scales and sensitivity analysis. *Global Change Biol.* 2003, 9, 697-717.

Papale, D.; Valentini, R. A new assesment of European forests carbon exchanges by eddy fluxes and artificial neural network spatialization. *Global Change Biology*, 2003, 9, 525-535.

Parton, W.J.; Ojima, D.S.; Cole, C.V.; Schimel, D.S. A general model for soil organic matter dynamics: sensitivity to litter chemistry, texture and management. *Soil Sci. Soc. Am. J.* 1993, *39*, 147 167.

Pataki, D.E.; Ehleringer, J.R.; Flanagan, L.B.; Yakir, D.; Bowling, D.R.; Still, C.J.; Buchmann, N.; Kaplan, J.O.; Berry, J.A. The application and interpretation of Keeling plots in terrestrial carbon cycle research. *Global Biogeochem.Cycles* 2003, *17*, 1022, doi:10.1029/2001GB001850.

Pielke, R.A. Sr. Influence of the spatial distribution of vegetation and soils on the prediction of cumulus convective rainfall. *Rev. Geophys.* 2001, *39*, 151-177.

Pietroniro, A.; Leconte, R. A review of Canadian remote sensing and hydrology, 1999-2003. *Hydro. Processes* 2005, *19*, 285-301.

Plummer, S.E. Perspectives on combining ecological process models and remotely sensed data. *Ecol. Model.* 2000, *129*, 169-186.

Ponton ,S; Flanagan, L.B.; Alstard, K; Johnson, B.G.; Morgenstern, K; Kljun, N; Black, T.A.; Barr, A.G. Comparison of ecosystem water-use efficiency among Douglas-fir forest, aspen forest and grassland using eddy covariance and carbon isotope techniques. *Global Change Biol.* 2006, *12*, 294-310.

Porporato, A.; D'odorico, P.; Laio, F.; Rodriguez-Iturbe, I. Hydrologic Controls on Soil Carbon and Nitrogen Cycles. I. Modeling Scheme. *Adv. Water Resour.* 2003, *26*, 45-58.

Prentice, I.C.; Cramer, W.; Harrison, S.P.; Leemans, R.; Monserud, R.A.; Solomon, A.M. A global biome model based on plant physiology and dominance, soil properties, and climate. *J. Biogeog.* 1992, *19*, 117-134.

Prentice; I.C.; Farquhar, G.D.; Fasham, M.J.R. et al. In The carbon cycle and atmospheric carbon dioxide; Houghton, J.T., et al. Eds.; Cambridge University Press: *Cambridge, UK*, 2001, pp.183-238.

Prentice, K.; Fung, I.Y. The sensitivity of terrestrial carbon storage to climate change, *Nature* 1990, 346(6279), 48-51.

Prince, S. D.; Goward, S. N. Global primary production: A remote sensing approach, *J. Biogeography*, 1995, 22, 815–835.

Raich, J. W.; Rastetter, E. B.; Melillo, J. M. et al. Potential net primary productivity in South-America — application of a global-model, *Ecological Applications*, 1991, 1, 399–429.

Rannik, U.; Kolari, P.; Vesala, T.; Hari, P. Uncertainties in Measurement and Modelling of Net Ecosystem Exchange of a Forest. *Agric. For. Meteor.* 2006, *138*, 244-257.

Ranson, K. J.; Kovacs, K.; Sun, G.; Kharuk, V. I. Disturbance recognition in the boreal forest using radar and Landsat-7. *Canadian Journal of Remote Sensing*, 2003 ,29, 271–285.

Reed, S.; Koren, V.; Smith, M.; Zhang, Z.; Moreda, F.; Seo, D.J. Overall distributed model intercomparison project results. *J. Hydro.* 2004, *298*, 27–60.

Reichstein, M.; Falge, E.; Baldocchi, D. et al. On the separation of net ecosystem exchange into assimilation and ecosystem respiration: review and improved algorithm, *Glob. Change Biol.* , 2005, 11, 1424–1439.

Reichstein, M., Tenhunen, J. D., Roupsard, O., Ourcival, J. M., Rambal,S., Dore, S., and Valentini, R.: Ecosystem respiration in two Mediterranean veergreen Holm Oak forests: drought effects and decomposition dynamics, *Funct. Ecol.* , 2002, 16, 27–39.

Richardson, A. D.; Hollinger, D.Y.B.; George, G. et al. A multi-site analysis of random error in tower-based measurements of carbon and energy fluxes, *Agricultural Forest Meteorology*, 2006, 136, 1-18.

Ritchie, J.C.; Rango, A. Remote sensing applications to hydrology: introduction. *Hydro. Sci. J.* 1996, *41*, 429–431.

Rodriguez-Iturbe, I. Ecohydrology: A hydrologic perspective of climate-soil-vegetation dynamics. *Water Resour. Res.* 2000, *36*, 3–9.

Rodriguez-Iturbe, I.; Porporato, A.; Laio, F.; Ridolfi, L. Plants in Water-Controlled Ecosystems: Active Role in Hydrologic Processes and Response to Water Stress - I. Scope and General Outline. *Adv. Water Resour.* 2001, *24*, 695-705.

Rouse, J.W.; Haas, R.H.; Schell, J.A. et al. Monitoring the vernal advancement and retrogradation (greenwave effect) of natural vegetation. In *NASA/GSFC Type III Final Report.* Greenbelt, 1974, MD: NASA

Roy, P.; Junchang, J.; Lewis, P. et al. Multi-temporal MODIS–Landsat data fusion for relative radiometric normalization, gap filling, and prediction of Landsat data. *Remote Sensing of Environment*, 2008, 112, 3112−3130.

Ruimy, A.; Jarvis, P. G.; Baldocchi, D. D.; Saugier, B. CO2 fluxes over plant canopies and solar radiation: A review, *Advances in Ecological Research*, 1995, 1-68.

Ruimy, A.; Kergoat, L.; Bondeau, A. Comparing global models of terrestrial net primary productivity (NPP): Analysis of differences in light absorption and light-use efficiency, *Global Change Biology*, 1999, 5, 56−64.

Running, S. W.; Baldocchi, D. D.; Turner, D. P. et al. A global terrestrial monitoring network integrating tower fluxes, flask sampling, ecosystem modeling and EOS satellite data, *Remote Sensing of Environment*, 1999, 70, 108−127.

Running, S. W.; Nemani, R. R.; Heinsch, F. A. et al. A continuous satellite-derived measure of global terrestrial primary productivity: Future science and applications, *Bioscience*, 2004, 56, 547–560.

Running, S. W.; Thornton, P. E.; Nemani, R.; Glassy, J. M. Global terrestrial gross and net primary productivity from the Earth Observing System. In O. E. Sala, R. B. Jackson, & H. A. Mooney (Eds.), *Methods in Ecosystem Science*, (pp. 44–57). , 2000, New York: Springer.

Running, S.W.; Coughlan, J.C. A general model of forest ecosystem processes for regional applications I. Hydrological balance, canopy gas exchange and primary production processes. *Ecol. Model*. 1998, 42, 125-154.

Running, S.W.; Nemani, R.R.; Heinsch, F.A.; Zhao, M.; Reeves, M.; Jolly, M. A continuous satellite-derived measure of global terrestrial primary productivity: Future science and applications. *Bioscience* 2004, 56, 547-560.

Saatchi, S.S.; Moghaddam, M. Estimation of crown and stem water content and biomass of boreal forests using polarimetric SAR imagery. *IEEE Transactions on Geoscience and Remote Sensing*, 2000, 38, 697–709.

Sass, G.Z.; Creed, I.F.: Bird's eye view of forest hydrology: novel approaches using remote sensing techniques. In D. Levia, D. Carlyle-Moses, & T. Tanaka (eds), *Forest Hydrology and Biogeochemistry: Synthesis of Past Research and Future Directions,*(pp.45-68) 2011, New York: Springer.

Schmid, H. P. Experimental design for flux measurements: matching scales of observations and fluxes, Agric For Meteorol. , 1997, 87, 179-200.

Schmugge, J.S.; Kustas, W.P.; Ritchie ,J.C.; Jackson, T.J.; Rango, A. Remote sensing in hydrology. *Adv. Water Resour*. 2002, 25, 1367–1385.

Schultz, G.A. Remote sensing applications to hydrology: runoff. *Hydro. Sci. J*. 1996, 41, 453–475.

Schwalm, C.R.; Ek, A.R. Climate change and site: relevant mechanisms and modeling techniques. *For. Ecol. Manage*. 2001, 150, 241-258.

Seaquist, J.W.; Olsson, L.; Ard" O, J. A remote sensing-based primary production model for grassland biomes. *Ecol. Model*. 2003, 169, 131–155.

Sellers, P.J., *et al*. Modeling the exchange of energy, water, and carbon between continentals and the atmosphere. *Science* 1997, 275, 502-509.

Sellers, P.J.; Mintz, Y.; Sud, Y.C.; Dalcher, A. A Simple Biosphere Model (SiB) for Use Within General-Circulation Models. *J. Atmos. Sci*. 1986, 43, 505-531.

Sellers, P.J.; Randall, D.A.; Collatz, G.J. et al. A revised land surface parameterization (SiB2) for atmospheric GCMs. Part I: model formulation. *J. Climat*. 1996, 9, 676-05.

Seneviratne, S.I.; Luthi, D.; Litschi, M.; Schar C. Land-atmosphere coupling and climate change in Europe. *Nature* 2006, 443, 205–209.

Seth, A.; Giorgi, F.; Dickinson, R.E. Simulating fluxes from heterogeneous land surfaces: Explicit subgrid method employing the biosphere-atmosphere transfer scheme (BATS). *J. Geophys. Res*., 1994, 99(D9), 18,651–18,667.

Shen, S.; Leclerc, M. Y. How large must surface layer inhomogeneities be before they influence the convective boundary layer structure? A case study, *Quart J Roy Meteorol Soc*. , 1995, 121, 1209-1228.

Snyder, P.K.; Delire, C.L.; Foley, J.A. Evaluating the influence of different vegetation biomes on the global climate. *Clim. Dyn*. 2004, 23, 279-302.

Sogachev, A.; Rannik, U.; Vesala, T. Flux footprints over complex terrain covered by heterogeneous forest, Agric. For. Meteorol. , 2004, 127, 142–158.

Stisen, S.; Jensen, K.H.; Sandholt, I.; Grimes D.I.F. A remote sensing driven distributed hydrological model of the Senegal River basin. *J. Hydro.* 2008, *354*, 131-148.

Stoy, P. C.; Katul, G.; Siqueira, G. Separating the effects of climate and vegetation on evapotranspiration along a successional chronosequence in the southeastern US, Global Change Biol. , 2006, 12, 2115-2135.

Su, Z. Remote sensing of land use and vegetation for mesoscale hydrological studies. *Inter. J. Remot. Sens.* 2000, *21*, 213–233.

Su, Z., The surface energy balance system (SEBS) for estimation of the turbulent heat fluxes. *Hydro. Earth Sci.* 2002, *6*, 85–99.

Suits, N.S.; Denning, A.S.; Berry, J.A.; Still, C.J.; Kaduk, J.; Miller, J.B.; Baker I.T. Simulation of carbon isotope discrimination of the terrestrial biosphere. *Global Biogeochem. Cycles* 2005, *19*, GB1017, doi:10.1029/2003GB002141.

Tans, P.P.; Fung, I.Y.; Takahashi, T. Observation Constraints on the global atmospheric $CO_2$ budget. *Science* 1990, *247*, 1431-1438.

Tarantola, A. Inverse Problem: Theory Methods for Data Fitting and Parameter Estimation. *Elsevier*, 1987, New York.

Tilford, S. Global habitability and Earth remote sensing. *Philos. Trans. R. Soc.* 1984 *A* 312(1519):115-18.

Treuhaft, R.N.; Asner, G.P.; Law, B.E.; Van Tuyl, S. Forest leaf area density profiles from the quantitative fusion of radar and hyperspectral data. *Journal of Geophysical Research*, 2002, 107, 4568–4580.

Treut, L.; Somerville, H. R.; Cubasch, U.; Ding, Y.; Mauritzen, C.; Mokssit, A.; Peterson T.; Prather, M. Historical Overview of Climate Change. In *Climate Change 2007: The Physical Science Basis. Contribution of Working Group I to the Fourth Assessment Report of the Intergovernmental Panel on Climate Change* Solomon, S., D. Qin, M. Manning, Z. Chen, M. Marquis, K.B. Averyt, M. Tignor and H.L. Miller Eds. Cambridge University Press, Cambridge, United Kingdom and New York, NY, USA, 2007; pp. 104-105.

Tucker, C.J.; Fung, I.Y.; Keeling, C.D.; Gammon, R. The relationship of global green leaf biomass to atmospheric CO2 concentrations. *Nature* , 1986 319:159–99.

Tucker, C.J. Red and photographic infrared linear combinations for monitoring vegetation. *Remote Sens. Environ.* 1979, 8(2):127–50.

Urbanski, S.; Barford, C.; Wofsy, S. Factors controlling CO2 exchange on timescales from hourly to decadal at Harvard Forest, *J. Geophys. Res.*, 2007, 112, G02020, doi:10.1029/2006JG000293.

Verseghy, D.L. Class-a Canadian Land Surface Scheme for Gcms.1. Soil Model. *Int. J. Climatol.* 1999, *11*, 111-133.

Verseghy, D.L.; McFarlane, N.A.; Lazare, M., CLASS: A Canadian land surface scheme for GCMs: II. Vegetation model and coupled runs. *Int. J. Climatol.* 1993, *13*, 347-370.

Verstraeten, W. W.; Veroustraete, F.; Feyen, J. Assessment of evapotranspiration and soil moisture content across different scales of observation, *Sensors*, 2008, *8*, 70–117.

Viterbo, P.; Beljaars, A.C.M.; An improved land surface parameterization scheme in the ECMWF model and its validation. *J. Climat.* 1995, *8*, 2716-2748.

Wang, J.; Price, K.P.; Rich, P.M. Spatial patterns of NDVI in response to precipitation and temperature in the central Great Plains. *Int. J. Remote Sens.* 2001, *22*, 3827-3844.

Wang, S.; Grant, R.F.; Verseghy, D.L.; Black, T.A. Modeling carbon-coupled energy and water dynamics of boreal aspen forest in a general circulation model land surface scheme. *Int. J. Climatol.* 2002, 22, 1249-1265.

Wang, S.; Grant, R.F.; Verseghy, D.L.; Black, T.A. Modeling carbon dynamics of boreal forest ecosystems using the Canadian land surface scheme. *Clim. Chang.* 2002, 55, 451-477.

Wang, X.-C.; Chen, R.F.; Gardner, G.B. Sources and transport of dissolved and particulate organic carbon in the Mississippi River estuary and adjacent coastal waters of the northern Gulf of Mexico, *Marine Chemistry*, 2004, 89, 241- 256.

Warren, C.R.; Adams, M.A. Distribution of N. Rubisco and photosynthesis in Pinus pinaster and acclimation to light. *Plant Cell Environ.* 2001, 24, 597-609.

Wen, X.; Yu, G.; Sun, X.; Li, Q. et al. Soil moisture effect on the temperature dependence ofecosystem respiration in a subtropical Pinus plantation of southeastern China, *Agric. For. Meteor.* , 2006, 137, 166-175.

Williams, M.S.; Law, B.E.; Anthoni, P.M.;Unsworth, W.H. Use of a simulation model and ecosystems flux data to examine carbon-water interactions in ponderosa pine. *Tree Physiol.* 2001, 21, 287-298.

Wilson, K.B.; Baldocchi, D.D.; Hanson, P.J. Leaf age affects the seasonal pattern photosynthetic capacity and net ecosystem exchange of carbon in a deciduous forest. *Plant Cell Environmen.* 2001, 24, 571-583.

Wilson, K.B.; Baldocchi, D.D.; Hanson, P.J. Spatial and seasonal variability of photosynthetic parameters and their relationship to leaf nitrogen in a deciduous forest. *Tree Physi.* 2000, 20, 565-578.

Wu, J. D.; Wang, D.; Bauer, M. E. Image-based atmospheric correction ofQuickBird imagery of Minnesota cropland. *Remote Sensing of Environment*, 2005,99, 315−325.

Xiao, X. M.; Zhang, Q. Y.; Hollinger, D.; Aber, J.; Moore, B. Modeling gross primary production of an evergreen needleleaf forest using MODIS and climate data, *Ecological Applications*, 2005, 15, 954−969.

Xiao, X.; Boles, S.; Liu, J. Y.; Zhuang, D. F.; Liu, M. L. Characterization of forest types in Northeastern China, using multi-temporal SPOT-4 VEGETATION sensor data, *Remote Sensing of Environment*, 2002, 82, 335−348.

Xiao, X.; Hollinger D.; Aber, J. D.; Goltz, M.; Davidson, E.; Zhang, Q. and Moore III B. Satellite-based Modeling of Gross Primary Production in an Evergreen Needle leaf Forest, Remote Sensing Envir. , 2004, 89, 519 - 534.

Zhang, L. M.; Yu, G. R.; Sun, X. M. et al.; Liu, Y. F.; Xin, D.; Guan D. X.; Yan, J. H. Seasonal variations of ecosystem apparent quantum yield ($\alpha$) and maximum photosynthesis rate (Pmax) of different forest ecosystems in China, *Agric. For. Meteor.* , 2006, 137, 176-187.

Zhang, Y.; Chen, W.; Cihlar, J. A process-based model for quantifying the impact of climate change on permafrost thermal regimes. *J. Geophys. Res.* 2003, 108(D22), 4695, doi:10.1029/2002JD003354.

Zhu, X.; Chen, J.; Gao, F.; Chen, X.; Masek J. An enhanced spatial and temporal adaptive reflectance fusion model for complex heterogeneous regions, *Remote Sensing of Environment*, 2010, doi:10.1016/j.rse.2010.05.032.

# 11

# Multi-Wavelength and Multi-Direction Remote Sensing of Atmospheric Aerosols and Clouds

Hiroaki Kuze

*Centre for Environmental Remote Sensing (CEReS), Chiba University*
*Japan*

## 1. Introduction

Aerosols are liquid and solid particles floating in the atmosphere. Aerosol particles are originated from both natural and anthropogenic origins (Seinfeld & Pandis, 1998). In regard to the radiation balance of the Earth's atmosphere, aerosols reflect solar radiation back to space (direct effect), thus reducing the influence of greenhouse gases, though some type of aerosol causes opposite effects due to absorption of radiation. At the same time, aerosol particles work as nuclei for cloud condensation (indirect effect). Knowledge on these radiative effects of aerosol and cloud, however, is still insufficient so that uncertainties remain in the prediction of future global warming trends (IPCC, 2007). In this respect, intensive efforts are needed to evaluate the optical/physical properties of aerosols and clouds by means of both ground- and satellite-based remote sensing observations.

In order to obtain better understanding of these particulate matters, what is obviously needed is the monitoring technique that enables the retrieval of their optical properties. In this chapter, we propose multi-wavelength and multi-directional remote sensing of atmospheric aerosols and clouds. The proposed method consists of the application of ground-based radiation measurement, lidar measurement, differential optical absorption spectroscopy (DOAS), and satellite observations using natural as well as artificial light sources. Such combinatory approach makes it possible to measure various aspects of radiation transfer through the atmosphere, especially the influence of tropospheric aerosols and clouds. Also, the data provided from the ground-based solar irradiance/sky radiance measurement and DOAS are valuable for precisely characterizing the optical property of aerosol particles near the ground level, including the information from both particulate scattering and gaseous absorption. Such ground data are also indispensable for the atmospheric correction of satellite remote sensing data in and around the visible range of the radiation spectrum. The multi-wavelength and multi-directional observation schemes treated in the present chapter are summarized in Table 1.

## 2. Differential optical absorption spectroscopy (DOAS)

The method of differential optical absorption spectroscopy (DOAS) provides a useful tool for monitoring atmospheric pollutants through the measurement of optical extinction (i.e., the sum of absorption and scattering) over a light path length of a few kilometres (Yoshii et al., 2003, Lee et al., 2009; Si et al., 2005; Kuriyama et al., 2011). The DOAS method in the

| Scheme | Wavelength | Direction | Aerosol | Trace gas |
|--------|-----------|-----------|---------|-----------|
| DOAS | UV, VIS, and NIR, with the resolution of array detector | Nearly horizontal measurement | Measurable through the spectral intensity | Measurable (e.g. $NO_2$ around 450 nm) |
| solar/skylight | UV and VIS, with the resolution of array detector | Solar direction/ any direction including the zenith | Measurable through the spectral intensity | Measurable (e.g. $H_2O$ around 720 nm) |
| Lidar | Fundamental and harmonics of Nd:YAG laser (1064, 532 and 355 nm) | Vertical and slant path observations | Profiling by solving the lidar equation | Not applicable unless tunable lasers are employed |
| Satellite | Spectral bands in UV, VIS, and NIR | Nadir or near-nadir directions | Evaluated and removed in the process of atmospheric correction | Spectral bands are usually too wide to retrieve trace gases |

Table 1. Various schemes of atmospheric observation discussed in this chapter.

visible spectral region is quite suitable for urban air pollution studies, since both nitrogen dioxide ($NO_2$) and aerosol, the most important pollutants originated from human activities, can directly be measured using a near horizontal light path in the lower troposphere.

Although conventional approach in the DOAS measurement is to install a light source, our group at the Centre for Environmental Remote Sensing (CEReS), Chiba University, has established a unique DOAS approach based on aviation obstruction lights (white flashlights) equipped at tall constructions such as smokestacks (Yoshii et al., 2003, Si et al., 2005; Kuriyama et al., 2011). Since those xenon lamps produce flash pulses every 1.5 s during the daytime, they can easily be recognized with the coverage of the whole visible spectral range. Thus, a simple setup consisting of an astronomical telescope and a compact spectro-radiometer can be employed for the measurement of $NO_2$. Also, the stable intensity of the light source makes it possible to retrieve aerosol, or suspended particulate matter (SPM) concentration in the lower troposphere, since the intensity variation of the detected light is mostly ascribable to the aerosol extinction over the light path (Yoshii et al., 2003).

As shown in Figs. 1 and 2, the principle of DOAS analysis of $NO_2$ concentration is based on matching high-pass filtered spectral (wavelength) features between the DOAS-observed optical thickness ($\Delta\tau$) and laboratory-observed molecular absorption spectrum ($\Delta\sigma$). Because of the Lambert-Beer's law, the optical thickness, $\tau$, is expressed as

$$\tau = -\ln(I / I_0),  \tag{1}$$

where $I$ and $I_0$ stand for the observed and reference spectrum, respectively. The reference spectrum can be obtained by either operating the DOAS spectrometer at a short distance

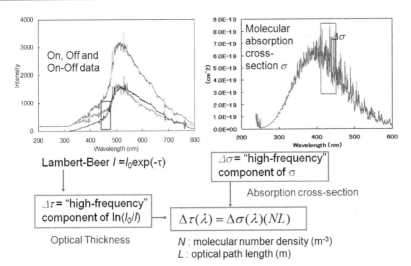

Fig. 1. Schematic flow of the DOAS analysis. The net radiation from the pulsed light source can be retrieved by subtracting the background due to sky radiation, and an appropriate portion of the spectrum is compared with the molecular cross-section spectrum obtained from laboratory measurement. Then, the "high-frequency" components of the observed optical thickness ($\Delta\tau$) and cross-section data ($\Delta\sigma$) are compared to derive the molecular number density along the optical path length, $L$.

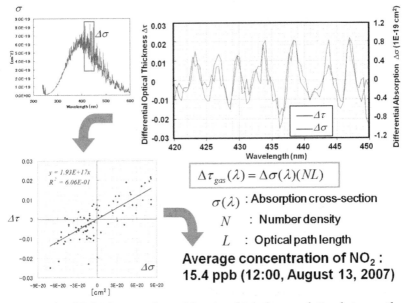

Fig. 2. An example of DOAS spectral matching, in which the correlation between the differential optical thickness from the DOAS data and differential absorption is examined to determine the average volume concentration of $NO_2$ molecules.

from the light source, or observing the spectrum under very clear atmospheric conditions with minimal aerosol loading. The optical thickness is generally proportional to the product of extinction coefficient, $\alpha$, and the light path length, $L$, i.e., $\tau = \alpha L$. In the case of molecular absorption, the extinction coefficient is equal to the absorption coefficient, which can be given as the product of absorption cross-section, $\sigma$, and the molecular number density, $N$, i.e., $\alpha = N\sigma$. Although molecular scattering (Rayleigh scattering) and aerosol scattering (Mie scattering) also exist, their contribution can be eliminated by applying the high-pass filtering to both $\tau$ ($\lambda$) and $\sigma$ ($\lambda$) (where $\lambda$ is wavelength), since the absorption feature of $NO_2$ is a rapidly varying function with wavelength (see insets in Figs. 1 and 2), while the wavelength dependence of Rayleigh or Mie scattering is much more moderate. Thus, after the high-pass filtering, one obtains

$$\Delta\tau = (NL)\Delta\sigma. \tag{2}$$

This indicates that the correlation analysis between the rapidly varying components of the optical thickness and $NO_2$ cross-section in an appropriate wavelength range can lead to the determination of the molecular number density, hence the volume concentration ratio, of $NO_2$ along the DOAS observation light path. An example of the retrieval of $NO_2$ in the Chiba city area is shown in Fig. 3. In this case, the DOAS result shows the average concentration over a light path length of 5.5 km. From Fig. 3, it is seen that the DOAS data show good temporal correlation with the ground sampling data from nearby sampling stations. Note that the temporal resolution (5 min) of the DOAS observation is much better than that of the ground sampling (1 h). The observation of DOAS spectra is limited to daytime, since the white flashlight (Xe light) is replaced with blinking red lamps during night time.

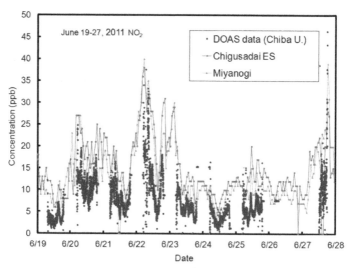

Fig. 3. Comparison of $NO_2$ volume concentration between the DOAS and conventional ground sampling measurements during June 19 - 27, 2011. The DOAS data are based on the measurement at CEReS, Chiba University, using an aviation obstruction flashlamp located around 5.5 km in the north direction. The ground sampling data are from two nearby sampling stations (Chigusadai Elementary School and Miyanogi stations) operated by the municipal government.

The analysis of light intensity detected with a DOAS spectrometer can yield information also on aerosol extinction along the light path. The wavelength dependence of each atmospheric component is exemplified in Fig. 4(a), where it is apparent that the contribution from aerosol extinction is much more significant than that from either $NO_2$ or molecular Rayleigh scattering. The optical thickness associated with aerosol extinction can generally be given as

$$\tau(\lambda) = B(\lambda / \lambda_0)^{-A}, \tag{3}$$

where $A = \alpha_{ang}$ is called the Angstrom exponent and $B$ the turbidity constant. The value of $A$ changes with the aerosol size distribution in such a way that a smaller value (~ 0.5) indicate the dominance of relatively coarse particles (such as sea salt or dust), while a large value (~ 2) that of relatively fine particles (such as ammonium sulfate or ammonium nitrate). The value of $B$, on the other hand, is equal to the aerosol optical thickness as wavelength $\lambda_0$,

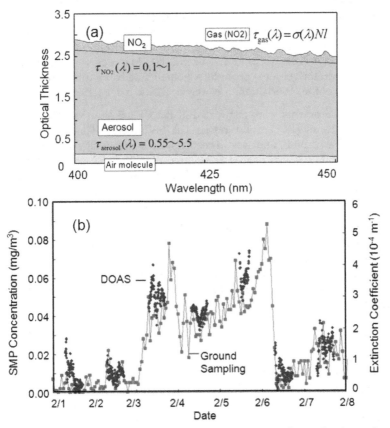

Fig. 4. Aerosol measurement from DOAS data: (a) comparison of contributions of gas ($NO_2$) absorption, aerosol extinction, and molecular extinction (Rayleigh scattering) to DOAS optical thickness, and (b) temporal change of SPM concentration from ground sampling and aerosol extinction coefficient from DOAS during February 1 to 7, 2011 observed in Chiba.

which is chosen to be 550 nm or some appropriate value within the observation wavelength range. Figure 4(b) shows the result of analysis based on eq. (4). As seen from Fig. 4(b) the temporal variation shows good agreement between the DOAS-derived aerosol optical thickness and the SPM mass concentration observed from the ground sampling.

## 3. Solar and skylight radiation measurement

For aerosols, network observation activities have been undertaken in terms of skyradiometer measurements (Takamura & Nakajima, 2004). Alternatively, the use of a compact, stand-alone spectroradiometer (EKO, MS-720) enables the spectral measurements of direct solar radiation (DSR), solar aureole (AUR) and scattered solar radiation (SSR) (Manago & Kuze, 2010; Manago et al., 2010). Since the instrument is powered by batteries with no PC requirement during measurements, it provides better portability compared to a skyradiometer. The wavelength coverage between 350 and 1050 nm with a resolution of 10 nm is useful for precise evaluation of the aerosol optical properties as well as that of the water vapor column amount. The wide dynamic-range measurement of both the direct and scattered solar radiation is attained by means of a thick diffuser and a stable photodiode array, in combination with the automatic exposure control equipped to the handy spectroradiometer (MS-720) (Manago & Kuze, 2011). In order to facilitate the radiation transfer calculation in the retrieval procedure, home-made baffle tubes are used to limit the field of view (FOV) of the observation to 20 deg (SSR) and 5 deg (DSR).

The radiation measurements were conducted at the CEReS site (35.62°N, 140.10°E) under clear-sky conditions, mostly around noon. The SSR measurements were made in 24 directions (north, east, south, and west directions, each with 6 elevation angles). The DSR and AUR components were measured before and after the SSR measurements. The total time required for a set of DSR, AUR, and SSR data was 30 - 40 minutes. Approximately 130 datasets were obtained during the observation period from August 2007 to March 2009.

Independent measurement of AOD was carried out with a sunphotometer (Prede, PSF-100). This instrument has four channels centred at 368, 500, 675, and 778 nm, each having the bandwidth of 5 nm. The wavelength dependence of AOT is analyzed with eq. (3) to obtain the Angstrom exponent. During the daytime the AOD is retrieved from the solar radiation intensity within a FOV of 1 deg at an interval of 10 s. From the sunphotometer measurement, $A = \alpha_{ang}$ and the coefficient $B$ (turbidity constant at the reference wavelength $\lambda_0 = 550$ nm in this case) can be retrieved.

In our ground observation with the battery-operated spectroradiometer, the direct solar irradiance, aureole radiance, and scattered solar radiance were measured in various directions as mentioned above. Even with these detailed measurements, however, usually it is not possible to determine the complete composition of aerosol particles. Thus, we rely on the three-component aerosol model (TCAM), in which three aerosol types of water soluble, oceanic, and soot components are considered as a basis set which is "quasi-complete" to describe the aerosol optical parameters, namely the wavelength dependence of extinction coefficient, single scattering albedo, asymmetry parameter, and scattering phase function. Figure 5 shows the wavelength dependence of the real and imaginary parts of the aerosol refractive index for the three aerosol components. It has been shown that most of the irradiance/radiance values are well reproduced by appropriately adjusting the total and relative contributions of these three basis components as well as the size distribution of each component (Manago et al, 2011). As seen from Fig. 5, the soot component shows remarkably high value of the imaginary part of the refractive index. This indicates that the absorption property is higher (single scattering albedo is lower) for aerosol with more contribution of soot particles. Figure 6 shows an example of the results of the irradiance and radiance observations. Figure 7 shows an example of aerosol optical parameters derived from the TCAM analysis of the data: Fig. 7(a) shows the wavelength dependence of the aerosol extinction coefficient (normalized to the value at 550 nm), (b) single scattering albedo, (c) asymmetry parameter, and (d) scattering phase function at wavelength 550 nm. In Sec. 5 below, we describe the application of these aerosol characteristics to the atmospheric correction of satellite remote sensing data.

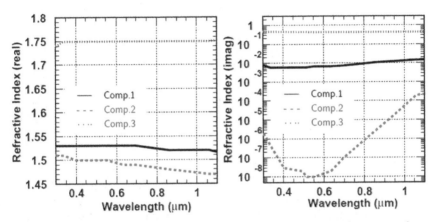

Fig. 5. Real and imaginary parts of the complex refractive index of the three aerosol components: component 1, 2 and 3 refer to the water soluble, oceanic, and soot aerosol types, respectively.

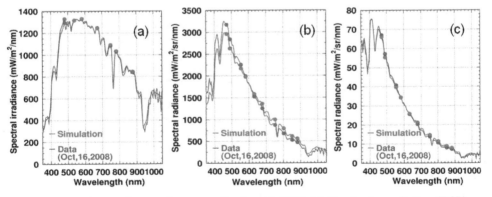

Fig. 6. Spectra observed around noon on October 16, 2008: (a) direct solar radiation (DSR), (b) aureole (AUR), and (c) scattered solar radiation (SSR). Acceptance angle of the instrument is 5 deg for DSR, 5-20 deg for AUR, and 20 deg for SSR. Simulation curves based on the TCAM best fitting are also shown with data points (circles) used for the fitting.

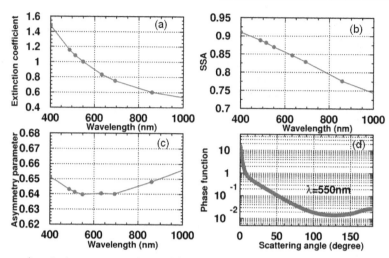

Fig. 7. Aerosol optical parameters derived from the TCAM analysis of the data shown in Fig. 6: (a) wavelength dependence of the aerosol extinction coefficient (normalized to the value at 550 nm), (b) single scattering albedo, (c) asymmetry parameter, and (d) scattering phase function at wavelength 550 nm.

## 4. Lidar measurement of aerosols and clouds

While the DOAS method and skylight/solar radiation measurement lead to the retrieval of atmospheric information integrated over optical paths, the lidar measurement makes it possible to measure aerosol and cloud distributions (profiles) along the optical path. Here we report the result of multi-wavelength lidar measurement conducted at CEReS. Conventionally lidar data have been analyzed by means of the Fernald method (Fernald, 1984), in which the lidar equation

$$P(R) = P_0 \frac{c\tau}{2} AK \frac{G(R)}{R^2} \beta(R) \exp\left[ -2\int_0^R \alpha(R')\, dR' \right] \quad (4)$$

is solved by starting the integration from the far-end boundary. In eq. (4), $P(R)$ is the power detected by the lidar system corresponding to a distance $R$, $P_0$ is the power of the emitted laser radiation, $c$ is the speed of light, $\tau$ is the time duration of the laser pulse, $A$ is the area of the lidar telescope, $G(R)$ is the function describing the overlap between the laser beam and telescope field of view, $\beta(R)$ is the backscattering coefficient, and $\alpha(R)$ is the extinction coefficient. Since both air molecules and aerosol particles contribute to the scattering and extinction, one needs to separate these two components in solving the lidar equation. This can be achieved by introducing the ratio between the extinction coefficient and the backscattering coefficient. Thus, for aerosols,

$$S_1(R) = \alpha_1(R) / \beta_1(R) = \sigma_1(R) / \left( \frac{d\sigma_1}{d\Omega} \right)_{\theta=\pi} \quad (5)$$

is assumed, whereas for air molecules,

$$S_2 = \alpha_2(R) / \beta_2(R) = 8.52 \ (\text{sr}) \tag{6}$$

is used as a constant. In eqs. (5) and (6), suffix 1 and 2 refer to aerosol and air molecule, respectively. In eq. (5), $\sigma_1(R)$ and $(d\sigma_1/d\Omega)_{\theta=\pi}$ indicate the total cross-section and backward differential cross-section of aerosol scattering, respectively. The parameter $S_1$ is often called the lidar ratio. In eq. (6), the range dependence of $S_2$ is omitted, since the composition of air molecules is stable througout the troposphere. Under these assumptions, the lidar equation can be analytically solved as

$$\alpha_1(R) = -\frac{S_1(R)}{S_2}\alpha_2(R) + \frac{S_1(R)\,X(R)\exp I(R)}{\dfrac{X(R_c)}{\dfrac{\alpha_1(R_c)}{S_1(R_c)} + \dfrac{\alpha_2(R_c)}{S_2}} + J(R)}. \tag{7}$$

Here

$$X(R) = R^2 P(R) \tag{8}$$

is the range-corrected signal, and functions $I(R)$ and $J(R)$ are defined as

$$I(R) = 2\int_R^{R_c}\left[\frac{S_1(R')}{S_2} - 1\right]\alpha_2(R')\,\mathrm{d}R' \tag{9}$$

and

$$J(R) = 2\int_R^{R_c} S_1(R')\,X(R')\exp I(R')\,\mathrm{d}R'. \tag{10}$$

In eqs. (7), (9) and (10), $R_c$ denotes the range of a far-end boundary, at which each integration is started. The reason that a far-end boundary value is assumed rather than a near-end boundary value is the stability of the numerical evaluation of eq. (7) (Fernald 1984).

Usually signals of a vertically looking lider are analyzed assuming that the aerosol property does not change with the altitude. Under this assumption, the range dependence in eq. (5) can be neglected. Even in this case, however, it is necessary to determine the value of lidar ratio as a function of wavelength [$S_1=S_1(\lambda)$] in order to analyze multi- wavenength lidar data. One way to accomplish this is to use ancillary data from a sunphotometer (Kinjo et al., 1999), since the wavelength dependence of optical thickness provides a constraint to the intagration of $\alpha_1(R, \lambda)$ from $R=0$ to $R =R_c$. Another approach is to employ the aerosol properties measured at the ground level. In the case of Fig. 8, for example, the $S_1$ values of 54.7, 53.0, 46.0 and 43.2 sr are used for $\lambda = 355$, 532, 756 and 1064 nm, respectively, as derived from the chemical analysis of ground sampling data taken monthly at CEReS (Fukagawa et al., 2006). In Fig. 8, panel (a) shows the temporal variation of the aerosol extinction profile measured for 1064 nm and relative humidity (RH) at the ground level, while panel (b) depicts that of the profile of the Angstrom exponent, $\alpha_{\text{ang}}$, as derived from the analysis of lidar data for the four wavelengths. The features in these panels indicate

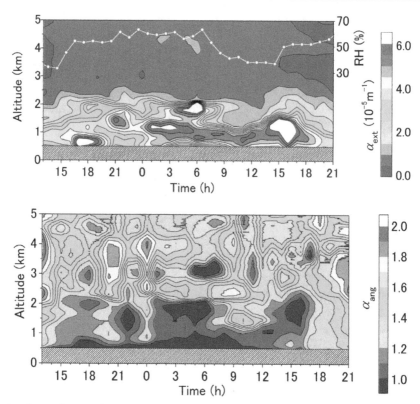

Fig. 8. Analysis of vertical looking multi-wavelength lidar data: (a) extinction profile for wavelength $\lambda$ = 1064 nm observed at CEReS on 17-18 November 2005, and Angstrom exponent derived from extinction coefficients observed for $\lambda$ = 355, 532, 756 and 1064 nm. The analysis is based on the Fernald method with lidar parameters $S_1$ = 54.7, 53.0, 46.0 and 43.2 sr for each lidar wavelength (based on sampling result at CEReS) and the reference altitude of $R_c$ = 5.5 km.

that relatively higher extinction near the ground level is observed when RH increases, and at the same time, smaller values of $\alpha_{ang}$ are observed. It is likely that both of these observations are due to the aerosol growth associated with the increase of RH.

## 5. Atmospheric correction of satellite remote sensing data

Images taken from satellite sensors are affected by both the ground reflectance and atmospheric conditions, which include the influence of scattering and absorption of air molecules and aerosol particles. The process of atmospheric correction, in which such atmospheric effects are precisely evaluated and removed, is indispensable for extracting the intrinsic information of the ground reflectance from satellite imagery (Tang et al., 2005; Kaufman et al., 1997). Although it is rather straightforward to make corrections on the Rayleigh scattering of air molecules, aerosol particles are quite variable both temporally and spatially. This is due to the variable origins of aerosols, consisting of relatively coarse particles of natural origins (such as sea-salt and soil particles) and relatively fine particles of anthropogenic origins (such as sulphate and soot particles).

In standard atmospheric correction, it is customary to assume some representative aerosol models such as maritime, rural, continental, or urban aerosol, to implement the radiative transfer calculation of a satellite scene. This approach has an obvious disadvantage that if the assumed aerosol properties are different from those of real aerosols included in the satellite scene, the resulting information on the ground reflectance is inaccurate. To overcome this difficulty, here we use the aerosol information derived from the ground observation implemented nearly simultaneously with the satellite overpass. Such ancillary information ensures better separation of the ground and atmospheric effects from satellite imagery. Figure 9 shows the schematic drawing of radiation components considered in the radiative transfer calculation (Kotchenova et al., 2006). In this scheme, the radiance originated from the target area is denoted as $L_{tar}$, which consists of the ground direct ($L_{gd}$) and ground indirect ($L_{gi}$ and $L_{gi}'$) components. The environmental radiance, $L_{env}$, is the component associated with the surface reflection that takes place in adjacent pixels. The atmospheric radiance, $L_{atm}$, consists of two terms, namely, the path radiance due to single scattering ($L_{ps}$) and that due to multiple scattering ($L_{pm}$).

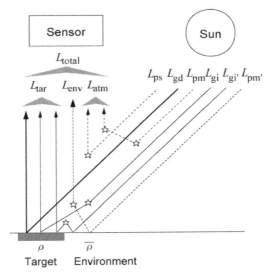

Fig. 9. Schematic drawing of radiation components considered in the radiative transfer calculation. See text for the explanation of radiance components shown in this figure.

In the present research, the ground measurement by means of a compact spectroradiometer was implemented in synchronous with the overpass of the satellite observation. The aerosol optical parameters were derived by analyzing both the direct solar radiation (DSR) and scattered solar radiation (SSR) through the Mie-scattering and radiative transfer calculations, as explained in Sec. 3 of this chapter. When the aerosol loading is relatively small (clear days), it is likely that the aerosol model resulting from this procedure can be applicable to the whole region of the Moderate Resolution Imaging Spectroradiometer (MODIS) image, and the atmospheric correction is applied to the image. Since this correction is based on the aerosol model from the simultaneous measurement, the resulting distribution of the surface reflectance is considered to be more reliable than the result that would be obtained by assuming usually available "standard" aerosol models such as urban, rural, or oceanic models. The surface reflectance map ($\rho_{clear}$ map) on such a "clear" day, in turn, can be used as a standard for that particular season of the year, and the atmospheric correction of MODIS data taken on more turbid days can be implemented on the basis of these standard $\rho_{clear}$ maps. This process leads to the derivation of the distribution of aerosol optical thickness ($\tau$ map).

For each of the visible bands of MODIS, a lookup table of the radiance at the top of the atmosphere, $L_{total}(\rho, \tau_{550})$, was constructed on the basis of the aerosol optical parameters and the geometric data describing the observational conditions of each image. Here, $\rho$ is the diffuse reflectance of each pixel, and $\tau_{550}$ is the aerosol optical thickness (AOT) at wavelength 550 nm. The reflectance property of the surface was assumed to be Lambertian, and the radiative transfer calculation was carried out using the 6S code (Kotchenova et al., 2006).

The atmospheric correction was applied to channels 1 through 4 covering wavelength range between 0.450 and 0.876 µm of the Terra/MODIS and Aqua/MODIS images. The ground resolution of the MODIS sensor is 0.5 km×0.5 km/pixel. The region of 600×600 pixels around Chiba University was extracted from each of the MODIS images, which were taken from the satellite data archiving system of CEReS, Chiba University. The ground observations using the spectroradiometer were carried out at CEReS around noon on nearly cloud-free days from 2007 to 2009 (around 130 days). In order to take the time lag of around 2 h between the satellite overpass (10:00 local time) and the spectroradiometer observation (12:00) into account, the sunphotometer data taken at CEReS were employed to examine the temporal stability of atmospheric conditions. If the AOT derived from the spectroradiometer was close to the AOT value measured with the sunphotometer at the time of satellite overpass, the data were employed in the atmospheric correction. Otherwise, the data were not included in the clear-day analysis lest the instability in the atmospheric condition might also degrade the regional stability (i.e. homogeneity) of the aerosol distribution. Figure 10 shows the wavelength dependence of the surface reflectance ($\rho_{clear}$ map) of the pixel including the location of Chiba University for various months in the year 2008.

Figure 11 shows the seasonal variation of the surface reflectance for the Chiba University pixel obtained from the analysis of MODIS band 4 centred at 550 nm. For the sake of comparison, our previous result obtained from the Landsat-5 analysis is also depicted (Todate et al., 2004). Note that the Landsat reflectance was obtained assuming a standard aerosol model (maritime), whereas the TCAM aerosol model is used in the present MODIS analysis. Pixels with vegetation and soil coverage are shown for the Landsat data, since the ground resolution associated with this sensor (30 m) is much better than the MODIS resolution of 500 m. From Fig. 11, it is seen that the surface reflectance decreases from November to December, due to the decrease in the vegetation coverage during winter. In winter the reflectance shows no critical dependence on the aerosol model assumed in the atmospheric correction because of the fact that the AOT tends to be small. In summer, on the contrary, the AOT generally increases so that the resulting value of surface reflectance varies in accordance with the aerosol model employed in the analysis.

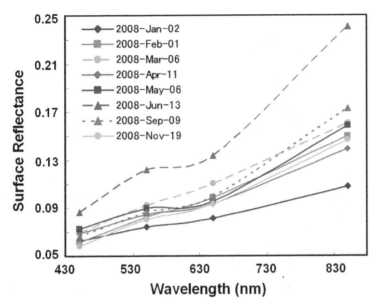

Fig. 10. Surface reflectance at Chiba University (2008)

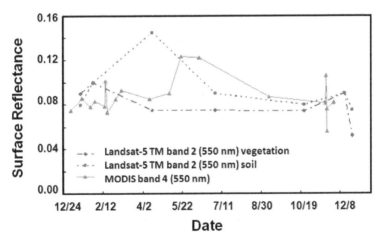

Fig. 11. Seasonal variation of surface reflectance at the MODIS pixel including the Chiba university campus ($\lambda$ =550 nm).

From the present TCAM analysis of MODIS data, monthly reflectance image ($\rho_{monthly}$) is generated for each month as a composite of pixels that exhibit the lowest reflectance. This process ensures the removal of cloud pixels that might contaminate the resulting $\rho$ map. These monthly $\rho$ maps, in turn, are employed in the radiative transfer analysis to derive the aerosol distribution ($\tau$ map) from images taken on relatively turbid days. Examples of the reflectance and aerosol distribution images are shown in Fig. 12.

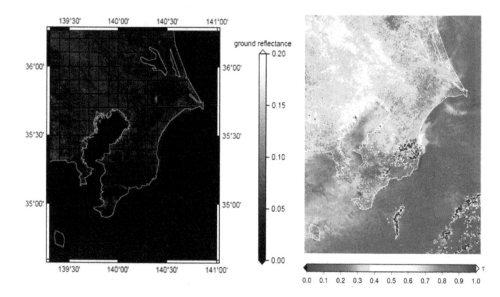

Fig. 12. Analysis of MODIS data in November 2007: (a) surface reflectance map (540 - 570 nm), and (b) aerosol optical thickness at 550 nm on 24 November 2007.

## 6. Conclusion

Optical properties of aerosols and clouds play an important role in the consideration of the Earth's radiation budget. In this chapter, we have described multi-wavelength and multi-directional remote sensing of the troposphere, putting emphasis on the visible part of the spectrum. The DOAS approach enables the direct observation of air pollutants by employing a nearly horizontal light path in the lowest part of troposphere, where the highest concentrations of pollutants such as $NO_2$ and aerosol (SPM) are found. The observation of direct solar radiation and scattered solar radiation (sky light), on the other hand, is useful for retrieving detailed aerosol optical properties under clear-sky conditions. Thus, the data can be quite useful for implementing precise atmospheric correction on satellite-observed imagery that includes the ground observation point. The multi-wavelength lidar observation provides an efficient tool to elucidate the vertical profiles of aerosol particles. By combining the lidar data with some appropriate ancillary data such as the ground-level characterization of aerosol properties, it becomes possible to derive useful information on temporal as well as spatial information on aerosol and cloud characteristics in the atmosphere.

## 7. Acknowledgment

We acknowledge the financial support of the Grant-in-Aid from the Ministry of Education, Culture, Sports, Science & Technology in Japan (#21510006). Also, contributions of a number of graduate students who participated in various researches presented in this chapter are gratefully acknowledged.

# 8. References

Fernald, F. G., Analysis of atmospheric lidar observation: some comments. Appl. Opt. Vol 23, (1984), pp. 652-653, ISSN 0003-6935

Fukagawa, S., Kuze, H., Bagtasa, G., Naito, S., Yabuki, M., Takamura, T., Takeuchi, N., Characterization of seasonal and long-term variation of tropospheric aerosols in Chiba, Japan. Atmospheric Environment, Vol. 40, No.12, (2006), pp. 2160-2169, ISSN 1352-2310

Intergovernmental Panel on Climate Change (IPCC), Climate Change 2007: The Physical Science Basis, Cambridge Univ. Press (2007) ISBN-13: 978-0521705967, Cambridge, U.K.

Kaufman, Y. J., Tanré, D., Gordon, H. R., Nakajima, T., Lenoble, J., Frouins, R., Grassl, H.,Herman, B. M., King, M. D., Teillet, P. M., Passive remote sensing of tropospheric aerosol and atmospheric correction for the aerosol effect. J. Geophys. Res., Vol.102, No.D14, (1997), pp. 16,815-16,830, ISSN 0148-0227

Kinjo, H., Kuze, H., Takeuchi, N., Calibration of the Lidar Measurement of Tropospheric Aerosol Extinction Coefficients. Jpn. J. Appl. Phys. Vol. 38, (1999), pp. 293-297, ISSN 0021-4922

Kotchenova, S.Y., Vermote, E.F., Matarrese, R., Klemm, Jr. F.J., Validation of a vector version of the 6S radiative transfer code for atmospheric correction of satellite data. Part I: Path radiance. Applied Optics Vol. 45, No. 26, No.10 (September 2006), pp.6762-6774, ISSN 0003-6935

Kuriyama, K., Kaba, Y., Yoshii, Y., Miyazawa, S., Manago, N., Harada, I., Kuze, H., Pulsed differential optical absorption spectroscopy applied to air pollution measurement in urban troposphere. J. Quantitative Spectroscopy & Radiative Transfer, Vol. 112, (2011), pp. 277-284, ISSN 0022-4073

Lee, H.; Kim, Y.J., Jung, J., Lee, C., Heue, K.P., Platt, U., Hu, M., 6 Zhu, T., Spatial and temporal variations in $NO_2$ distributions over Beijing, China measured by imaging differential optical absorption spectroscopy. J. Environ. Management, Vol. 90, No.5, (April 2009), pp. 1814-1823, ISSN 0301-4797

Manago, N., Kuze, H., Determination of tropospheric aerosol characteristics by spectral measurements of solar radiation using a compact, stand-alone spectroradiometer. Appl. Opt. Vol. 49, (2010), pp. 1446-1458, ISSN 1539-4522

Manago, N., Miyazawa, S., Bannu, Kuze, H., Seasonal variation of tropospheric aerosol properties by direct and scattered solar radiation spectroscopy. J. Quantitative Spectroscopy & Radiative Transfer, Vol. 112, Issue 2, (January 2011), pp. 285-291, ISSN 0022-4073

Miyazawa, S., Manago, N., Kuze, H., Precice atmospheric correction of MODIS data using aerosol optical parameters derived from simultaneous ground measurement, in Proceedings of the 49th Annual Meeting of the Remote Sensing Society of Japan, (2010), pp.245-246 (in Japanese)

Seinfeld, H. and Pandis, S.N. (1998). Atmospheric Chemistry and Physics, from Air Pollution to Climate Change, John Wiley, ISBN 0-471-17815-2, New York, U.S.A.

Si, F., Kuze, H., Yoshii, Y., Nemoto, M., Takeuchi, N., Kimura, T., Umekawa, T., Yoshida, T., Hioki, T., Tsutsui, T., Kawasaki, M., Measurement of regional distribution of atmospheric $NO_2$ and aerosol particles with flashlight long-path optical monitoring. Atmos. Environ. Vol. 39, (2005), pp. 4959-4968, ISSN 1352-2310

Takamura, T., Nakajima, T., Overview of SKYNET and its activities. Optica Pura y Aplicada Vol. 37, (2004), pp. 3303–3308, ISSN 0030-3917

Tang, J., Xue, Y., Yu, T., Guan,Y. Aerosol optical thickness determination by exploiting the synergy of TERRA and AQUA MODIS. Remote Sensing of Environment Vol. 94, (2005), pp. 327–334, ISSN 0034-4257

Todate, Y., Minomura, M., Kuze, H., Takeuchi, N., On the retrieval of aerosol information on the land surface from Landsat-5/TM data, in Proceedings of the 36th Annual Meeting of the Remote Sensing Society of Japan, (2004), pp.57 - 58. (*in Japanese*)

Yoshii, Y., Kuze, H., Takeuchi, N., Long-path measurement of atmospheric $NO_2$ with an obstruction flashlight and a charge-coupled-device spectrometer. Appl. Opt., Vol. 42, (2003), pp. 4362-4368, ISSN 0003-6935

# Permissions

The contributors of this book come from diverse backgrounds, making this book a truly international effort. This book will bring forth new frontiers with its revolutionizing research information and detailed analysis of the nascent developments around the world.

We would like to thank Boris Escalante-Ramírez, for lending his expertise to make the book truly unique. He has played a crucial role in the development of this book. Without his invaluable contribution this book wouldn't have been possible. He has made vital efforts to compile up to date information on the varied aspects of this subject to make this book a valuable addition to the collection of many professionals and students.

This book was conceptualized with the vision of imparting up-to-date information and advanced data in this field. To ensure the same, a matchless editorial board was set up. Every individual on the board went through rigorous rounds of assessment to prove their worth. After which they invested a large part of their time researching and compiling the most relevant data for our readers. Conferences and sessions were held from time to time between the editorial board and the contributing authors to present the data in the most comprehensible form. The editorial team has worked tirelessly to provide valuable and valid information to help people across the globe.

Every chapter published in this book has been scrutinized by our experts. Their significance has been extensively debated. The topics covered herein carry significant findings which will fuel the growth of the discipline. They may even be implemented as practical applications or may be referred to as a beginning point for another development. Chapters in this book were first published by InTech; hereby published with permission under the Creative Commons Attribution License or equivalent.

The editorial board has been involved in producing this book since its inception. They have spent rigorous hours researching and exploring the diverse topics which have resulted in the successful publishing of this book. They have passed on their knowledge of decades through this book. To expedite this challenging task, the publisher supported the team at every step. A small team of assistant editors was also appointed to further simplify the editing procedure and attain best results for the readers.

Our editorial team has been hand-picked from every corner of the world. Their multi-ethnicity adds dynamic inputs to the discussions which result in innovative outcomes. These outcomes are then further discussed with the researchers and contributors who give their valuable feedback and opinion regarding the same. The feedback is then collaborated with the researches and they are edited in a comprehensive manner to aid the understanding of the subject.

Apart from the editorial board, the designing team has also invested a significant amount of their time in understanding the subject and creating the most relevant covers. They scrutinized every image to scout for the most suitable representation of the subject and create an appropriate cover for the book.

The publishing team has been involved in this book since its early stages. They were actively engaged in every process, be it collecting the data, connecting with the contributors or procuring relevant information. The team has been an ardent support to the editorial, designing and production team. Their endless efforts to recruit the best for this project, has resulted in the accomplishment of this book. They are a veteran in the field of academics and their pool of knowledge is as vast as their experience in printing. Their expertise and guidance has proved useful at every step. Their uncompromising quality standards have made this book an exceptional effort. Their encouragement from time to time has been an inspiration for everyone.

The publisher and the editorial board hope that this book will prove to be a valuable piece of knowledge for researchers, students, practitioners and scholars across the globe.

# List of Contributors

Ellen Eigemeier, Janne Heiskanen, Miina Rautiainen, Matti Mõttus, Veli-Heikki Vesanto, Titta Majasalmi and Pauline Stenberg
University of Helsinki, Finland

Jorge Alberto Bustamante, Regina Alvalá and Celso von Randow
Brazilian National Institute for Space Researches, Brazil

Wenjiang Huang, Juhua Luo, Jingcheng Zhang, Jinling Zhao, Chunjiang Zhao, Jihua Wang, Guijun Yang, Muyi Huang, Linsheng Huang and Shizhou Du
Beijing Research Center for Information Technology in Agriculture, Beijing, China

José Maria Filippini Alba, Victor Faria Schroder and Mauro Ricardo R. Nóbrega
Brazilian Agricultural Reasearch Corporation (Embrapa), Brazil

A. Gelfan, E. Muzylev and Z. Startseva
Water Problem Institute of Russian Academy of Sciences, Moscow, Russia

A. Uspensky
State Research Center of Space Hydrometeorology Planeta, Moscow, Russia

P. Romanov
City College of City University of New York, New York, NY, USA

Ignacio Melendez-Pastor, Encarni I. Hernández, Jose Navarro-Pedreño and Ignacio Gómez
Department of Agrochemistry and Environment, University Miguel Hernández of Elche, Spain

Yida Fan, Siquan Yang, Shirong Chen, Haixia He, Sanchao Liu, Wei Wu, Lei Wang, Juan Nie, Wei Wang, Baojun Zhang, Feng Xu, Tong Tang, Zhiqiang Lin, Ping Wang, Qi Wen and Wei Zhang
National Disaster Reduction Center of China, China

Z. Servadio and N. Villeneuve
Laboratoire Géosciences Réunion, Université de la Réunion, Institut de Physique du Globe de Paris, CNRS, UMR 7154, Géologie des Systèmes Volcaniques, Saint Denis, France

Z. Servadio
Institut de Recherche pour le Développement, US 140, BP172, 97492 Sainte-Clotilde Cedex, France

**P. Bachèlery**
Clermont Université, Université Blaise Pascal, Laboratoire Magmas et Volcans, CNRS, UMR 6524, Observatoire de Physique du Globe de Clermont-Ferrand, BP 10448, F-63000 Clermont-Ferrand, France
IRD, R 163, LMV, F-63038 Clermont-Ferrand, France

**Long S. Chiu and Si Gao**
Department of Atmospheric, Oceanic and Earth Sciences, George Mason University, USA

**Chung-Lin Shie**
UMBC/JCET, NASA/GSFC, USA

**Baozhang Chen**
Institute of Geographic Sciences and Nature Resources Research, Chinese Academy of Sciences, Beijing, P.R. China

**Hiroaki Kuze**
Centre for Environmental Remote Sensing (CEReS), Chiba University, Japan

Printed in the USA
CPSIA information can be obtained
at www.ICGtesting.com
JSHW011503221024
72173JS00005B/1183